Springer Tracts in Natural Philosophy

Volume 40

Edited by C. Truesdell

Springer

New York
Berlin
Heidelberg
Barcelona
Budapest
Hong Kong
London
Milan
Paris
Santa Clara
Singapore
Tokyo

Nanny Fröman Per Olof Fröman

Phase-Integral Method

Allowing Nearlying Transition Points

With adjoined papers by:
A. Dzieciol, N. Fröman, P.O. Fröman, A. Hökback
S. Linnaeus, B. Lundborg, and E. Walles

With 17 Illustrations

Springer

Nanny Fröman
Department of Theoretical Physics
University of Uppsala
Box 803
S-751 08 Uppsala
Sweden

Per Olof Fröman
Department of Theoretical Physics
University of Uppsala
Box 803
S-751 08 Uppsala
Sweden

Mathematical Subject Classification (1991): 810xx

Library of Congress Cataloging in Publication Data
Fröman, Nanny.
  Phase-integral method : allowing nearlying transition
points / Nanny Fröman, Per Olof Fröman.
    p.   cm. — (Springer tracts in natural philosophy ; v. 40)
  Includes bibliographical references and indexes.
  ISBN-13:978-1-4612-7511-4   e-ISBN-13:978-1-4612-2342-9
  DOI: 10.1007/978-1-4612-2342-9

    1. WKB approximation.   2. Wave equation.   I. Fröman, Per Olof.
  II. Title.   III. Series.
  QC20.7.W53F76   1995
  530.1′24—dc20   95-12919
                    Printed on acid-free paper.

            © 1996 Springer-Verlag New York, Inc.
        Softcover reprint of the hardcover 1st edition 1996

Production managed by Terry Kornak; manufacturing supervised by Jeffrey Taub.
Photocomposed copy prepared from LATEX files.

            9 8 7 6 5 4 3 2 1

    ISBN-13:978-1-4612-7511-4

Preface

The efficiency of the phase-integral method developed by the present authors has been shown both analytically and numerically in many publications. With the inclusion of supplementary quantities, closely related to new Stokes constants and obtained with the aid of comparison equation technique, important classes of problems in which transition points may approach each other become accessible to accurate analytical treatment.

The exposition in this monograph is of a mathematical nature but has important physical applications, some examples of which are found in the adjoined papers. Thus, we would like to emphasize that, although we aim at mathematical rigor, our treatment is made primarily with physical needs in mind.

To introduce the reader into the background of this book, we start by describing the phase-integral approximation of arbitrary order generated from an unspecified base function. This is done in Chapter 1, which is reprinted, after minor changes, from a review article. Chapter 2 is the result of research work that was pursued during more than two decades, interrupted at times. It started in the sixties, when we were still using a phase-integral approximation, which in our present terminology corresponds to a special choice of the base function. At the time our primary aim was to derive expressions for the supplementary quantities needed in order to obtain an accurate connection formula for a real potential barrier, when the energy lies in the neighborhood of the top of the barrier. In 1972 we published analytic expressions, without derivation, for such quantities up to the fifth order of the phase-integral approximation. These results were then used in a number of applications. After our derivation in 1974 of the more general and flexible phase-integral approximation generated from an unspecified base function, described in Chapter 1, we had to generalize the comparison equation technique. We then also decided to elaborate that technique into a general scheme for obtaining the Stokes constants, pertaining to an arbitrary order of the phase-integral approximation used, and valid also when transition points, not specified *a priori*, approach each other. We wished to formulate the comparison equation technique, developed chiefly by Cherry, Erdélyi, and others, in such a way that, when the resulting solution sufficiently far from transition points is expanded in terms of a "small" bookkeeping parameter, the result is the phase-integral approximation of arbitrary order generated from an unspecified base function, with expressions for phase and amplitude that remain valid also when transition points approach each other. In addition, our goal was to perform the rather lengthy calculations once and for all, in order to obtain formulas that can readily be particularized to specific situations encountered in various applications.

Thus, the comparison equation technique described in Chapter 2 is developed in general terms with the aim of facilitating its application to particular problems as much as possible. The exposition displays both the power and the limitations of comparison equation technique. Although the general ideas that we exploit can be found in papers by earlier authors, our aim required a nontrivial adaptation of comparison equation technique to phase-integral technique with a number of new steps needed in the procedure. The resulting phase-integral formulas, involving supplementary quantities, are new. The accomplishment of our aims turned out to be a long toil; we had to rewrite the manuscript numerous times over the years, in order to make the exposition clear step by step.

To illustrate the application of the results in Chapters 1 and 2, we have added a number of adjoined papers (Chapters 3 through 11), which were originally intended to be published in scientific journals. The papers in Chapters 3 through 8 are essentially of a mathematical nature, but with a physical aim, whereas those in Chapters 9 through 11 are of a more physical nature.

In Chapters 3 and 4, the formulas obtained in Chapter 2 are applied to the case in which there is one transition zero; in Chapter 5 they are applied to the case in which there are two transition zeros, and in Chapters 6 and 7 they are applied to the case in which there are two or three transition points (one first- or second-order pole and one or two simple zeros of the square of the base function). In Chapter 8 the phase-amplitude method for numerical solution of Schrödinger-like differential equations is combined with comparison equation technique in order to master an important special problem.

Chapters 9 through 11 concern physical applications in which transition points may lie close together. Numerical results illustrate the accuracy that can be achieved by means of the phase-integral method when supplementary quantities are included.

We are much indebted to the Editor, Professor C. Truesdell, for his generous help with the publication of this book. We are also very happy to have enjoyed the friendship of Charlotte and Clifford Truesdell during many years.

We are grateful to Miss Ebba Johansson, who has typed several early versions of some of the chapters in this book, and to Mrs. Maud Högberg, who has typed several later versions as well as other chapters. For their unfailing and generous help we would like to give them our heartfelt thanks.

Uppsala   Nanny Fröman
July 1995   Per Olof Fröman

Contents

1
Phase-Integral Approximation of Arbitrary Order Generated from an Unspecified Base Function

Abstract. We begin with a brief review of the so-called WKB approximation, its deficiencies in higher order, and attempts by several authors to remedy them. It is then shown that these deficiencies do not appear in the phase-integral approximation generated from an *a priori* unspecified base function, which was originally devised by the present authors in 1974 and is presented here in a way that clarifies the role of the "small" parameter in the differential equation. The advantage of this approximation versus the WKB approximation in higher order is also discussed. In a discussion of relations between solutions of the Schrödinger equation and the q-equation, the Ermakov–Lewis invariant is considered. In the concluding section we mention other items that constitute the phase-integral method beside the phase-integral approximation in question.

1.1 Introduction

The mathematical approximation method that since the breakthrough of quantum mechanics has usually been called the WKB method, has in reality been known for a very long time.[1] The method describes various kinds of wave motion in an inhomogeneous medium, in which the properties change only slightly over one wavelength; in addition, the method provides the connection between classical mechanics and quantum mechanics. To a surprisingly large extent it can already be found in an investigation by Carlini (1817) on the motion of a planet in an unperturbed elliptic orbit. After that the method was independently developed and used by many people. However, the important connection formulas were missing until Rayleigh (1912), very implicitly, and Gans (1915), somewhat more explicitly, derived

---

[1]Chapter 1 is reprinted after minor changes from *Forty More Years of Ramifications* (Discourses in Mathematics and Its Applications, No. 1, 1991), by permission of the Department of Mathematics of Texas A & M University.

one of them, later rediscovered independently by Jeffreys (1925), who also derived another connection formula (although not in quite correct form), and by Kramers (1926). A review of the history of the method from 1817 to 1926 is given by Fröman and Fröman (1985).

Obviously unaware of the existence of the work by the previous authors, Brillouin (1926a,b), Wentzel (1926) and Kramers (1926) introduced analogous considerations in quantum mechanics. Brillouin established, for a system of particles, the connection between the Schrödinger equation of quantum mechanics and the Hamilton–Jacobi equation of classical mechanics, while Wentzel and Kramers introduced for the radial Schrödinger equation the main results obtained in the course of the development described above. Kramers also pointed out that in the first order of the approximation it is sometimes convenient to replace $\ell(\ell+1)$ by $(\ell+1/2)^2$, where ℓ is the orbital angular momentum quantum number. These results turned out to be extremely useful in applications of the new quantum theory and became known as the WKB method. However, Brillouin, Wentzel, and Kramers contributed hardly anything new to the mathematical approximation method that had already been found by previous authors, as described in more detail in the above-mentioned historical review (Fröman and Fröman, 1985). In brief, one can say that the so-called WKB method consists of the use of Carlini's approximation, which he derived in arbitrary-order approximation, and Jeffreys' connection formulas, which he derived in the first-order approximation. For the radial Schrödinger equation it also involves Kramers' modification of the first-order approximation by the replacement of $\ell(\ell+1)$ by $(\ell+1/2)^2$.

Since the publication of the papers by Brillouin (1926a,b), Wentzel (1926) and Kramers (1926) the method has uncritically been called "the WKB method" by most writers in theoretical physics, although this is not a very appropriate name (as should be obvious from what has been said above concerning the early development of the theory).

New contributions to the mathematical method were later made by other authors, especially Zwaan (1929), Kemble (1935), Furry (1947), Olver (1961), and Heading (1962). In the beginning of the sixties there existed an extensive literature on the subject, but there still were various kinds of essential deficiencies in the so-called WKB method.

Our interest in the problems associated with the mathematical foundations of the approximation method arose about 1960 in connection with a quantal decay problem for which a precise knowledge of the properties of the so-called WKB approximation, especially the connection formulas, was needed. The existing literature did not provide what we were looking for, and the lack of a satisfactory and rigorous treatment allowing for the physical aspects grew more and more embarrassing. Thus we were actually compelled to find our own way out of the impasse. Using as a starting point the ideas introduced by Zwaan (1929) and by Kemble (1935), we developed in the years 1960 to 1964 a new rigorous method for handling the connection problems appearing for the so-called WKB approximation of the first order and modifications of it. The study of connection problems was thereby transferred to the study of a certain matrix, the F-matrix, the elements of which were given by convergent series. This method, which was also powerful in intricate and complicated applications, was presented in a book by Fröman and Fröman (1965). It provided a sound basis for handling the connection problems of the first-order WKB approximation and led to satisfactory estimates of the accuracy of that approximation. Soon after the publication of our book, Olver (1965a,b) published related estimates, which he had derived quite independently of our work.

In parallel to the work with our above mentioned book, we made several attempts to generalize the method. In this connection a very important result was revealed. It turned out that the usual WKB expansion could be transformed (N. Fröman, 1966b) so that the sum of the terms of odd order in the expansion parameter could be eliminated. This transformation is exact when the whole series is retained. If the series is truncated at a finite power of the expansion parameter, however, a new approximation is obtained, which differs from the WKB approximation of corresponding order. The distinguishing property lies in the fact that the new approximation also displays in higher order a simple relationship between the preexponential factor and the integrand of the integral in the exponent. An important consequence of this simple analytical form is that the Wronskian of two approximate solutions is a constant, as it should be (since the Wronskian of two exact solutions is constant). This, in turn, implied that the F-matrix method for mastering connection problems (Fröman and Fröman, 1965) could be applied to the new approximation after minor generalizations. Certain test calculations showed that the higher-order approximations yielded extremely accurate results. A new line of development was therefore opened: Many problems that could be treated analytically to within an accuracy of a given percent with the first-order WKB approximation and the F-matrix method for mastering connection problems could now be treated to within an accuracy that in many cases was several powers of ten better. In view of this fact, we considered it very important to continue our work along the same lines, to develop the theory further, and to study its application to actual physical problems.

Our immediate concern then became to overcome, for the new approximation, the difficulties associated with the Coulomb or centrifugal barrier singularity of the radial Schrödinger equation. Because of our previous extensive study of the connection problems for the first-order WKB approximation, we clearly realized that the serious difficulties associated with the solution of the connection problems for the higher orders of the WKB approximation were due to a large extent to the complicated connection between phase and amplitude. Though the new arbitrary-order approximation derived by N. Fröman (1966b) did not suffer from this deficiency, it was not sufficiently general to be useful at singularities such as appear, in general, in the radial Schrödinger equation. We first tried to apply the ideas used by Beckel and Nakhleh (1963) for the WKB approximation (Fröman and Fröman, 1974b, pp. 124–126), but we soon realized that this could not be the appropriate way to proceed. Finally, we solved the problems associated with this remaining deficiency of the higher-order approximations by devising a more general kind of phase-integral approximation of arbitrary order (Fröman and Fröman, 1974a, 1974b, pp. 126–131), in which there appears an unspecified function, the base function, which we now denote by $Q(z)$. The presence of this *a priori* unspecified function, which is said to generate the phase-integral approximation in question, makes the new approximation very flexible, a property which can be taken advantage of in several different respects. In particular, the function $Q(z)$ can in general be chosen so that the approximation remains valid in the neighborhood of a first- or second-order pole of the coefficient function in the differential equation. For a particular choice of $Q(z)$ (the one which is in fact the simplest) the approximation becomes identical to the more special phase-integral approximation previously devised by N. Fröman (1966b, 1970).

In this chapter we first describe in some detail the deficiencies of the so-called WKB approximation (i.e., the Carlini approximation) in higher order and the successive steps in the research work that led to the construction of the arbitrary-order phase-integral approximation generated from the unspecified base function $Q(z)$. We thereby strive to present the derivation of this approximation in a way that clarifies the role of the "small" parameter in the differential equation. We then discuss essential properties of the new approximation.

1.2   The So-Called WKB Approximation, Its Deficiencies in Higher Order, and Early Attempts to Remedy These Deficiencies

In this section the WKB approximation is first derived in a well known way, and the conditions for its validity, as they appear by the derivation, are given. The deficiencies of the WKB approximation of higher order, and early attempts to remove these deficiencies, are then discussed in some detail.

1.2.1   Derivation of the WKB Approximation

We shall consider the differential equation

$$\frac{d^2\psi}{dz^2} + R(z)\psi = 0, \tag{1.2.1}$$

where $R(z)$ is assumed to be an analytic function of z in the region of the complex z-plane under consideration. To derive the WKB approximation for the solution of (1.2.1) we shall assume $R(z)$ to be large, and to account for this assumption in an explicit way we shall treat the differential equation

$$\frac{d^2\psi}{dz^2} + \frac{R(z)}{\lambda^2}\psi = 0, \tag{1.2.2}$$

where λ is a "small" parameter that we shall at the end set equal to unity to obtain the solution of the original differential equation (1.2.1). As is well known, the formal solution of (1.2.2), known as the WKB expansion, which had already been introduced by Carlini (1817), is obtained by inserting

$$\psi = \exp\left(\frac{i}{\lambda}\int^z \sum_{\nu=0}^{\infty} y_\nu(z)\lambda^\nu dz\right) \tag{1.2.3}$$

into the differential equation (1.2.2), dividing both sides of the resulting equation by ψ, and setting the coefficient of each power of λ equal to zero. This yields

$$y_0 = \pm R^{1/2}(z) \tag{1.2.4a}$$

and the recurrence formula

$$\frac{dy_{\nu-1}}{dz} = -i\sum_{\mu=0}^{\nu} y_\mu y_{\nu-\mu}, \qquad \nu = 1, 2, 3, \ldots, \tag{1.2.4b}$$

from which $y_1, y_2, y_3, y_4, \ldots$ can be determined successively. On determining these functions from the recurrence formula, one finds that each odd-order function y_{2n+1} can be written as the derivative with respect to z of an expression containing only the even-order functions $y_0, y_2, \ldots, y_{2n}$. Furthermore, the odd-order functions are not affected by the double sign in (1.2.4a), whereas all the even-order functions $y_0, y_2, y_4, \ldots$ will have the double sign $\pm$. Thus we obtain two formal solutions of (1.2.2), one belonging to the plus sign of the even-order terms, the other belonging to the minus sign of the same terms. It is well known that the infinite series in (1.2.3) is in general not convergent, but only asymptotic. Truncating this series after the term $y_\nu(z)\lambda^\nu$, we obtain the WKB approximation of order ν, which obviously is of the general form

$$\psi = \exp[u(z) \pm i\,w(z)], \tag{1.2.5}$$

where $u(z)$ is the truncated sum of the odd-order terms, and $\pm i\,w(z)$ is the truncated sum of the even-order terms.

If z is real and lies in a classically allowed region (i.e., a region where $R(z)/\lambda^2$ is positive) one improves only the phase of ψ by going from an odd order to the next even order of the WKB approximation, but one improves both the phase and the amplitude by going to the next odd order. Furthermore, the first-order approximation is often useful, while the amplitude for the zeroth order is too crude. Therefore, the odd orders of the WKB approximation are more satisfactory than the even orders. By truncating the series in (1.2.3) one finds that, except for a constant factor, the first-order WKB approximation is

$$\psi = \frac{\exp\left(i \int^z \frac{1}{\lambda} y_0 dz\right)}{\sqrt{\frac{1}{\lambda} y_0}}, \tag{1.2.6a}$$

the third-order WKB approximation is

$$\psi = \frac{\exp\left[i \int^z \frac{1}{\lambda}(y_0 + y_2 \lambda^2) dz\right]}{\sqrt{\frac{1}{\lambda} y_0 \exp\left(\frac{y_2 \lambda^2}{y_0}\right)}}, \tag{1.2.6b}$$

and the fifth-order WKB approximation is

$$\psi = \frac{\exp\left[i \int^z \frac{1}{\lambda}(y_0 + y_2 \lambda^2 + y_4 \lambda^4) dz\right]}{\sqrt{\frac{1}{\lambda} y_0 \exp\left\{\frac{y_2 \lambda^2}{y_0} + \left[\frac{y_4 \lambda^4}{y_0} - \frac{1}{2}\left(\frac{y_2 \lambda^2}{y_0}\right)^2\right]\right\}}}. \tag{1.2.6c}$$

On account of the double sign of $y_0, y_2, y_4, \ldots$ the exponent in the numerator of these expressions will have a double sign [cf. (1.2.5)], and each one of the expressions (1.2.6a,b,c) thus represents two linearly independent functions.

From (1.2.4a,b) it follows that

$$y_1/y_0 = -\frac{1}{2} i \frac{d}{dz}(1/y_0) \tag{1.2.7a}$$

and

$$y_2/y_0 = \frac{1}{2}\varepsilon_0 \tag{1.2.7b}$$

where

$$\varepsilon_0 = R^{-3/4}(z)\frac{d^2}{dz^2}R^{-1/4}(z). \tag{1.2.8}$$

According to (1.2.4a) and (1.2.7b) the two linearly independent third-order WKB functions (1.2.6b) for the solution of the differential equation (1.2.1); i.e., (1.2.2) with λ set equal to unity; are (except for arbitrary constant factors)

$$\psi = \frac{\exp\left(\pm i \int_{z_0}^z \left(1 + \frac{1}{2}\varepsilon_0\right) R^{1/2} dz\right)}{\left[R^{1/2} \exp\left(\frac{1}{2}\varepsilon_0\right)\right]^{1/2}}, \tag{1.2.9}$$

where z_0 is a conveniently chosen point in the region under consideration. One may expect the derivation of (1.2.9) to be justified if $|y_2| \ll |y_1| \ll |y_0|$; i.e., if [see (1.2.4a) and (1.2.7a,b)]

$$\frac{1}{2}|\varepsilon_0| \ll \frac{1}{2}\left|\frac{d}{dz}R^{-1/2}\right| \ll 1. \qquad (1.2.10)$$

If the further condition

$$\frac{1}{2}\left|\int_{z_0}^{z} \varepsilon_0 R^{1/2}dz\right| \ll 1 \qquad (1.2.11)$$

is fulfilled, one can instead of (1.2.9) use the well known first-order WKB approximation

$$\psi = R^{-1/4} \exp\left(\pm i \int_{z_0}^{z} R^{1/2}dz\right). \qquad (1.2.12)$$

Similar derivations of the first-order WKB approximation are given in many textbooks on quantum mechanics; but, with few exceptions, it is said that the condition for its validity is only the last inequality in (1.2.10). Therefore, we want to emphasize that the magnitude of the function ε_0; which appears to the left in the chain of local inequalities (1.2.10), as well as in the global condition (1.2.11); is very important. We shall now shed further light on this fact.

The formula (1.2.12) indicates that, if one uses the first-order WKB approximation, the general solution of the differential equation (1.2.1); i.e., (1.2.2) with $\lambda = 1$; is given by the approximate formula

$$\psi = AR^{-1/4} \exp\left(i \int_{z_0}^{z} R^{1/2}dz\right) + BR^{-1/4} \exp\left(-i \int_{z_0}^{z} R^{1/2}dz\right), \quad (1.2.13)$$

where A and B are arbitrary constants. It is very natural that the validity of this approximate solution should depend on an integral such as the one in the condition (1.2.11). In fact, at any point, let us say z_0, one can choose the constants A and B in (1.2.13) such that the approximate solution (1.2.13), as well as its first derivative, coincides exactly with the corresponding quantity for a given solution of the differential equation (1.2.1). The question is then how far one can move away from the point z_0 before the approximate solution becomes bad. According to the rather qualitative treatment now under discussion, the answer is that one must not move so far that the condition (1.2.11) is violated. In a rigorous treatment with the aid of the F-matrix method it turns out that the conditions that we have now obtained for the validity of the approximate solution (1.2.13) are not sufficient. The reason for this is that there is an interference between the two terms in (1.2.13) that cannot be neglected in a rigorous treatment of the problem. This interference is connected with the Stokes phenomenon in the theory of asymptotic expansions, a consequence of which is that the question of the accuracy of (1.2.13) and the question whether in a *classically forbidden region* (i.e., in a region of the real z-axis where $R(z) < 0$) it is meaningful to keep both terms in (1.2.13) cannot be answered in a satisfactory way by the use of the elementary qualitative discussion given above (Fröman and Fröman, 1965; N. Fröman, 1966a).

1.2.2   Deficiencies of the WKB Approximation in Higher Order

The WKB approximation of higher order has two serious deficiencies, one of which is due to its unsatisfactory form, the other due to its lack of flexibility.

First, the form of the expression (1.2.6b,c) for the third- and fifth-order WKB approximation shows that in higher order the preexponential factor is complicated and not tied to the integrand in the exponent by the simple relation which must be fulfilled for the exact solution and which can be described as follows: For the particular case in which $R(z)$ is real on the real z-axis, every solution ψ of the differential equation (1.2.1), which is represented by a single term, and which does not have a constant phase on the real axis (i.e., not equal to a [possibly complex] constant times a real function of z), can (except for an arbitrary constant factor) be written in the form

$$\psi = q^{-1/2}(z) \exp \left(\pm i \int^z q(z)\, dz \right) \qquad (1.2.14)$$

with $q(z)$ real on the real axis. For the general case in which $R(z)$ is not necessarily real on the real z-axis, the requirement that there exist for the differential equation (1.2.1) two solutions of the form $\psi = A(z) \exp[\pm i\phi(z)]$, where the functions $A(z)$ and $\phi(z)$ need not necessarily be real on the real z-axis, implies that $A(z) = \mathrm{const} \times [\phi(z)/dz]^{-1/2}$ and hence that (except for constant factors) the two solutions are of the form (1.2.14), where, however, the function $q(z)$ need not be real on the real z-axis. These assertions are proved in the Appendix. We remark that when $R(z)$ is not real on the real z-axis, or when z lies in the complex z-plane outside the real z-axis, the functions $A(z)$ and $\phi(z)$ are *not in the usual sense* an amplitude and a phase, respectively, and the right-hand member of (1.2.14) is *not in the usual sense* written as the product of an amplitude and a phase factor; however, it may still be convenient to use the names "amplitude" and "phase" factor. As a consequence of the complicated relation between the "amplitude" $\exp[u(z)]$ and the derivative with respect to z of the "phase" $\pm w(z)$ for the WKB functions (1.2.5) of higher order, the Wronskian of the two linearly independent WKB functions (1.2.5) is constant only for the first-order approximation, in distinction to the Wronskian of two exact solutions, which is always constant, as is also the Wronskian of the two functions (1.2.14), which is equal to $-2i$ for any function $q(z)$. That the

Wronskian must be constant is thus a strong reason for using approximate solutions of the form (1.2.14) also when $R(z)$ is not necessarily real on the real z-axis. From a physical point of view it is worth noticing that in quantum mechanics the constancy of the Wronskian is related to the principle of current conservation, which prevails if $R(z)$ is real and nonsingular on the real z-axis.

Because of the inconvenient structure of the higher orders of the WKB approximation, the connection problems (i.e., the Stokes phenomenon) cannot be handled efficiently and rigorously. While the F-matrix method (Fröman and Fröman, 1965) provides an efficient tool for mastering the connection problems of the first-order WKB approximation, this method becomes practically inapplicable to the higher orders of the WKB approximation because of their complicated analytical form, yielding a nonconstant Wronskian. For a discussion of the difficulties and complications that, in general, appear if one tries to use the WKB approximation in higher order for solving various physical problems, we refer to the analysis in a paper by Dammert and P.O. Fröman (1980). To the best of our knowledge, it is only in particularly simple situations that one can actually treat a physical connection problem satisfactorily by means of the WKB approximation in higher order (Fröman and Fröman, 1977).

Secondly, as regards the lack of flexibility, we note that there are important commonly appearing cases when the WKB approximation does not give a satisfactory approximate solution of the differential equation (1.2.1) in the whole range of the independent variable of interest in the problem considered. As a well known example of a situation in which this deficiency appears, we mention the radial Schrödinger equation for the case when there is a Coulomb singularity in the potential at the origin or the angular momentum quantum number ℓ is different from zero. The difficulty is reflected in the fact that the original half-integer WKB quantization condition does not, in general, give the correct bound-state energy levels of the radial Schrödinger equation. For the first-order WKB approximation, Kramers (1926) noticed that the replacement of $\ell(\ell+1)$ by $(\ell+1/2)^2$ in that quantization condition was needed in order that correct energy levels would be reproduced in the case of an attractive Coulomb field. Furthermore, for the radial Schrödinger equation with a Coulomb potential or a centrifugal potential, the first-order WKB wave function does not have the correct behavior at the origin, unless this replacement is made. This was pointed out by Young and Uhlenbeck (1930) for attractive Coulomb fields and by Yost *et al.* (1936) for repulsive Coulomb fields. The justification for the replacement of $\ell(\ell+1)$ by $(\ell+1/2)^2$ in the first-order WKB approximation was discussed in more detail in a paper by Kemble (1935), p. 560, and in the well known book by Kemble (1937), pp. 107–108. In connection with the radial Schrödinger equation for a particle moving in a potential with a Coulomb singularity at the origin, Langer (1937) eliminated the deficiency of the first-order WKB approximation at the origin in a new way: he introduced the logarithm of the distance from the center of force as a new independent variable, introducing at the same time a new dependent variable such that no first derivative appeared in the transformed differential equation; which he then solved by means of the first-order WKB approximation. When he finally went back to the original variables, the replacement of $\ell(\ell + 1)$ by $(\ell+1/2)^2$ appeared automatically in the expression for the first-order WKB solution. There are authors who justify Langer's transformation procedure by asserting that the WKB approximation is valid when the independent variable ranges from $-\infty$ to $+\infty$, but is not valid when it ranges from 0 to $+\infty$ (as in a radial problem). Because in the derivation of the WKB approximation, only local but not global considerations are made, this assertion is obviously not relevant. The correct justification for Langer's transformation procedure is that the WKB approximation of the original differential equation is not valid at the origin, whereas the WKB approximation of the transformed differential equation is actually valid at the point corresponding to the origin. The replacement of $\ell(\ell+1)$ by $(\ell+1/2)^2$ in the first-order WKB approximation has often been referred to as the Langer modification of the first-order WKB approximation, although the replacement in question had been introduced and justified in various ways by earlier authors. For some further comments on the early history of the replacement of $\ell(\ell + 1)$ by $(\ell+1/2)^2$ in the first-order WKB approximation and related questions we refer to pp. 451–453 in an article by Hull and Breit (1959). For the WKB approximation of higher order the difficulty now discussed could, however, not be removed by the simple replacement of $\ell(\ell + 1)$ by $(\ell+1/2)^2$. Thus, in connection with the radial Schrödinger equation, it was not understood how to modify the WKB approximation in higher order to make it also valid in the neighborhood of the Coulomb or centrifugal barrier singularity at the origin.

In the sixties, considerable interest was focused on the study of modifications of the WKB approximation in higher order with the purpose of obtaining an approximate solution of the radial Schrödinger equation that remains valid at the origin. Beckel and Nakhleh (1963) pointed out the fact mentioned above that for the WKB aproximation of higher order the difficulties at the origin are not eliminated by the replacement of $\ell(\ell+1)$ by $(\ell+1/2)^2$ in the expression for the WKB approximation of higher order. By an analysis of the differential equations obeyed by the second-order WKB functions they found that these functions will satisfy the original differential equation approximately near the origin if $\ell(\ell+1)$ is replaced by a root K of the equation $K^2 + 1/(64K) = \ell(\ell + 1)$. The Beckel–Nakhleh (1963) idea of how to improve the WKB approximation in higher order close to the origin was adapted to the asymptotic approximation obtained by truncating the series in formula (1.2.20) up to the ninth order by Fröman and Fröman (1974b), pp. 124–126, who concluded, however, that the resulting kind of approximation was not to the point. One obtains a treatment built on a more satisfactory theoretical basis by adhering to the above mentioned transformation procedure introduced by Langer (1937) and using the WKB approximation in higher order for solving the transformed differential equation. This point of view was adopted by Choi and Ross (1962) for a phase shift problem and by Krieger and Rosenzweig (1967) for a bound-state problem, but the connection problems were disregarded. When these authors finally went back to the original variables, they obtained modified WKB expressions for the physical quantities considered (the phase shift and the quantization condition). In these expressions the contributions from the higher-order terms of the WKB approximation, as distinct from the first-order term, could not be obtained merely by the replacement of $\ell(\ell+1)$ by any other quantity in the corresponding unmodified WKB expressions, since new additional terms appeared in the modified expressions. Numerical calculations of energy eigenvalues performed at our department around 1970 by Nordlund [unpublished; quoted by Fröman and Fröman (1974b)] indicated that the transformation procedure is superior to the previously mentioned modification procedure, which is based on the Beckel–Nakhleh idea. By replacing, in the above-mentioned transformation procedure, the logarithm by an unspecified function (i.e., by using a generalized Langer transformation) Krieger (1969) derived a modified WKB quantization condition containing the unspecified transformation function. In a paper by Engelke and Beckel (1970) this function was chosen to be a special function depending on two parameters (viz., an arbitrary parameter and a trivial scaling factor) and reducing to the Langer logarithm for a particular choice of the values of the parameters. In the sixties, correct higher-order WKB expressions for the solutions of a radial phase shift problem (Choi and Ross, 1962) and a radial eigenvalue problem (Krieger and Rosenzweig, 1967) had thus been given; however, there did not exist an explicit asymptotic expression of arbitrary order for the wave function that was of simple and satisfactory form and sufficiently flexible to remove the difficulties of the WKB approximation when applied, for instance, to the radial Schrödinger equation. Such an asymptotic expression was derived in two independent ways by Fröman and Fröman (1974a,b) and will be described in §1.3.

1.2.3 Phase-Integral Approximation of Arbitrary Order, Freed from the First Deficiency

The important separation of the wave function into amplitude and phase factor, with the correct relation between phase and amplitude [cf. (1.2.14)], can be traced back to papers by Rayleigh (1912) and Fowler et al. (1921). In a numerical method developed in the thirties this relation was used on the real z-axis by Milne (1930), by Wilson (1930) and especially by Young (1931, 1932) who pointed out its great fundamental importance in quantum mechanics. Adding to Milne's work, Wheeler (1937) showed the great practical advantage of what he called the "amplitude-phase method" for efficient numerical solution of the radial Schrödinger equation. The reason for the success of this method is that the amplitude and the phase can be obtained as nonoscillating functions in space. More recently, there have also appeared several publications in which the above-mentioned phase-amplitude relation is used for numerical solution of the Schrödinger equation with the aid of modern computers. Although the connection between phase and amplitude was thus used at an early stage for numerical solution of the Schrödinger equation, there are very few publications before 1966 in which there are traces of higher-order asymptotic approximations of the differential equations (1.2.1) or (1.2.2) that have the correct relation between phase and amplitude. If actually pursued to higher order, the procedure used by Messiah (1959) in Section 7 of Chapter 6 would have yielded such a higher-order asymptotic approximation [cf. (1.2.14)] instead of the WKB approximation of higher order [cf. (1.2.5)]. The distinction between these two kinds of approximations does not, however, appear in Messiah's treatment, as he restricts himself to deriving the first-order approximation only. Broer (1963) used the same procedure as Messiah (1959) and pursued it by deriving the first two higher-order approximations, but in his terminology he did not distinguish the resulting higher-order approximations from the higher-order WKB approximations. This new kind of higher-order asymptotic approximations also appears implicitly in a paper by Bertocchi, Fubini, and Furlan (1965), but the advantage of using these approximations for solving the connection problems is neither indicated, nor made use of, by these authors.[2] That there is actually an essential difference between the two types of approximations (cf. [1.2.5] and [1.2.14]) was thus not noticed or pointed out in the treatment by Bertocchi, Fubini and Furlan (1965); see the discussion in the paper by Dammert and P.O. Fröman (1980). It was only in 1966 that asymptotic approximations of arbitrary order, with the correct relation between phase and amplitude (cf. [1.2.14]), were derived in a systematic way by N. Fröman (1966b). Using the recurrence formula (1.2.4b), she showed that in the infinite series occurring in (1.2.3) the sum of the odd-order terms can simply be expressed in terms of the sum of the even-order terms; she thus obtained a formal solution of the time-independent Schrödinger equation containing only the even-order terms. Her derivation proceeds as follows. She first noted that for odd values of ν $(= 2N + 1)$ the recurrence formula (1.2.4b) can be

---

[2]Higher-order asymptotic approximations of the form (1.2.14) are actually obtained if the expansion of $a(x, \lambda)$ on p. 604 in the paper by Bertocchi, Fubini, and Furlan (1965) is truncated and inserted into their Eq. (2.11), but instead of using the advantages of these approximations in their reasoning they use the higher-order WKB approximations, which are obtained when the expansions for $a(x, \lambda)$ and $b(x, \lambda)$ on p. 604 in their paper are truncated and inserted into Eq. (2.5) with Eqs. (2.6) and (2.9) of the same paper.

written

$$\frac{dy_{2N}}{dz} = -2i \sum_{n=0}^{N} y_{2n} y_{2N+1-2n} \qquad (N = 0, 1, 2, 3, \ldots). \qquad (1.2.15)$$

Multiplying both members of (1.2.15) by λ^{2N} and summing over all possible values of N, she then obtained

$$\sum_{N=0}^{\infty} \frac{dy_{2N}}{dz} \lambda^{2N} = -\frac{2i}{\lambda} \sum_{N=0}^{\infty} \sum_{n=0}^{N} y_{2n} y_{2N+1-2n} \lambda^{2N+1}. \qquad (1.2.16)$$

Noting that the right-hand member of (1.2.16) can be written as the product of two infinite series, she finally obtained

$$\sum_{N=0}^{\infty} \frac{dy_{2N}}{dz} \lambda^{2N} = -\frac{2i}{\lambda} \left(\sum_{n=0}^{\infty} y_{2n} \lambda^{2n} \right) \left(\sum_{n=0}^{\infty} y_{2n+1} \lambda^{2n+1} \right) \qquad (1.2.17)$$

and hence

$$\frac{i}{\lambda} \sum_{n=0}^{\infty} y_{2n+1} \lambda^{2n+1} = -\frac{1}{2} \frac{\sum_{n=0}^{\infty} \frac{dy_{2n}}{dz} \lambda^{2n}}{\sum_{n=0}^{\infty} y_{2n} \lambda^{2n}}$$

$$= \frac{d}{dz} \ln \left[\left(\sum_{n=0}^{\infty} y_{2n} \lambda^{2n} \right)^{-1/2} \right], \qquad (1.2.18)$$

that is,

$$\exp \left(\frac{i}{\lambda} \int^{z} \sum_{n=0}^{\infty} y_{2n+1} \lambda^{2n+1} dz \right) = \text{const} \times \left(\sum_{n=0}^{\infty} y_{2n} \lambda^{2n} \right)^{-1/2} \qquad (1.2.19)$$

This result made it possible to eliminate the sum of the odd-order terms occurring in the formal solution (1.2.3), which can thus be written

$$\psi = \frac{\exp \left(\frac{i}{\lambda} \int^{z} \sum_{n=0}^{\infty} y_{2n} \lambda^{2n} dz \right)}{\left(\frac{1}{\lambda} \sum_{n=0}^{\infty} y_{2n} \lambda^{2n} \right)^{1/2}} \qquad (1.2.20)$$

apart from a constant factor that may be accounted for by an appropriate choice of the constant lower limit of integration in the exponent of (1.2.20). An exact transformation of the infinite WKB series (1.2.3) thus yielded the expansion (1.2.20). By truncating at $n = N$ the sum of the even-order terms, which occur in the amplitude as well as in the phase of the formal solution (1.2.20), and by recalling the sign ambiguity of the even-order terms $y_{2n} \lambda^{2n}$ that was mentioned to occur between (1.2.4b) and (1.2.5), N. Fröman obtained two approximate solutions of the order $2N + 1$. Thus, she obtained a phase-integral approximation of arbitrary (odd) order that, in contrast to the corresponding higher-order WKB approximation, is of the form (1.2.14) and hence is remedied from the first one of the two deficiencies mentioned previously. For the new phase-integral approximation thus obtained, the connection problems could be solved with the aid of the F-matrix method (Fröman and Fröman, 1965). The derivation of this phase-integral approximation can also be performed in an alternative way, if one abstains from the WKB Ansatz (1.2.3), where everything is placed in the exponent; and instead, in a generalized sense, defines both a phase and an amplitude for the wave function in the complex z-plane (with the appropriate relation between phase and amplitude, as displayed in [1.2.14] and [1.2.20]). For the functions y_{2n} (or, more precisely, for the functions $y_{2n}\lambda^{2n}/y_0$), N. Fröman (1966b) gave a general recurrence formula and derived explicit expressions for the first few orders of the phase-integral approximation. Starting from this recurrence formula, Campbell (1972) used symbolic formula manipulation on a computer to calculate explicit expressions up to the 21st order. The new approximation is the particular case of the phase-integral approximation generated from an unspecified base function (to be described in §1.3), which one obtains by choosing the base function to be equal to $R^{1/2}(z)$.

Although both (1.2.3) and (1.2.20) are exact formal solutions of the differential equation (1.2.2), it is evident that we obtain different approximations on truncating the series in (1.2.3) and the one in (1.2.20). When the series in (1.2.3) is truncated, we obtain the well known WKB approximation of various orders. When the series in (1.2.20) is truncated, we obtain the asymptotic approximation of various orders just described, displaying for every order of approximation the same relation between phase and amplitude as the exact wave function does. The orders of this kind of asymptotic approximation are designated by odd integers in order to retain the correspondence with the orders of the odd-order WKB approximation. Thus, truncating the series in (1.2.20) after the term $y_{2N}\lambda^{2N}$, we obtain an approximation of the order $2N + 1$. The first order of this approximation is identical to the first-order WKB approximation, while the higher orders of this approximation have the same phase as the corresponding higher orders of the WKB approximation, but differ in amplitude from the latter approximation (see [1.2.6a,b,c] and [1.2.20]). Well away from the zeros and singularities of $R(z)$, the two kinds of approximations agree in general very closely in numerical terms, the relative error being of the order λ^{2N+1} for both kinds of approximations of order $2N + 1$. Thus, the accuracy of the higher-order approximation obtained by truncating the series in (1.2.20) and the WKB approximation (1.2.5) of the corresponding odd order can, in general, be expected to be essentially the same. However, the former approximation is to be preferred, since it is free from the first of the two previously mentioned deficiencies.

In the paper in which (1.2.20) was originally derived, N. Fröman (1966b) used the terminology "higher order approximations of the JWKB-type" for the resulting truncated approximations. We now find this terminology inappropriate and unfortunate, as the resulting approximations differ in essential respects from the higher-order JWKB approximations. In a historical survey of ordinary linear differential equations with a large parameter and turning points, McHugh (1971, p. 280) characterized the higher-order approximation introduced by N. Fröman (1966b) in a better way, by saying that the distinctive feature of this higher-order approximation is the symmetry between the integrand of the exponential integral and the factor of the exponential.

Although, in higher orders, the asymptotic approximation obtained from (1.2.20) by truncation has considerable advantages as regards simplicity and suitability for the solution of connection problems, compared to the WKB approximation (1.2.5), it is not flexible enough for the treatment of important and commonly appearing physical problems; for instance, in connection with the radial Schrödinger equation when the effective potential has a first- or second-order pole at the origin. An asymptotic approximation that is also free from the second of the previously mentioned deficiencies was derived by Fröman and Fröman (1974a,b) and will be described in §1.3.

1.3 Phase-Integral Approximation of Arbitrary Order, Generated from an Unspecified Base Function

The derivation of a phase-integral approximation without the deficiencies of the WKB approximation described in §1.2 was originally made in two different ways, which can be characterized as the direct procedure (Fröman and Fröman, 1974a) and the transformation procedure (Fröman and Fröman, 1974b, pp. 126–131).

1.3.1 Direct Procedure

Let us consider the one-dimensional Schrödinger equation, and let us for a moment consider Planck's constant as a "small" expansion parameter; let us solve the Schrödinger equation asymptotically either in the WKB form (1.2.3) or in the phase-integral form (1.2.20), where λ is now Planck's constant. These expansions break down, however, at points where the effective potential has certain singularities; therefore, they can not be used even in the simple and frequent case of a radial Schrödinger equation with a Coulomb singularity or a centrifugal barrier, if one wants a solution that remains valid in the neighborhood of the origin. For this situation there is no appropriate "small" parameter λ in the Schrödinger equation. However, to be able to use the powerful tool of asymptotic expansion we need a "small" parameter λ; hence, we shall introduce such a parameter in a convenient way in the Schrödinger equation. We shall later see that our way of introducing this expansion parameter into the Schrödinger equation is equivalent to the introduction of an *a priori* unspecified *base function* $Q(z)$ which appears in, and can be said to generate, the resulting asymptotic approximation. Our purpose is to obtain a solution that is as accurate as possible for fixed values of the parameters. If there is already in the differential equation a "small", convenient parameter, and if we want to investigate the behavior of the solution as this parameter tends to zero, we can do this simply by choosing the base function appropriately.

After these remarks concerning the "small" parameter, we shall now return to the original differential equation (1.2.1) and derive the arbitrary-order phase-integral approximation, generated from an unspecified base function, by an improved version of one of the direct procedures used in the original paper by Fröman and Fröman (1974a), where this approximation was first given.

As we have explained in some detail in connection with (1.2.14) and in the Appendix of this chapter, it is convenient to represent two solutions of the differential equation (1.2.1); i.e.,

$$\frac{d^2\psi}{dz^2} + R(z)\psi = 0, \tag{1.3.1}$$

in the form of the functions (1.2.14); i.e.,

$$\psi = q^{-1/2}(z) \exp\left(\pm i \int^z q(z)\,dz\right), \tag{1.3.2}$$

which have a constant Wronskian (equal to $-2i$).

Inserting (1.3.2) into the differential equation (1.3.1), we obtain for $q(z)$ the differential equation (1.A.7); i.e.,

$$q^{-3/2}\frac{d^2}{dz^2}q^{-1/2} + R(z)/q^2 - 1 = 0, \tag{1.3.3}$$

which in the present context will be called the original q-equation in order to distinguish it from the auxiliary q-equation (1.3.6) below. Let us now assume that, in some way or another, one has determined a function $Q(z)$, which will be called the *base function*, such that $Q(z)$ is an approximate solution of (1.3.3). We achieve this result by requiring that the quantity ε_0 defined by the left-hand member of (1.3.3) with $q(z)$ replaced by $Q(z)$; i.e.,

$$\varepsilon_0 = Q^{-3/2}(z)\frac{d^2}{dz^2}Q^{-1/2}(z) + R(z)/Q^2(z) - 1, \tag{1.3.4}$$

be small compared to unity. We take this smallness explicitly into account by considering ε_0 to be proportional to λ^2, where λ is a "small" parameter. This is attained if $Q(z)$ is proportional to $1/\lambda$ and $R(z) - Q^2(z)$ is independent of λ; i.e., if $R(z)$ is replaced by $Q^2(z)/\lambda^2 + [R(z) - Q^2(z)]$. Therefore, we consider, instead of the original differential equation (1.3.1), the auxiliary differential equation

$$\frac{d^2\psi}{dz^2} + \left(\frac{Q^2(z)}{\lambda^2} + [R(z) - Q^2(z)]\right)\psi = 0. \qquad (1.3.5)$$

We have thus introduced the "small" parameter λ, which at the end is to be put equal to unity, and which plays only the role of a bookkeeping parameter. By comparing (1.3.1) and (1.3.5) one realizes that the introduction of the "small" parameter λ $(= 1)$ is formally equivalent to the assumption that the function $R(z)$ in (1.3.1) be split into a "larger" part $Q^2(z)$ and a "smaller" part $R(z) - Q^2(z)$. The old question concerning which parameter is actually to be considered as small in the Schrödinger equation can now be answered. There is in the equation from the beginning no quantity that can be used as such a "small" parameter in every situation that occurs. Instead, the parameter has to be introduced in a convenient way for every actual problem. One can often use Planck's constant h, but this cannot be done, for instance, in connection with the radial Schrödinger equation and the well known replacement of $\ell(\ell + 1)$ by $(\ell + 1/2)^2$, introduced when one wants the first-order wave function to be valid at the origin.

Inserting (1.3.2) into the auxiliary differential equation (1.3.5), or, equivalently, replacing in (1.3.3) $R(z)$ by $Q^2(z)/\lambda^2 + R(z) - Q^2(z)$, we obtain for $q(z)$ the differential equation

$$q^{+1/2}\frac{d^2}{dz^2}q^{-1/2} - q^2 + Q^2(z)/\lambda^2 + R(z) - Q^2(z) = 0, \qquad (1.3.6)$$

which will be called the auxiliary q-equation. Introducing instead of z the variable

$$\zeta = \int^z Q(z)\,dz, \qquad (1.3.7)$$

we can, with the aid of the identity

$$q^{+1/2}\frac{d^2}{dz^2}q^{-1/2} = Q^2\left[\left(\frac{q}{Q}\right)^{+1/2}\frac{d^2}{d\zeta^2}\left(\frac{q}{Q}\right)^{-1/2}\right.$$

$$\left. + Q^{-3/2}\frac{d^2}{dz^2}Q^{-1/2}\right], \qquad (1.3.8)$$

write (1.3.6) in the form

$$1 - \left(\frac{q\lambda}{Q(z)}\right)^2 + \varepsilon_0\lambda^2 + \left(\frac{q\lambda}{Q(z)}\right)^{+1/2}\frac{d^2}{d\zeta^2}\left(\frac{q\lambda}{Q(z)}\right)^{-1/2}\lambda^2 = 0, \qquad (1.3.9)$$

where ε_0 is defined by (1.3.4). It is convenient to rewrite (1.3.9) in the form

$$\left(\frac{q\lambda}{Q(z)}\right)^2 - \left(\frac{q\lambda}{Q(z)}\right)^4 + \varepsilon_0 \left(\frac{q\lambda}{Q(z)}\right)^2 \lambda^2 + \frac{3}{4}\left[\frac{d}{d\zeta}\left(\frac{q\lambda}{Q(z)}\right)\right]^2 \lambda^2$$

$$-\frac{1}{2}\frac{q\lambda}{Q(z)}\frac{d^2}{d\zeta^2}\left(\frac{q\lambda}{Q(z)}\right)\lambda^2 = 0. \tag{1.3.10}$$

To obtain a formal solution of (1.3.10) we set

$$q = \frac{1}{\lambda}Q(z)\sum_{n=0}^{\infty} Y_{2n}\lambda^{2n}, \tag{1.3.11}$$

where Y_0 is assumed to be different from zero, and Y_{2n} $(n = 0,1,2,\ldots)$ is considered to be independent of λ. Inserting the expansion (1.3.11) into (1.3.10), and setting the coefficients of successive powers of λ equal to zero, we get $Y_0 = \pm 1$ and the recurrence formula

$$\sum_{\alpha+\beta=n} Y_{2\alpha}Y_{2\beta} - \sum_{\alpha+\beta+\gamma+\delta=n} Y_{2\alpha}Y_{2\beta}Y_{2\gamma}Y_{2\delta}$$

$$+ \sum_{\alpha+\beta=n-1}\left[\varepsilon_0 Y_{2\alpha}Y_{2\beta} + \frac{3}{4}\frac{dY_{2\alpha}}{d\zeta}\frac{dY_{2\beta}}{d\zeta}\right.$$

$$\left.-\frac{1}{4}\left(Y_{2\alpha}\frac{d^2Y_{2\beta}}{d\zeta^2} + \frac{d^2Y_{2\alpha}}{d\zeta^2}Y_{2\beta}\right)\right] = 0, \quad n \geq 1, \tag{1.3.12}$$

where the summation indices α, β, γ and δ are nonnegative integers. As we have already introduced the double sign $\pm$ in the exponent of (1.3.2), it does not restrict us to choose

$$Y_0 = +1. \tag{1.3.13a}$$

Noting that in (1.3.12) the terms containing Y_{2n} occur only in the first two sums, where they yield $2Y_0Y_{2n}$ and $4Y_0^3Y_{2n}$, respectively, we can rewrite (1.3.12) in such a form that Y_{2n} (for $n \geq 1$) is expressed explicitly in terms of quantities $Y_{2\nu}$ with indices $\nu \leq n-1$. Recalling (1.3.13a), we thus obtain the recurrence formula

$$Y_{2n} = \frac{1}{2}\sum_{\substack{\alpha+\beta=n-1\\0\leq\alpha,\beta\leq n-1}}\left[\varepsilon_0 Y_{2\alpha}Y_{2\beta} + \frac{3}{4}\frac{dY_{2\alpha}}{d\zeta}\frac{dY_{2\beta}}{d\zeta}\right.$$

$$\left.-\frac{1}{4}\left(Y_{2\alpha}\frac{d^2Y_{2\beta}}{d\zeta^2} + \frac{d^2Y_{2\alpha}}{d\zeta^2}Y_{2\beta}\right)\right] + \frac{1}{2}\sum_{\substack{\alpha+\beta=n\\0\leq\alpha,\beta\leq n-1}} Y_{2\alpha}Y_{2\beta}$$

$$-\frac{1}{2} \sum_{\substack{\alpha+\beta+\gamma+\delta=n \\ 0\leq\alpha,\beta,\gamma,\delta\leq n-1}} Y_{2\alpha}Y_{2\beta}Y_{2\gamma}Y_{2\delta}, \quad n \geq 1. \qquad (1.3.13b)$$

Starting with (1.3.13a), i.e., $Y_0 = 1$, and using the recurrence formula (1.3.13b), one can successively obtain the functions $Y_2, Y_4, Y_6, \ldots$. It is easily seen that every function Y_{2n} $(n > 0)$ becomes expressed purely in terms of ε_0, defined by (1.3.4), and derivatives of ε_0 with respect to ζ; i.e., in terms of the quantities

$$\varepsilon_\nu = d^\nu \varepsilon_0 / d\zeta^\nu, \quad 0 \leq \nu \leq 2n - 2 \quad (\text{excluding } \nu = 2n - 3), \qquad (1.3.14)$$

with ζ defined by (1.3.7). The first few functions Y_{2n} are

$$Y_0 = 1 \qquad (1.3.15a)$$

$$Y_2 = \frac{1}{2}\varepsilon_0 \qquad (1.3.15b)$$

$$Y_4 = -\frac{1}{8}(\varepsilon_0^2 + \varepsilon_2). \qquad (1.3.15c)$$

The expressions for Y_{2n} were originally obtained up to Y_4 by N. Fröman (1966b), up to Y_8 by Karlsson and Yngve, quoted in Fröman and Fröman (1970), and in a slightly different form by Yngve (1971), and up to Y_{20} by Campbell (1972), who used a symbolic (algebraic) computing system; since evaluation of Y_{2n} by hand quickly becomes rather complicated as n increases. These expressions indicate that for $2n \geq 2$ the quantity ε_ν with largest index ν that appears in Y_{2n} is ε_{2n-2}, which means that the highest derivatives of $R - Q^2$ and Q^2 appearing in Y_{2n} are $d^{2n-2}(R - Q^2)/dz^{2n-2}$ and $d^{2n}Q^2/dz^{2n}$, respectively.

The choice of the unspecified base function $Q(z)$ shows itself only in the expressions (1.3.4) for ε_0 and (1.3.7) for ζ as functions of $R(z)$ and $Q(z)$, while the expressions for the functions Y_{2n} in terms of ε_0 and derivatives of ε_0 with respect to ζ do not depend explicitly on $R(z)$ and the choice of the base function $Q(z)$. Therefore, the expressions for Y_{2n} can be determined once and for all. We would also like to remark that at the zeros and poles of $Q^2(z)$ the functions $Q(z)$ and $Q^{1/2}(z)$ may have branch-points, whereas the functions ε_0, Y_{2n} and $q(z)/Q(z)$ are all single-valued functions of z, when $R(z)$ and $Q^2(z)$ are single-valued. A further number of general properties of Y_{2n} were pointed out by Skorupski (1980, 1988a).

It is often of great practical importance to note that one can write

$$Y_{2n} = Z_{2n} + \frac{d}{d\zeta}U_{2n}, \qquad (1.3.16)$$

where this splitting is uniquely determined by the requirement that in every term in the expression for Z_{2n} the indices ν of the quantities ε_ν be as small as possible. Thus, from (3.15a–c) one finds that

$$Z_0 = 1 \qquad (1.3.17a)$$

$$Z_2 = \frac{1}{2}\varepsilon_0 \tag{1.3.17b}$$

$$Z_4 = -\frac{1}{8}\varepsilon_0^2 \tag{1.3.17c}$$

and

$$U_0 = 0 \tag{1.3.18a}$$

$$U_2 = 0 \tag{1.3.18b}$$

$$U_4 = -\frac{1}{8}\varepsilon_1. \tag{1.3.18c}$$

The expressions for Z_{2n} up to Z_8 were originally obtained by P.O. Fröman (1974), and the expressions for both Z_{2n} and U_{2n} up to $2n = 16$ were obtained by Campbell (1979) by means of a symbolic (algebraic) computing system. These expressions indicate that for $2n \geq 4$ the quantity ε_ν with the largest index ν is ε_{n-2} in Z_{2n} and ε_{2n-3} in U_{2n}.

Formal solutions of the auxiliary differential equation (1.3.5) are obtained from (1.3.2) with $q(z)$ obtained from (1.3.11), where Y_{2n} are given by (1.3.15a,b,c,...), (1.3.14), (1.3.4) and (1.3.7). The corresponding approximate solutions (1.3.2) of the original differential equation (1.3.1), obtained by truncating the infinite series in (1.3.11) with $\lambda = 1$ as

$$q(z) = Q(z) \sum_{n=0}^{N} Y_{2n}, \tag{1.3.19}$$

will be called the *phase-integral functions* of the order $2N+1$ generated from the unspecified base function $Q(z)$. One obtains the corresponding phase-integral approximation of the order $2N + 1$ by forming linear combinations of these phase-integral functions. It is often natural to choose the square of the base function (i.e., $Q^2(z)$) to be equal to the function $R(z)$ in the original differential equation (1.3.1), but in many cases it is preferable to choose $Q^2(z)$ to differ from $R(z)$.

During attempts in the sixties to remove the breakdown of the WKB approximation at the origin for the radial Schrödinger equation, the terminology "modification of the WKB approximation" was often used. At that time our minds were dominated by the idea of modifying already existing approximations. Therefore, when the arbitrary-order phase-integral approximation generated from an unspecified base function was originally derived (Fröman and Fröman, 1974a,b), and also subsequently, the terminology "modified phase-integral approximation" was used to distinguish the resulting phase-integral approximation, when $Q^2(z)$ differs from $R(z)$, from the more special phase-integral approximation, corresponding to $Q^2(z) = R(z)$, which was originally derived by N. Fröman (1966b). We now find this terminology unfortunate, since (once one has actually derived the general approximation and become accustomed to its use) it is natural to consider any reasonable choice of the base function $Q(z)$ to be on the same footing with any other such choice. From the present point of view it is therefore quite unnatural to call a phase-integral approximation with $Q^2(z)$ chosen to be different from $R(z)$ "modified phase-integral approximation"; the choice of $Q^2(z)$ should not affect the name of the approximation. Therefore, the names "unmodified approximation" and "modified approximation" should be avoided.

It should also be remarked that in the early papers concerning the arbitrary-order phase-integral approximation generated from an unspecified base function, the function $R(z)$ in (1.3.1) was denoted by $Q^2(z)$, whereas the base function was denoted by $Q_{\mathrm{mod}}(z)$. From the present point of view, these old notations are not convenient and have now been abandoned.

The zeros and poles of the function $Q^2(z)$ are in our terminology called *transition zeros* and *transitions poles*, respectively; we can refer to both of these as *transition points*. These points are of crucial importance in the analysis of connection problems. With $Q^2(z) = R(z)$, the transition points lying on the real z-axis have a well defined physical meaning, e.g., classical turning points in quantum mechanics, and cutoff points or resonance points in the theory of electromagnetic wave propagation in inhomogeneous media. Although the phase-integral approximation may break down when z approaches such points, for computational purposes it is important to know how the phase-integral quantities behave in the neighborhood of these points, as one often needs this knowledge in order to be able to choose convenient paths of integration in various physical applications involving the phase-integral approximation of higher order.

Detailed studies of the behavior of the phase-integral approximation in the neighborhood of zeros and singularities of $Q^2(z)$ were made by Skorupski (1980, 1988a). He also utilized the phase-integral approximation (in the asymptotic region) in connection with numerical integration of the wave equation (Skorupski, 1988b).

A very useful approximate formula for Y_{2n} that is valid in the region around a first-order pole or an arbitrary-order zero of $Q^2(z)$, was obtained by Skorupski (1980, 1988a) for the choice $Q^2(z) = R(z)$; in addition the formula is also valid for reasonable choices of $Q^2(z)$ differing from $R(z)$. Skorupski found that in the region under consideration, the term in the expression for Y_{2n}, $2n \geq 2$, that contains ε_ν with the largest index ν, dominates. He thus obtained the approximate formula

$$Y_{2n} = \frac{(-1)^{n-1}}{2^{2n-1}} \varepsilon_{2n-2}, \qquad 2n \geq 2, \tag{1.3.20}$$

which is sufficiently accurate in qualitative considerations.

1.3.2   Transformation Procedure

We have in §1.3.1 derived the phase-integral approximation generated from an unspecified base function by a direct procedure. Another way of obtaining the same result is to use a transformation procedure (Fröman and Fröman, 1974b) which will now be described briefly. We start by performing the most general transformation that preserves the reduced form of the differential equation (1.3.1), which means that there appears no first derivative in the transformed differential equation. Thus, we introduce a new independent variable

$$\tilde{z} = \tilde{z}(z), \tag{1.3.21}$$

where $\tilde{z}(z)$ is an unspecified function of z, and a new wave function $\tilde{\psi}$ defined by

$$\psi = \left(\frac{d\tilde{z}(z)}{dz}\right)^{-1/2} \tilde{\psi}. \tag{1.3.22}$$

Inserting (1.3.22) into (1.3.1), we get

$$\frac{d^2\tilde{\psi}}{d\tilde{z}^2} + \tilde{R}(\tilde{z})\tilde{\psi} = 0 \tag{1.3.23}$$

where

$$\tilde{R}(\tilde{z}) = \frac{R(z) + \left(\frac{d\tilde{z}(z)}{dz}\right)^{+1/2} \frac{d^2}{d\tilde{z}^2} \left(\frac{d\tilde{z}(z)}{dz}\right)^{-1/2}}{\left(\frac{d\tilde{z}(z)}{dz}\right)^2}. \tag{1.3.24}$$

We then solve the transformed differential equation (1.3.23) by the procedure described in §1.2.3. In the phase-integral solution thus obtained, one transforms back to the original variables z and ψ and obtains a formal solution of the differential equation (1.3.1). In this solution the unspecified transformation function $\tilde{z}(z)$ reflects itself in the appearance of the unspecified base function $Q(z)$, the square of which is related to $\tilde{z}(z)$ according to the formula

$$Q^2(z) = R(z) + \left(\frac{d\tilde{z}(z)}{dz}\right)^{+1/2} \frac{d^2}{dz^2} \left(\frac{d\tilde{z}(z)}{dz}\right)^{-1/2}. \tag{1.3.25}$$

The use of the transformation procedure can be justified when the asymptotic solution of the original differential equation obtained in §1.2.3 is unsatisfactory, while it is possible to find a transformation function $\tilde{z}(z)$ such that the corresponding asymptotic solution of the transformed differential equation is free from objection. Using the transformation procedure one finds that the resulting solution is of such general form that it contains the a priori unspecified base function $Q(z)$. In practice it may be very difficult, if not impossible, to find a convenient transformation function $\tilde{z}(z)$; fortunately, this is not necessary, since besides $R(z)$ the only function related to

$\tilde{z}(z)$ that appears in the *final* formulas for the phase-integral approximation is the base function $Q(z)$, which one can choose conveniently without
knowing explicitly a corresponding transformation function $\tilde{z}(z)$. Therefore, a convenient function $Q(z)$ can be obtained directly from a criterion
on $Q(z)$, and it is not necessary to obtain an explicit expression for the
transformation function $\tilde{z}(z)$. Thus, the final formulas for the arbitrary-
order phase-integral approximation in question are also applicable to situations in which a useful, generalized Langer transformation $\tilde{z}(z)$ cannot
in practice be found in explicit form, whereas one may directly be able
to find a useful expression for $Q(z)$, and then make direct use of general,
final formulas, which have been worked out once and for all. It is thus seen
that, if one elaborates sufficiently well the generalization of the Langer
transformation procedure (Fröman and Fröman, 1974b); *the transformation function $\tilde{z}(z)$ appears only in intermediate steps, whereas in the final
formulas only the function $Q(z)$ appears, beside $R(z)$.* The final formulas
for the arbitrary-order phase-integral approximation, obtained by the use
of a generalization and detailed elaboration of the transformation procedure used by Langer (1937), are identical to those obtained by the quite
different and conceptually much simpler direct approach in §1.3.1, which
leads to the same result without transformations back and forth. Both the
direct procedure and the transformation procedure give consistent derivations of the arbitrary-order phase-integral approximation, containing the *a
priori* unspecified base function $Q(z)$, which can be chosen conveniently so
as to make the approximation as satisfactory and as easily manageable as
possible. The direct procedure has the advantage of much greater concep-
tual simplicity; however, it is important to realize that the transformation
procedure is equivalent to the use of the direct procedure.

1.4   Advantage of Phase-Integral Approximation Versus WKB Approximation in Higher Order

One obtains the WKB approximation by writing the solution ψ of the
differential equation (1.3.1) as an exponential function with the exponent
as an indefinite integral of an unknown function which is then expanded in
a well known way, as described in the beginning of §1.2.

The phase-integral approximation derived in §1.3 is instead obtained by
first formally introducing in the differential equation an unspecified base
function $Q(z)$ and a "small" bookkeeping parameter λ ($= 1$), then writing
ψ according to (1.3.2) as the product of an exponential factor (with the
phase in the form of an integral) and an amplitude factor related in a simple
way to the phase, the derivative of which is finally determined by means of
series expansion in terms of λ. As in the discussion below (1.2.14) in §1.2,
the concepts "phase" and "amplitude" are used here in a generalized sense;

i.e., even when the quantities in question are complex functions and hence do not physically represent the phase and amplitude of a wave.

The phase-integral approximation derived in §1.3 does not suffer from the first of the deficiencies of the WKB approximation discussed in §1.2; i.e., the lack of the simple and correct relation between phase and amplitude, which is seen when one compares (1.2.6b,c) with (1.2.14). The F-matrix method (Fröman and Fröman, 1965) can therefore be used for handling the phase-integral approximation. This method is rigorous and permits an exhaustive treatment of the connection problems appearing in the solution of physical problems; the use of an arbitrary order of the phase-integral approximation does not introduce any essential complication in the F-matrix method. We do not enter here upon the difficulties and complications that would in general appear if one tried to use the WKB approximation in higher order, instead of the phase-integral approximation of corresponding higher order, for solving various physical problems. We would only like to remark that Dammert and P.O. Fröman (1980) have made a concrete and detailed analysis of a calculation performed by means of the WKB approximation in higher order, for the purpose of making comparisons that are illuminating at some crucial points with a treatment performed by means of the phase-integral approximation of corresponding higher order. This analysis clearly demonstrates the great simplification and advantage one achieves when one treats connection problems by using the phase-integral approximation instead of the WKB approximation in higher order. Further important consequences of the form of the phase-integral approximation, displaying the crucial feature of a purely local amplitude and an integral occurring only in the phase, have been noted by Fulling, who also generalized the phase-integral approach to systems of differential equations and pointed out that the phase-integral expansion is a better approximation than the WKB expansion and formally more appropriate for the application in quantum field theory (Fulling, 1979, 1981, 1982, 1983a,b). The necessity of making a clear distinction between the two kinds of approximations is thus clear.

The phase-integral approximation derived in §1.3 is also freed from the second deficiency of the WKB approximation in higher order discussed in §1.2, i.e., its lack of flexibility. It is an essential advantage of the phase-integral approximation of arbitrary order versus the WKB approximation in higher order that the former approximation contains the unspecified base function $Q(z)$, which one can take advantage of in several ways. A criterion for the determination of the base function is that the function ε_0, defined by (1.3.4), be in some sense "small" in the region of the complex z-plane relevant for the problem under consideration. However, this criterion does not determine the base function $Q(z)$ uniquely; it turns out that, within certain limits, the results are not very sensitive to the choice of $Q(z)$, when the approximation is used in higher orders. An inconvenient, but possible, choice of $Q(z)$ introduces in the first-order approximation an unnecessarily large error that is, however, in general already corrected in the third-order approximation. The function $Q^2(z)$ can be chosen in many important cases to be identical to $R(z)$, and then the phase-integral approximation generated from an unspecified base function, described in §1.3.1, reduces to the more special approximation obtained by N. Fröman (1966b). In other important cases; for instance, when one wants to include the immediate neighborhood of a first- or second-order pole of $R(z)$ in the region of validity of the phase-integral approximation; the function $Q^2(z)$ is in general chosen to be similar to $R(z)$ except in the neighborhood of the pole (see [1.4.1] to [1.4.4]). However, it may sometimes be convenient to choose $Q^2(z)$ to be quite different from $R(z)$ throughout the whole relevant region of the complex z-plane. Such a case, concerning Mathieu functions, has been treated by N. Fröman (1979), Eqs. (14) and (58). By choosing the base function appropriately she obtained very accurate results for parameter values such that the choice of $Q^2(z)$ equal to $R(z)$ would not work at all.

The freedom one has in the choice of the base function $Q(z)$ will now be illuminated in a more concrete way. For a radial Schrödinger equation the usual choice of $Q^2(z)$ is

$$Q^2(z) = R(z) - \frac{1}{4z^2}. \tag{1.4.1}$$

However, the replacement of (1.4.1) by

$$Q^2(z) = R(z) - \frac{1}{4z^2} - \frac{\text{const}}{z}, \tag{1.4.2}$$

where the coefficient of $1/z$ should be comparatively small, does not destroy the great accuracy of the results usually obtained with the phase-integral approximation in higher orders. There is thus a whole set of base functions that may be used; there are various ways in which one can take advantage of this nonuniqueness to make the choice of the base function well adapted to the particular physical problem under consideration. For instance, by adapting the choice of $Q^2(z)$ to the analytical form of $R(z)$ one can sometime achieve the result that the integrals occurring in the phase-integral approximation can be evaluated analytically. To give an example, we assume that $R(z)$ contains only $\exp(z)$ but not z itself. In this case it is convenient to replace the choice (1.4.2) by the choice

$$Q^2(z) = R(z) - \frac{1}{4(e^z - 1)^2} - \frac{\text{const}}{e^z - 1}. \tag{1.4.3}$$

To give another example we assume that $R(z)$ has poles at $z = 0$ and $z = \pi$, and that the analytical expression for $R(z)$ contains only $\text{ctg}z$, but not z itself. In this case it is convenient to choose

$$Q^2(z) = R(z) - \frac{1}{4}\text{ctg}^2z - \text{const} \times \text{ctg}z. \tag{1.4.4}$$

By a convenient choice of $Q^2(z)$; for instance, by a convenient choice of the unspecified coefficient in (1.4.2), (1.4.3) or (1.4.4); one can sometimes attain the result that, for example, eigenvalues or phase shifts are obtained exactly for some particular parameter value, in every order of the phase-integral approximation. By making this exactness fulfilled in the limit of a parameter value, for which the phase-integral result without this adaptation would deteriorate, one can actually extend the region of validity of the phase-integral treatment. When the differential equation contains one or more parameters, the accurate calculation of the wave function may require different choices of the base function $Q(z)$ for different ranges of the parameters. For instance, the accurate evaluation of Coulomb wave functions requires a different choice of $Q(z)$ when the angular momentum quantum number is large than when it is small. To illustrate this fact we consider the Coulomb wave functions $F_\ell(\eta, z)$ and $G_\ell(\eta, z)$, the differential equation of which is given by (1.3.1) with

$$R(z) = 1 - \frac{2\eta}{z} - \frac{\ell(\ell+1)}{z^2}. \qquad (1.4.5)$$

For sufficiently large absolute values of both ℓ and η we obtain an accurate phase-integral approximation (valid also close to $z = 0$) if we choose

$$Q^2(z) = 1 - \frac{2\eta}{z} - \frac{(\ell+1/2)^2}{z^2}. \qquad (1.4.6)$$

If the absolute value of ℓ is too small, then the corresponding phase-integral approximation is not very good. It can be considerably improved (except close to $z = 0$) if instead one chooses

$$Q^2(z) = 1 - \frac{2\eta}{z}, \qquad (1.4.7)$$

where the absolute value of η is still assumed to be sufficiently large.

1.5   Relations Between Solutions of the Schrödinger Equation and the q-Equation

To get a qualitative understanding of the behavior of the exact solutions of the q-equation (1.3.3), it is often very useful to express $q(z)$ in terms of two linearly independent solutions of the Schrödinger equation (1.3.1). Such an insight into the properties of the possible functions $q(z)$ is indispensable, for instance, in connection with numerical integration of the q-equation (1.3.3); since a judicious choice of starting values, as well as of paths of integration, is of crucial importance for the efficiency of the integration procedure.

In this section we shall give a number of relationships that demonstrate the connection between solutions of the q-equation and the Schrödinger equation.

1.5.1 Solutions of the Schrödinger Equation and Solutions of the q-Equation Expressed in Terms of Each Other

Still refraining from imposing any reality condition on $R(z)$, we will consider once again the Schrödinger equation (1.3.1). Let $\psi_1(z)$ and $\psi_2(z)$ be two unspecified, given, linearly independent solutions. Their Wronskian

$$W = \psi_1(z)d\psi_2(z)/dz - \psi_2(z)d\psi_1(z)/dz \qquad (1.5.1)$$

is a constant different from zero. Considering two so far unspecified linear combinations $\bar{\psi}_1(z)$ and $\bar{\psi}_2(z)$ of $\psi_1(z)$ and $\psi_2(z)$:

$$\bar{\psi}_1(z) = A_1\psi_1(z) + A_2\psi_2(z) \qquad (1.5.2a)$$

$$\bar{\psi}_2(z) = B_1\psi_1(z) + B_2\psi_2(z), \qquad (1.5.2b)$$

we try to write these linear combinations as

$$\bar{\psi}_1(z) = q^{-1/2}(z) \exp\left(+i\int_{z_0}^{z} q(z)dz\right) \qquad (1.5.3a)$$

$$\bar{\psi}_2(z) = q^{-1/2}(z) \exp\left(-i\int_{z_0}^{z} q(z)dz\right), \qquad (1.5.3b)$$

where z_0 is a given, unspecified constant, and $q(z)$ is an unknown function. For the Wronskian of the functions $\bar{\psi}_1(z)$ and $\bar{\psi}_2(z)$ one obtains from (1.5.2a,b)

$$\bar{\psi}_1(z)d\bar{\psi}_2(z)/dz - \bar{\psi}_2(z)d\bar{\psi}_1(z)/dz = W(A_1B_2 - A_2B_1) \qquad (1.5.4a)$$

and from (1.5.3a,b)

$$\bar{\psi}_1(z)d\bar{\psi}_2(z)/dz - \bar{\psi}_2(z)d\bar{\psi}_1(z)/dz = -2i. \qquad (1.5.4b)$$

Hence

$$A_1B_2 - A_2B_1 = -2i/W. \qquad (1.5.5)$$

The expression in the left-hand member of (1.5.5) is thus different from zero, which is in fact the condition necessary and sufficient for the functions $\bar{\psi}_1(z)$ and $\bar{\psi}_2(z)$ in (1.5.2a,b) to be linearly independent. Since $\bar{\psi}_1(z_0) = \bar{\psi}_2(z_0)$ according to (1.5.3a,b), we obtain, with the aid of (1.5.2a,b),

$$A_1\psi_1(z_0) + A_2\psi_2(z_0) = B_1\psi_1(z_0) + B_2\psi_2(z_0). \qquad (1.5.6)$$

In order that it be possible to represent the linear combinations (1.5.2a,b) according to (1.5.3a,b), the constants A_1, A_2, B_1 and B_2 must thus fulfill the conditions (1.5.5) and (1.5.6). From (1.5.3a,b) we obtain

$$\bar{\psi}_1(z)\bar{\psi}_2(z) = \frac{1}{q(z)}, \qquad (1.5.7a)$$

$$\frac{1}{2i} \ln \frac{\bar{\psi}_1(z)}{\bar{\psi}_2(z)} = \int_{z_0}^{z} q(z)dz. \tag{1.5.7b}$$

With the aid of (1.5.4b) and (1.5.7a) we obtain

$$\frac{d}{dz}\left(\frac{1}{2i} \ln \frac{\bar{\psi}_1(z)}{\bar{\psi}_2(z)}\right) = \frac{1}{\bar{\psi}_1(z)\bar{\psi}_2(z)} = q(z), \tag{1.5.8}$$

and therefore (1.5.7b) is a consequence of (1.5.4b) and (1.5.6), and hence of (1.5.3a,b). Inserting (1.5.2a,b) into (1.5.7a), we get

$$q(z) = \frac{1}{[A_1\psi_1(z) + A_2\psi_2(z)][B_1\psi_1(z) + B_2\psi_2(z)]}. \tag{1.5.9}$$

As a consequence of the assumption that $\psi_1(z)$ and $\psi_2(z)$ are solutions of the Schrödinger equation (1.3.1), one can directly verify that the expression (1.5.9) for $q(z)$ satisfies the q-equation (1.3.3), provided that the condition (1.5.5) is fulfilled. For any two given linearly independent solutions $\psi_1(z)$ and $\psi_2(z)$ of the Schrödinger equation (1.3.1), we have thus in a unique way determined the solution (1.5.9) of the q-equation (1.3.3) and the conditions (1.5.5) and (1.5.6) on the coefficient A_1, A_2, B_1 and B_2 in (1.5.2a,b) that must be fulfilled in order that it be possible to write the linear combinations (1.5.2a,b) according to (1.5.3a,b).

Let us continue to assume that $\psi_1(z)$ and $\psi_2(z)$ are any two given linearly independent solutions (with the Wronskian W) of the Schrödinger equation (1.3.1), but now let $q(z)$ be any given solution of the q-equation (1.3.3). The two functions $\bar{\psi}_1(z)$ and $\bar{\psi}_2(z)$ defined by (1.5.3a,b), where z_0 is a so far unspecified constant, are then also linearly independent solutions of the Schrödinger equation (1.3.1). This implies that there must exist constants A_1, A_2, B_1 and B_2 such that $\bar{\psi}_1(z)$ and $\bar{\psi}_2(z)$ can be expressed in terms of $\psi_1(z)$ and $\psi_2(z)$ according to (1.5.2a,b). The results derived in the present section from (1.5.1), (1.5.2a,b), and (1.5.3a,b) still apply, and therefore we see that $q(z)$ must be given by (1.5.9), where the constants A_1, A_2, B_1 and B_2 must fulfill the condition (1.5.5). There is no further condition on these constants, if we consider (1.5.6) to be an equation for the determination of z_0. Defining

$$\alpha = A_1 B_1 \tag{1.5.10a}$$

$$\beta = A_2 B_2 \tag{1.5.10b}$$

$$\gamma = A_1 B_2 + A_2 B_1 \tag{1.5.10c}$$

where, according to (1.5.5),

$$\alpha\beta - \gamma^2/4 = 1/W^2, \tag{1.5.11}$$

we can write (1.5.9) as

$$1/q(z) = \alpha[\psi_1(z)]^2 + \beta[\psi_2(z)]^2 + \gamma\psi_1(z)\psi_2(z). \tag{1.5.12}$$

Since $\psi_1(z)$ and $\psi_2(z)$ are assumed to be solutions of the Schrödinger equation (1.3.1), one obtains in a straightforward way the formula

$$\left(\alpha[\psi_1(z)]^2 + \beta[\psi_2(z)]^2 + \gamma\psi_1(z)\psi_2(z)\right)^{-1/2} \frac{d^2}{dz^2}\left(\alpha[\psi_1(z)]^2 + \beta[\psi_2(z)]^2\right.$$

$$\left. + \gamma\psi_1(z)\psi_2(z)\right)^{+1/2} = W^2(\alpha\beta - \gamma^2/4)\left(\alpha[\psi_1(z)]^2\right.$$

$$\left. + \beta[\psi_2(z)]^2 + \gamma\psi_1(z)\psi_2(z)\right)^{-2} - R(z) \qquad (1.5.13)$$

by means of which one can verify that the function $q(z)$ given by (1.5.12) is a solution of the q-equation (1.3.3), provided that the condition (1.5.11) is fulfilled. Given two unspecified linearly independent solutions $\psi_1(z)$ and $\psi_2(z)$ of the Schrödinger equation (1.3.1), we have thus shown that any given solution $q(z)$ of the q-equation (1.3.3) can be written in the form (1.5.12) with uniquely determined constants α, β, and γ that fulfill the condition (1.5.11); this result can be found in a paper by Eliezer and Gray (1976). The general solution $q(z)$ of the q-equation (1.3.3) can thus be expressed in terms of $\psi_1(z)$ and $\psi_2(z)$ according to the formula

$$1/q(z) = \alpha[\psi_1(z)]^2 + \beta[\psi_2(z)]^2 \pm 2(\alpha\beta - 1/W^2)^{1/2}\psi_1(z)\psi_2(z), \quad (1.5.14)$$

where α and β are arbitrary constants.

1.5.2   Ermakov–Lewis Invariant

If we now allow $\psi(z)$ to be any given solution of the Schrödinger equation (1.3.1) and $q(z)$ to be any solution of the q-equation (1.3.3), we can define the Ermakov–Lewis invariant I (introduced first by Ermakov [1880], pp. 22–23, and later, independently, by Lewis [1967]) as follows:

$$I = \frac{1}{2}\left[\left(q^{-1/2}(z)\frac{d\psi(z)}{dz} - \psi(z)\frac{dq^{-1/2}(z)}{dz}\right)^2 + \left(\frac{\psi(z)}{q^{-1/2}(z)}\right)^2\right] \qquad (1.5.15)$$

$$= \frac{1}{2}q(z)\left[\left(q^{-1}(z)\frac{d\psi(z)}{dz} - \frac{1}{2}\psi(z)\frac{dq^{-1}(z)}{dz}\right)^2 + [\psi(z)]^2\right]. \qquad (1.5.15')$$

As a direct consequence of the Schrödinger equation (1.3.1) and the q-equation (1.3.3), it follows from the definition (1.5.15) that $dI/dz = 0$; i.e., that I is independent of z. If, as previously, $\psi_1(z)$ and $\psi_2(z)$ are two given linearly independent solutions of the Schrödinger equation (1.3.1), with the Wronskian denoted by W according to (1.5.1), we can write

$$\psi(z) = D_1\psi_1(z) + D_2\psi_2(z), \qquad (1.5.16)$$

where D_1 and D_2 are constants uniquely determined by $\psi(z)$, and $1/q(z)$ is given by (1.5.12), where α, β, and γ are constants uniquely determined by $q(z)$ which fulfill the relation (1.5.11). To calculate the invariant I we insert (1.5.12) and (1.5.16) into (1.5.15'), getting

$$I = \frac{1}{2}W^2(\alpha D_2^2 + \beta D_1^2 - \gamma D_1 D_2)$$

$$- \frac{1}{2}W^2(\alpha\beta - \gamma^2/4 - 1/W^2)\frac{[D_1\psi_1(z) + D_2\psi_2(z)]^2}{\alpha[\psi_1(z)]^2 + \beta[\psi_2(z)]^2 + \gamma\psi_1(z)\psi_2(z)}.$$
$$(1.5.17)$$

Inserting (1.5.11) into (1.5.17), we obtain for the Ermakov–Lewis invariant the expression

$$I = \frac{1}{2}W^2(\alpha D_2^2 + \beta D_1^2 - \gamma D_1 D_2), \tag{1.5.18}$$

where α, β, and γ are still related to W according to (1.5.11). A corresponding formula has been given by Floyd (1986).

1.6   Phase-Integral Method

In this section we shall briefly mention a few items that, beside the phase-integral approximation described in detail above and the F-matrix method mentioned in passing, are essential parts of a *phase-integral method*, which has proved to be very efficient and accurate in a great variety of important applications in various fields of theoretical physics, as well as in applied mathematics. The most essential feature of this method is the combination of the arbitrary-order phase-integral approximation generated from an unspecified base function (derived in §1.3 of the present chapter) with the F-matrix technique for solving connection problems (Fröman and Fröman, 1965). If all transition points relevant for the problem under consideration are well isolated, then the F-matrix technique is always sufficient for mastering the global problem; the Stokes constants, i.e., certain F-matrix elements, are then independent of the order of the approximation used. When transition points lie close together, this is in general no longer the case, and if formulas for F-matrix elements are needed, results obtainable by means of the comparison equation technique, described in Chapter 2 of the present book, are to be included in the phase-integral method. For an example of such supplementary quantities, which are of decisive importance when transition points (associated with a potential barrier) approach each other, we refer to a paper concerning the energy levels of a double oscillator (Fröman et al., 1972). Another essential constituent of the phase-integral method is the derivation of formulas, not involving wave functions, for a number of quantities (such as normalization factors, probability densities at the origin, expectation values, matrix elements, and Franck–Condon factors) that, by definition, involve integration of wave functions. In these cases, derivatives of an integral of $q(z)$, with respect to parameters appearing in $q(z)$, enter in intermediate steps in the derivation of the formulas.

Such a parameter, which we shall denote by κ, may be a parameter appearing from the beginning in $q(z)$, as for instance the energy E, or it may be an auxiliary parameter that one introduces into $q(z)$. By partial differentiation with respect to κ of the explicit expression for $q(z)$ in the right-hand member of (1.3.11), one obtains a complicated expression that, when integrated along a closed contour encircling two transition zeros, was brought to a simple form by P.O. Fröman (1974). On the other hand, partial differentiation of the q-equation (1.3.3) with respect to κ yields a complicated differential equation which cannot easily be solved for $\partial q / \partial \kappa$. After some manipulation one can, however, write the differential equation in the form (N. Fröman, 1974)

$$\frac{\partial q}{\partial \kappa} = \frac{1}{2q} \frac{\partial R}{\partial \kappa} - \frac{\partial}{\partial z} \left[\frac{1}{2q} \frac{\partial}{\partial z} \left(\frac{1}{2q} \frac{\partial q}{\partial \kappa} \right) \right]. \tag{1.6.1}$$

This formula is satisfied formally exactly by the series expansion (1.3.19) for $q(z)$ with $N = \infty$ (i.e., the infinite expansion (1.3.11) with $\lambda = 1$), and it is satisfied to within quantities of the order of the first term omitted, if the series is truncated at $n = N$ (i.e., the expression (1.3.19) for $q(z)$ appearing in the phase-integral approximation of the finite order $2N + 1$) is substituted for $q(z)$ in (1.6.1). It is a common situation that $\partial q / \partial \kappa$ appears in the integrand of an integral along a closed contour Λ on which $q(z)$ is a single-valued function of z. This implies that on Λ the last term in the right-hand member of (1.6.1) is the derivative with respect to z of a single-valued function of z. Therefore, one obtains from (1.6.1)

$$\int_\Lambda \frac{\partial q}{\partial \kappa} dz = \int_\Lambda \frac{1}{2q} \frac{\partial R}{\partial \kappa} dz. \tag{1.6.2}$$

Since, in higher-order approximation, the function $q(z)$ depends on κ in a much more complicated way than does $R(z)$, the right-hand member of (1.6.2) is much simpler than the left-hand member. This result, obtained from the q-equation (1.3.3), is an important consequence of the separation of the wave function into amplitude and phase factor, with the correct relation between phase and amplitude.

Appendix: Phase-Amplitude Relation

Considering the differential equation (1.2.1); that is,

$$\frac{d^2 \psi}{dz^2} + R(z)\psi = 0, \tag{1.A.1}$$

we shall argue for the representation of ψ either as the product of two factors, which on the real z-axis represent an amplitude and a phase factor, if $R(z)$ is real there, or as a linear combination of two such terms.

Assuming for a moment that $R(z)$ is real on the real z-axis, we note that if the function $\psi = A(z)\exp[+i\phi(z)]$, with $A(z)$ and $\phi(z)$ real on the real z-axis, is a solution of (1.A.1), so also is the function $\psi = A(z)\exp[-i\phi(z)]$; and thus both functions

$$\psi = A(z)\,\exp[\pm i\phi(z)] \qquad (1.A.2)$$

are solutions of (1.A.1). Moreover, in the general case, when $R(z)$ need not be real on the real z-axis, it is very natural to look for two linearly independent solutions of the form (1.A.2). Therefore, abandoning the assumption that $R(z)$ be real on the real z-axis, and letting z lie anywhere in the complex z-plane, we can write two solutions of the differential equation (1.A.1) in the form of (1.A.2), where $A(z)$ and $\phi(z)$ need not now be real on the real z-axis. Inserting (1.A.2) into (1.A.1), we obtain

$$d^2A/dz^2 + \left[R(z) - (d\phi/dz)^2\right]A \pm 2iA\,d\phi/dz\,\frac{d}{dz}\ln\left[A(d\phi/dz)^{1/2}\right] = 0,$$
$$(1.A.3)$$

from which it follows that

$$A\,d\phi/dz\,\frac{d}{dz}\ln\left[A(d\phi/dz)^{1/2}\right] = 0 \qquad (1.A.4a)$$

and

$$d^2A/dz^2 + \left[R(z) - (d\phi/dz)^2\right]A = 0. \qquad (1.A.4b)$$

Assuming $A(z)$ and $d\phi(z)/dz$ to be different from zero, we obtain from (1.A.4a)

$$A = \text{const} \times (d\phi/dz)^{-1/2} \qquad (1.A.5a)$$

and hence from (1.A.4b)

$$\left(\frac{d\phi}{dz}\right)^{+1/2}\frac{d^2}{dz^2}\left(\frac{d\phi}{dz}\right)^{-1/2} - \left(\frac{d\phi}{dz}\right)^2 + R(z) = 0. \qquad (1.A.5b)$$

Introducing for $d\phi(z)/dz$ the notation $q(z)$, we find from (1.A.2) and (1.A.5a,b) that the differential equation (1.A.1) has the two solutions

$$\psi = q^{-1/2}(z)\,\exp\left(\pm i\int^z q(z)dz\right), \qquad (1.A.6)$$

where $q(z)$, which is not necessarily real on the real z-axis, is a solution of the differential equation

$$q^{1/2}\frac{d^2}{dz^2}q^{-1/2} - q^2 + R(z) = 0, \qquad (1.A.7)$$

that is,

$$\frac{d^2}{dz^2}q^{-1/2} + R(z)q^{-1/2} = \frac{1}{(q^{-1/2})^3}, \qquad (1.A.7')$$

which is the q-equation (1.3.3) written in slightly different form. The two functions (1.A.6) are linearly independent since their Wronskian (which is equal to $-2i$) is different from zero. It should be emphasized that the exact function $q(z)$ introduced here is sometimes not even approximately the same as the approximate function $q(z)$ in (1.3.19).

Assuming the differential equation (1.A.1) to have two solutions of the form (1.A.2), we shall now demonstrate in a different way that (except for an arbitrary constant factor) these solutions must be of the form (1.A.6). Without introducing any reality conditions on the functions $R(z)$, $A(z)$, and $\phi(z)$, we note that the Wronskian of two linearly independent solutions of (1.A.1) is constant and that the Wronskian of the two functions (1.A.2) is equal to $-2i\,A^2(z)d\phi(z)/dz$ and is thus constant only if $A(z)$ is given by the formula (1.A.5a). Writing $q(z)$ instead of $d\phi(z)/dz$, we see that the requirement that the Wronskian of the two functions (1.A.2) be constant implies that the two functions (except for an arbitrary constant factor) can be written in the form of the right-hand member of (1.A.6). For these two functions to be solutions of (1.A.1) the further condition (1.A.7), i.e., the q-equation, must be fulfilled.

To elucidate (1.A.6) further, we assert that for any two linearly independent solutions ψ_1 and ψ_2 of the differential equation (1.A.1) one can define a function $q(z)$ (which depends on the choice of ψ_1 and ψ_2) such that ψ_1 and ψ_2 (except for a common constant factor) can be written in the form of the right-hand member of (1.A.6) with the plus and minus sign, respectively. To prove this we note that the Wronskian W of the linearly independent functions ψ_1 and ψ_2 is a constant that is different from zero; i.e., $\psi_1\,d\psi_2/dz - \psi_2\,d\psi_1/dz = W \neq 0$. From this equation it follows that $d[\ln(\psi_1/\psi_2)]/dz = -W/(\psi_1\psi_2)$ and hence $\psi_1/\psi_2 = \exp[-W\int^z(\psi_1\psi_2)^{-1}dz]$. Defining $q(z) = iW/(2\psi_1\psi_2)$, we obtain $\psi_1 = (iW/2)^{1/2}q^{-1/2}\exp(+i\int^z qdz)$ and $\psi_2 = (iW/2)^{1/2}q^{-1/2}\exp(-i\int^z qdz)$. Thus, the assertion is proved.

In (1.A.6) we have expressed two solutions of the differential equation (1.A.1) in terms of the function $q(z)$ that must be a solution of the differential equation (1.A.7). Alternatively, we shall now instead express two solutions of the differential equation (1.A.1) in terms of the "amplitude" $A(z)$ for which we shall also obtain a differential equation. Solving (1.A.5a) with respect to $d\phi/dz$, we obtain

$$d\phi/dz = C/A^2, \qquad (1.A.8)$$

where C is an arbitrary constant. In order that the solutions (1.A.2) be linearly independent, $\phi(z)$ cannot be constant; hence the constant C in (1.A.8) is different from zero. By changing the definition of $A(z)$ by multiplication with an arbitrary constant factor, one realizes from (1.A.8) that C can, without restriction, be set equal to any constant (for instance, unity). Inserting into (1.A.2) the expression for $\phi(z)$ obtained from (1.A.8), we obtain

$$\psi = A(z) \exp\left(\pm i \int^z \frac{C}{A^2(z)} dz\right). \tag{1.A.9}$$

Inserting (1.A.8) into (1.A.4b), one obtains for the "amplitude" $A(z)$ in (1.A.9) the differential equation

$$d^2A/dz^2 + [R(z) - C^2/A^4]A = 0, \tag{1.A.10}$$

that is,

$$d^2A/dz^2 + R(z)A = C^2/A^3, \tag{1.A.10'}$$

which one can also obtain by inserting into (1.A.7') the expression $q^{-1/2}(z) = (d\phi/dz)^{-1/2} = A/C^{1/2}$ obtained from (1.A.8).

References

Beckel, C.L. and Nakhleh, J., 1963, *J Chem Phys* **39**, 94–97.

Bertocchi, L., Fubini, S., and Furlan, G., 1965, *Nuovo Cimento* **35**, 599–632.

Brillouin, L., 1926a, *C R Acad Sci (Paris)* **183**, 24–26.

Brillouin, L., 1926b, *J Physique Radium (Série VI)* **7**, 353–368.

Broer, L.J.F., 1963, *Appl Sci Res* **B10**, 110–118.

Campbell, J.A., 1972, *J Comp Phys* **10**, 308–315.

Campbell, J.A., 1979, *J Phys A: Math Gen* **12**, 1149–1154.

Carlini, F., 1817, "Ricerche sulla convergenza della serie che serve alla soluzione del problema di Keplero." Appendice all' *Effemeridi Astronomiche di Milano per l'anno 1818*, pp. 3–48, (Review: *Giornale di Fisica, Chimica, Storia Naturale, Medicina ed Arti* **10**, 458–460, 1817.)

Choi, S. and Ross. J., 1962, *Proc Nat Acad Sci* **48**, 803–806.

Dammert, Ö. and Fröman, P.O., 1980, *J Math Phys* **21**, 1683–1687.

Eliezer, C.J. and Gray, A., 1976, *SIAM J Appl Math* **30**, 463–468.

Engelke, R. and Beckel, C.L., 1970, *J Math Phys* **11**, 1991–1994.

Ermakov, V., 1880, *Univ Izv Kiev* **20**, No. 9, 1–25.

Floyd, E.R., 1986, *J Acoust Soc Am* **80**, 877–887.

Fowler, R.H., Gallop, E.G., Lock, C.N.H., and Richmond, W.H., 1921, *Phil Trans Roy Soc London* **A221**, 295–387.

Fröman, N., 1966a, *Ark Fys* **31**, 381–408.

Fröman, N., 1966b, *Ark Fys* **32**, 541–548.

Fröman, N., 1970, *Ann Phys (NY)* **61**, 451–464.

Fröman, N., 1974, *Phys Lett* **48A**, 137–139.

Fröman, N., 1979, *J Phys A: Math Gen* **12**, 2355–2371.

Fröman, N. and Fröman, P.O., 1965, *JWKB Approximation, Contributions to the Theory*. North-Holland, Amsterdam. Russian translation: MIR, Moscow, 1967.

Fröman, N. and Fröman, P.O., 1970, *Nucl Phys* **A147**, 606–626.

Fröman, N. and Fröman, P.O., 1974a, *Ann Phys (NY)* **83**, 103–107. (Review: *Zentralblatt für Mathematik und ihre Grenzgebiete, Mathematics Abstracts* **279**, 190–191, 1974.)

Fröman, N. and Fröman, P.O., 1974b, *Nuovo Cimento* **20B**, 121–132.

Fröman, N. and Fröman, P.O., 1977, *J Math Phys* **18**, 96–99.

Fröman, N. and Fröman, P.O., 1985, "On the history of the so-called WKB-method from 1817 to 1926," pages 1–7 of *Semiclassical Descriptions of Atomic and Nuclear Collisions*, edited by J. Bang and J. de Boer. North-Holland, Amsterdam.

Fröman, N., Fröman, P.O., Myhrman, U., and Paulsson, R., 1972, *Ann Phys (NY)* **74**, 314–323.

Fröman, P.O., 1974, *Ann Phys (NY)* **88**, 621–630.

Fulling, S.A., 1979, *J Math Phys* **20**, 1202–1209.

Fulling, S.A., 1981, "The local asymptotics of continuum eigenfunction expansions," pages 181–187 of *Spectral Theory of Differential Operators*, edited by I.W. Knowles and R.T. Lewis. North-Holland, Amsterdam.

Fulling, S.A., 1982, *SIAM J Math Anal* **13**, 891–912.

Fulling, S.A., 1983a, *SIAM J Math Anal* **14**, 780–795.

Fulling, S.A., 1983b, *J Phys A: Math Gen* **16**, 2615–2631.

Furry, W.H., 1947, *Phys Rev* **71**, 360–371.

Gans, R., 1915, *Ann Physik (Vierte Folge)* **47**, 709–736.

Heading, J., 1962, *An Introduction to Phase-Integral Methods*. Methuen's Monographs on Physical Subjects, London and New York.

Hull, M.H., Jr. and Breit, G., 1959, "Coulomb wave functions," pages 408–465 of *Encyclopedia of Physics*, **41/1**, edited by S. Flügge. Springer-Verlag, Berlin.

Jeffreys, H., 1925, *Proc Lond Math Soc (Second Series)* **23**, 428–436.

Kemble, E.C., 1935, *Phys Rev* **48**, 549–561.

Kemble, E.C., 1937, *The Fundamental Principles of Quantum Mechanics*. McGraw-Hill, New York and London. Reissue: Dover Publications, New York, 1958.

Kramers, H.A., 1926, *Z f Physik* **39**, 828–840.

Krieger, J.B., 1969, *J Math Phys* **10**, 1455–1458.

Krieger, J.B. and Rosenzweig, C., 1967, *Phys Rev* **164**, 171–173.

Langer, R.E., 1937, *Phys Rev* **51**, 669–676.

Lewis, H.R., Jr., 1967, *Phys Rev Lett* **18**, 510–512 and 636.

McHugh, J.A.M., 1971, *Archive for History of Exact Sciences* **7**, 277–324.

Messiah, A., 1959, *Mécanique Quantique, Tome* **1**. Dunod, Paris. English translation: North-Holland, Amsterdam, 1961.

Milne, W.E., 1930, *Phys Rev* **35**, 863–867.

Olver, F.W.J., 1961, *Proc Cambr Phil Soc* **57**, 790–810.

Olver, F.W.J., 1965a, *J Res NBS* **69B**, 271–290.

Olver, F.W.J., 1965b, *J Res NBS* **69B**, 291–300.

Rayleigh, Lord, (J.W. Strutt), 1912, *Proc Roy Soc London* **A86**, 207–226.

Skorupski, A.A., 1980, *Reports on Mathematical Physics* **17**, 161–187.

Skorupski, A.A., 1988a, *J Math Phys* **29**, 1814–1823.

Skorupski, A.A., 1988b, *J Math Phys* **29**, 1824–1831.

Wentzel, G., 1926, *Z f Physik* **38**, 518–529.

Wheeler, J.A., 1937, *Phys Rev* **52**, 1123–1127.

Wilson, H.A., 1930, *Phys Rev* **35**, 948–956.

Yngve, S., 1971, *J Math Phys* **12**, 114–117.

Yost, F.L., Wheeler, J.A., and Breit, G., 1936, *Phys Rev* **49**, 174–189.

Young, L.A., 1931, *Phys Rev* **38**, 1612–1614.

Young, L.A., 1932, *Phys Rev* **39**, 455–457.

Young, L.A. and Uhlenbeck, G.E., 1930, *Phys Rev* **36**, 1154–1167.

Zwaan, A., 1929, *Intensitäten im Ca-Funkenspektrum*. Academisch Proefschrift. Joh Enschedé en Zonen, Harlem.

2
Technique of the Comparison Equation Adapted to the Phase-Integral Method

Abstract. For a cluster consisting of an unspecified number of transition points we develop a convenient comparison equation technique for obtaining the wave function within the cluster and the Stokes constants that determine the phase-integral solution outside the cluster. The resulting formulas remain valid even if the transition points approach each other. We perform the rather lengthy calculations in a general way and obtain formulas which can easily be particularized to the specific situations encountered in applications. The final analytic formulas for the Stokes constants are expressed in terms of phase integrals, in which the contributions to the integrands in successive orders of approximation are to be integrated along contours enclosing pairs of transition points. With the aid of the comparison equation technique we thus obtain *supplementary quantities* which, when incorporated into the general formulas and the connection formulas of the phase-integral method, are of decisive importance when the transition points lie close to each other, but which become very small when those points recede sufficiently far away from each other.

It is shown explicitly that, after appropriate asymptotic expansion in terms of a "small" bookkeeping parameter, the comparison equation solution constructed here yields locally, when there are no transition points in the region under consideration (i.e., sufficiently far away from transition points) the arbitrary-order phase-integral approximation generated from an unspecified base function, which has been described in Chapter 1 of the present monograph.

The comparison equation technique used is based on general ideas that are already known; however, their adaptation to our purpose requires a number of new steps in the procedure, and the resulting phase-integral formulas are new.

2.1 Background

We shall be concerned with the solution of the differential equation

$$\frac{d^2\psi}{dz^2} + R(z)\psi = 0, \tag{2.1.1}$$

where $R(z)$ is an analytic function of z. This differential equation which could possibly result from separation of variables, describes large classes of important problems in various fields of physics, not only in quantum mechanics.

The phase-integral functions, in terms of which the solution will be expressed, are of the general form

$$f_1(z) = q^{-1/2}(z) \exp\left(+i\int^z q(z)\,dz\right) \tag{2.1.2a}$$

$$f_2(z) = q^{-1/2}(z) \exp\left(-i\int^z q(z)\,dz\right) \tag{2.1.2b}$$

(see Chapter 1). For the $(2N+1)$th-order approximation

$$q(z) = \sum_{n=0}^{N} q^{(2n+1)}(z) \tag{2.1.3}$$

where

$$q^{(2n+1)}(z) = Y_{2n}Q(z); \tag{2.1.4}$$

the function $Q(z)$ is an *a priori* unspecified base function, and the first few functions Y_{2n} are given by

$$Y_0 = 1 \tag{2.1.5a}$$

$$Y_2 = \frac{1}{2}\,\varepsilon_0 \tag{2.1.5b}$$

$$Y_4 = -\frac{1}{8}\left(\varepsilon_0^2 + \frac{d^2\varepsilon_0}{d\zeta^2}\right), \tag{2.1.5c}$$

where

$$\varepsilon_0 = Q^{-3/2}(z)\frac{d^2}{dz^2}Q^{-1/2}(z) + \frac{R(z) - Q^2(z)}{Q^2(z)} \tag{2.1.6}$$

and

$$\zeta = \int^z Q(z)\,dz. \tag{2.1.7}$$

All the functions Y_{2n} can be expressed in terms of ε_0 and the derivatives of ε_0 with respect to ζ. Thus, the choice of the base function $Q(z)$ affects ε_0 and ζ but not the analytic form of Y_{2n}, expressed in terms of ε_0 and derivatives of ε_0 with respect to ζ.

It is well known that transition points exert a decisive influence on the phase-integral solution of a wave equation such as (2.1.1). By *transition points* we mean zeros and poles of the square of the base function; i.e., of $Q^2(z)$. With the aid of comparison equation technique the wave function can be obtained in the neighborhood of such transition points. This may be necessary, for instance, if there is a boundary condition close to a transition point; but in general one does not need the wave function in the neighborhood of transition points. In the simple case of an isolated transition point, one can derive the connection formulas or, more generally, the Stokes constants for tracing a solution around the transition point, without using a comparison equation. On the other hand, when two or possibly more transition points lie close together, so that there is no region between them, in which the phase-integral approximation is valid, a convenient kind of comparison equation technique is most appropriate for obtaining the phase-integral solution outside the region containing the closely lying transition points in question. For instance, when barrier transmission takes place near the top of a barrier, two transition points lie close to each other, and comparison equation technique is, at present, practically indispensable for obtaining the Stokes constants and hence the connection formulas that determine the solution outside the region in which the close-lying transition points are located.

We shall refer to the phase-integral quantities, resulting from comparison equation technique, as *supplementary quantities*, since they appear in the solution together with the phase-integral functions generated from the unspecified base function $Q(z)$. When incorporated into the formulas of the phase-integral method, they are of decisive importance when the relevant transition points lie close to each other, but they become very small when those points recede sufficiently far away from each other. In the example of barrier transmission [10] the supplementary quantity is the quantity σ (given by the approximate expressions (10) and (10a,b,c) in [18]), which is of essential importance when the energy is near the top of the barrier. This supplementary quantity σ is the same as the quantity $-\phi/2$ in Chapter 5 of the present monograph. Another important example, which occurs, for instance, in phase shift problems, is when $Q^2(z)$ has a first- or second-order pole and one or two zeros in the relevant region of the complex z-plane. One can then specialize formulas to be given in §2.2 to the case of the Coulomb equation as comparison equation and derive supplementary quantities, which can be used to construct formulas, valid when transition points approach each other and containing formulas, derived previously, as limiting cases corresponding to the situation that the pole and the zero or zeros in question recede sufficiently far from each other; see [17].

The derivation of supplementary quantities (or corresponding Stokes constants) for differing situations of close-lying zeros and poles of $Q^2(z)$ forms an important part of our development of the phase-integral method.

The particular comparison equation technique that will be described in

§2.2 is developed in general terms with the aim of making its application to particular problems as much facilitated as possible. The exposition displays in a general way both the power and the practical limitations of comparison equation techniques. Here it is constructed so that, when the resulting solution is expanded appropriately in terms of a "small" parameter λ, the result is the phase-integral approximation of arbitrary odd order generated from an unspecified base function [5, 6, 11, 12, 15], with the appropriate values of the phase and the amplitude. Although the general ideas of the comparison equation technique can be found in papers by earlier authors, our present aim requires a nontrivial adaptation of the comparison equation technique to the phase-integral technique; a number of new steps are needed in the procedure. The resulting phase-integral formulas are new.

In §2.2 we derive a formal solution of the differential equation (2.1.1) in a region where there is an unspecified number of zeros and poles of the function $Q^2(z)$. In §2.3 we show that, as a particular case of the treatment (*viz.* the case when there are neither zeros nor poles of $Q^2(z)$ in the region under consideration) one obtains asymptotically (as $\lambda \to 0$) the arbitrary-order phase-integral functions (2.1.2a,b) generated from the unspecified base function $Q(z)$. On the one hand, this fact throws light on the connection between the solutions obtained by comparison equation techniques and the phase-integral approximation generated from an unspecified base function, first obtained in [11, 12]; see the review of that approximation given in Chapter 1 of the present monograph. On the other hand, the lengthy and cumbersome calculations needed illuminate the fact that comparison equation technique is inconvenient, unless one is really interested in exploiting its power beyond the power of the much simpler and now well developed technique in which one handles the arbitrary-order phase-integral approximation generated from an unspecified base function with the aid of connection formulas pertaining to an isolated transition point. This is worth mentioning, since one often encounters papers, in which comparison equation technique is used with the erroneous belief that the result of that technique always goes beyond what is obtained from the use of the existing arbitrary-order phase-integral connection formulas. To illustrate this fact we point out that in the case of a bound-state problem when $R(z)$ has two first-order zeros but no poles in the region considered, and when $Q^2(z)$ is a function with those properties, the comparison equation solution with a parabolic well as the "comparison potential" yields the same arbitrary-order quantization condition that can be obtained in a much simpler way with the aid of one of the connection formulas. In the case of the bound-state problem the merit of the comparison equation treatment with a parabolic "comparison potential" is to yield the wave function in the neighborhood of the turning points; however, this information is seldom needed. In particular, such information is in general not needed for the calculation of expectation values and matrix elements [7, 8, 13, 14, 16]. It should also be noted that the comparison equation solution with a parabolic barrier as the "comparison potential" yields the same transmission coefficient (except for the phase) as can be obtained with the aid of F-matrix analysis [9]. However, in barrier transmission problems the phase of the wave function is often extremely important, and with the aid of the comparison equation solution we can, as mentioned above, determine a supplementary quantity, which allows us to construct an arbitrary-order connection formula for a barrier, valid uniformly for all energies, even those in the neighborhood of the top of the barrier.

It is with the crucial problems in mind, where comparison equation technique is indispensable, that we derive the general final formulas presented in §2.2. Since comparison equation technique, taken to its full power, is too complicated to be worked out satisfactorily from first principles in every particular application where it is needed, our aim is to perform the rather lengthy calculations once and for all in order to obtain formulas that can readily be particularized to the specific cases encountered in the applications. After appropriate series expansions of parameters and variables in terms of the "small" parameter λ, we obtain the $(2N + 1)$th-order approximation of the comparison equation solution which then, after further formal expansion in powers of λ, yields the $(2N + 1)$th-order phase-integral approximation generated from the unspecified base function $Q(z)$, unless z is too close to a transition point. The treatment in §2.2 provides the basis for working out general results of a standard nature for important model problems, characterized by close-lying transition points. Such model problems occur frequently and appear in several fields of theoretical physics. Having specified the model problem and chosen the comparison equation, one can use the formulas derived in §2.2 rather straightforwardly as is done, e.g., in Chapter 6.

The final analytic formulas for the supplementary quantities are expressed in terms of phase-integrals in which the contributions $q^{(1)}(z)$, $q^{(3)}(z)$, ... to the function $q(z)$ in successive orders of approximation are to be integrated along contours enclosing a pair of two transition points.

It should be emphasized that comparison equation technique allows us only to show that there exists a solution that is approximated by the resulting formulas. As has been stressed by N. Fröman on pp. 390–391 in [4], the knowledge of an approximate solution, valid through a turning point, does not mean that one has solved the connection problem, formulated as follows: Given a linear combination of the phase-integral functions $f_1(z)$ and $f_2(z)$ representing a certain exact solution on one side of a turning point, find the correct linear combination representing the same exact solution on the other side of the turning point. In particular the knowledge of the above-mentioned solution does not mean that one has been able to circumvent the inherent one-directional nature of the connection formulas pertaining to a simple turning point. The connection problem is an intricate separate problem. It must first be formulated quite precisely and then solved in a rigorous way.

2.2   Comparison Equation Technique

We shall now describe in detail the comparison equation technique which we shall use. Consider the differential equation (2.1.1), i.e.,

$$\frac{d^2\psi}{dz^2} + R(z)\psi = 0. \tag{2.2.1}$$

Our aim is to derive a formal solution from which the arbitrary-order phase-integral approximation generated from the unspecified base function $Q(z)$ can be obtained when z lies sufficiently far away from the zeros and poles of $Q^2(z)$. To achieve this goal we formally assume that $R(z)$ in (2.2.1) is the sum of a "large" term $Q^2(z)$ and a "smaller" term $R(z) - Q^2(z)$, where $Q^2(z)$ is an unspecified function (the square of the base function) that can be conveniently chosen. To take explicit account of the magnitudes of the two terms constituting $R(z)$ in that way, we introduce a "small" bookkeeping parameter λ and replace the original differential equation (2.2.1) by the differential equation

$$\frac{d^2\psi}{dz^2} + \left(\frac{Q^2(z)}{\lambda^2} + [R(z) - Q^2(z)]\right)\psi = 0, \tag{2.2.2}$$

which agrees with (2.2.1) for $\lambda = 1$. The differential equation (2.2.2) will be solved by comparison equation technique. The parameter λ is used explicitly in the calculations, but at the end, in the resulting formulas, we set $\lambda = 1$; hence, we obtain formulas that apply to the original differential equation (2.2.1). From the solution thus obtained one can derive the phase-integral approximation of arbitrary order generated from the unspecified base function $Q(z)$ when z moves away from the zeros and poles of $Q^2(z)$. Beside the property of being valid in the neighborhood of transition points, the most important merit of the solution obtained by comparison equation technique is the possibility of determining the supplementary quantities in the phase-integral formulas and of constructing approximate solutions and connection formulas, valid also when there are zeros and poles of $Q^2(z)$ that may lie close to each other in the region of the complex z-plane considered.

We found it very important to start from a convenient version of comparison equation technique for solving the differential equation. As regards our choice, we refer to papers by Cherry [1], Miller and Good [19], Erdélyi [2, 3], Zauderer [20, 21] and to further references given in these papers. However, as already mentioned, the adaptation of that technique to the phase-integral technique, so that the comparison equation solution will asymptotically (as $\lambda \to 0$) yield the appropriate phase-integral approximation generated from the unspecified base function $Q(z)$, is not trivial.

We shall consider the general case in which there is an unspecified number of zeros of $Q^2(z)$, as well as an unspecified number of poles of $Q^2(z)$, in the region of the complex z-plane under consideration. The consideration of the poles requires some discussion, which we restrict to the particularly important case of poles of the first or second order, since our whole treatment is restricted to this case.

To solve (2.2.2) by convenient comparison equation technique we set

$$\psi = \left(\frac{d\varphi}{dz}\right)^{-1/2} \Phi(\varphi), \qquad (2.2.3)$$

where $\Phi(\varphi)$ is an analytic function of φ and λ that will be specified below, and φ is an analytic function of z and λ which will be determined fiom the differential equation (2.2.2). According to (2.2.3) it is natural to require that $d\varphi/dz$ must everywhere be finite and different from zero; hence the mapping of the whole region of the complex z-plane under consideration on the complex φ-plane must be conformal. Inserting (2.2.3) into (2.2.2), we get

$$\frac{d^2\Phi}{d\varphi^2} + \left[\frac{Q^2(z)/\lambda^2 + R(z) - Q^2(z)}{(d\varphi/dz)^2} + \left(\frac{d\varphi}{dz}\right)^{-3/2}\frac{d^2}{dz^2}\left(\frac{d\varphi}{dz}\right)^{-1/2}\right]\Phi = 0.$$
$$(2.2.4)$$

We now assume that $\Phi(\varphi)$ is a known solution of the differential equation

$$\frac{d^2\Phi}{d\varphi^2} + \frac{\Pi(\varphi)}{\lambda^2}\Phi = 0, \qquad (2.2.5)$$

which will be used as our comparison equation. The function $\Pi(\varphi)$ depends in general not only on φ but also on λ, although this latter dependence is not indicated in the notation. In order that the solutions of the differential equation (2.2.2) be expressible in terms of the solutions of the differential equation (2.2.5), the function $\Pi(\varphi)$ must in some sense be similar to the function $Q^2(z) + [R(z) - Q^2(z)]\lambda^2$. The necessity and the precise meaning of this assumption for the success of the technique to be described here will be seen below.

For a moment we let t'_j denote a second-order pole of $Q^2(z) + [R(z) - Q^2(z)]\lambda^2$, which corresponds to a second-order pole $\varphi(t'_j)$ of $\Pi(\varphi)$. Since $d\varphi/dz$ is everywhere finite and different from zero, it follows that for sufficiently small values of $|z - t'_j|$ the function $\varphi(z) - \varphi(t'_j)$ is approximately proportional to $z - t'_j$. Therefore, in order that the regular solution of (2.2.2), when expressed in terms of the regular solution of (2.2.5) according to (2.2.3), has the correct behavior at the pole t'_j it is (according to the theory of linear, second-order differential equations) necessary that

$$\lim_{\varphi\to\varphi(t'_j)}[\varphi-\varphi(t'_j)]^2\Pi(\varphi) = \lim_{z\to t'_j}(z-t'_j)^2(Q^2(z)+[R(z)-Q^2(z)]\lambda^2). \quad (2.2.6)$$

Using (2.2.5), and assuming that $\Phi(\varphi)$ is not identically equal to zero, we obtain from (2.2.4) the differential equation

$$-\Pi(\varphi) + \frac{Q^2(z) + [R(z) - Q^2(z)]\lambda^2}{(d\varphi/dz)^2} + \left(\frac{d\varphi}{dz}\right)^{-3/2} \frac{d^2}{dz^2} \left(\frac{d\varphi}{dz}\right)^{-1/2} \lambda^2 = 0,$$

(2.2.7)

which can also be written in the form

$$-\Pi(\varphi) \left(\frac{d\varphi}{dz}\right)^4 + Q^2(z) \left(\frac{d\varphi}{dz}\right)^2 + \left[[R(z) - Q^2(z)] \left(\frac{d\varphi}{dz}\right)^2\right.$$

$$\left. + \frac{3}{4} \left(\frac{d^2\varphi}{dz^2}\right)^2 - \frac{1}{2} \frac{d\varphi}{dz} \frac{d^3\varphi}{dz^3}\right] \lambda^2 = 0.$$

(2.2.8)

We now assume the function $\Pi(\varphi)$ to be given by the expression

$$\Pi(\varphi) = \frac{S(\varphi)}{T(\varphi)}$$

(2.2.9a)

with

$$S(\varphi) = \sum_{n=0}^{\infty} A_n \varphi^n$$

(2.2.9b)

and

$$T(\varphi) = \sum_{q=0}^{\infty} B_q \varphi^q,$$

(2.2.9c)

where the coefficients A_n and B_q are independent of φ but may depend on λ. In applications to physical problems the coefficients A_n and B_q are in general equal to zero when the values of n and q are sufficiently large; however, it is formally simpler to use infinite instead of finite series in the expression for $\Pi(\varphi)$. It should be remarked that there is some freedom in the choice of the coefficients A_n and B_q, from which one can benefit in the treatment of concrete applications. In fact, by introducing in (2.2.5) along with (2.2.9a–c) a new independent variable φ, which depends linearly on the original variable φ, one realizes that the expressions (2.2.9b,c) can be simplified in a way which corresponds to the introduction of two conditions on the coefficients A_n and B_q. The fact that it is possible to introduce two such conditions can be used to simplify the calculations in particular applications.

Inserting (2.2.9a–c) into (2.2.8), we obtain the differential equation

$$-\sum_{n=0}^{\infty} A_n \varphi^n \left(\frac{d\varphi}{dz}\right)^4 + Q^2(z) \sum_{q=0}^{\infty} B_q \varphi^q \left(\frac{d\varphi}{dz}\right)^2$$

$$+ \left(\sum_{q=0}^{\infty} B_q \varphi^q\right) \left[[R(z) - Q^2(z)] \left(\frac{d\varphi}{dz}\right)^2\right.$$

(2.2.10)

$$+\frac{3}{4}\left(\frac{d^2\varphi}{dz^2}\right)^2 - \frac{1}{2}\frac{d\varphi}{dz}\frac{d^3\varphi}{dz^3}\right]\lambda^2 = 0$$

from which φ is to be determined as a function of z and λ. To this purpose we write

$$\varphi = \sum_{\beta=0}^{\infty}\varphi_{2\beta}\lambda^{2\beta}, \tag{2.2.11}$$

or in truncated form

$$\varphi = \sum_{\beta=0}^{N}\varphi_{2\beta}\lambda^{2\beta}, \tag{2.2.11'}$$

where the quantities $\varphi_{2\beta}$ depend on z but are independent of λ, and we assume A_n to depend on λ according to the formal series

$$A_n = \sum_{\beta=0}^{\infty}A_{n,2\beta}\lambda^{2\beta}. \tag{2.2.12}$$

Inserting (2.2.12) into (2.2.9b), we obtain

$$S(\varphi) = \sum_{\beta=0}^{\infty}S_{2\beta}(\varphi)\lambda^{2\beta} \tag{2.2.13a}$$

where

$$S_{2\beta}(\varphi) = \sum_{n=0}^{\infty}A_{n,2\beta}\varphi^n. \tag{2.2.13b}$$

It turns out that, although it is in general necessary to assume the coefficients A_n to depend on λ, this is not the case regarding the coefficients B_q. Therefore, we shall assume B_q to be independent of λ. Inserting (2.2.11) and (2.2.12) into (2.2.10) and setting the coefficients of the successive powers of λ equal to zero in the resulting equation, we get from the coefficient of λ^0

$$\left(\frac{d\varphi_0}{dz}\right)^2\left[\left(\frac{d\varphi_0}{dz}\right)^2\sum_{n=0}^{\infty}A_{n,0}\varphi_0^n - Q^2(z)\sum_{q=0}^{\infty}B_q\varphi_0^q\right] = 0 \tag{2.2.14}$$

and from the coefficient of λ^{2N} $(N \geq 1)$

$$\sum_{n=0}^{\infty}\sum_{\beta_0+\cdots+\beta_{n+4}=N}A_{n,2\beta_0}\varphi_{2\beta_1}\cdots\varphi_{2\beta_n}\frac{d\varphi_{2\beta_{n+1}}}{dz}\cdots\frac{d\varphi_{2\beta_{n+4}}}{dz}$$

$$-Q^2(z)\sum_{q=0}^{\infty}B_q\sum_{\beta_1+\cdots+\beta_{q+2}=N}\varphi_{2\beta_1}\cdots\varphi_{2\beta_q}\frac{d\varphi_{2\beta_{q+1}}}{dz}\frac{d\varphi_{2\beta_{q+2}}}{dz}$$

$$-\sum_{q=0}^{\infty}B_q\sum_{\beta_1+\cdots+\beta_{q+2}=N-1}\varphi_{2\beta_1}\cdots\varphi_{2\beta_q}$$

$$\times \left([R(z) - Q^2(z)] \frac{d\varphi_{2\beta_{q+1}}}{dz} \frac{d\varphi_{2\beta_{q+2}}}{dz} + \frac{3}{4} \frac{d^2\varphi_{2\beta_{q+1}}}{dz^2} \frac{d^2\varphi_{2\beta_{q+2}}}{dz^2} \right.$$

$$\left. - \frac{1}{4} \frac{d\varphi_{2\beta_{q+1}}}{dz} \frac{d^3\varphi_{2\beta_{q+2}}}{dz^3} - \frac{1}{4} \frac{d^3\varphi_{2\beta_{q+1}}}{dz^3} \frac{d\varphi_{2\beta_{q+2}}}{dz} \right) = 0, \quad N \geq 1, \quad (2.2.15)$$

where all the indices β_i are nonnegative. We have thus obtained a system of differential equations from which the functions $\varphi_{2\beta}$ can be determined successively.

2.2.1 Differential Equation for φ_0

Disregarding the possibility that $d\varphi_0/dz$ is equal to zero, which can be justified since $d\varphi/dz \neq 0$ and φ_0 is the dominating part of φ, and defining

$$P^2(\varphi_0) = \frac{S_0(\varphi_0)}{T(\varphi_0)} \qquad (2.2.16a)$$

with (cf. [2.2.13b])

$$S_0(\varphi_0) = \sum_{n=0}^{\infty} A_{n,0} \varphi_0^n \qquad (2.2.16b)$$

and (cf. [2.2.9c])

$$T(\varphi_0) = \sum_{q=0}^{\infty} B_q \varphi_0^q, \qquad (2.2.16c)$$

we obtain from (2.2.14) the formula

$$\frac{d\varphi_0}{dz} = Q(z)/P(\varphi_0), \qquad (2.2.17)$$

where the signs of $Q(z)$ and $P(\varphi_0)$ are to be chosen such that $Q(z)$ and $P(\varphi_0)$ behave similarly.

Thus, from the first-order solution of the comparison equation (i.e., the solution to the zeroth order in λ), we obtain the relation (2.2.17) characterizing the mapping of the complex z-plane onto the complex φ_0-plane.

2.2.2 Determination of the Coefficients $A_{n,0}$ and B_q

When discussing the expression (2.2.3) for the comparison equation solution of (2.2.2) we have already required the mapping of the region of the complex z-plane under consideration onto the corresponding region of the complex φ-plane to be conformal. Since this has to be true for all sufficiently small values of λ, and in particular for $\lambda = 0$, the mapping of the z-plane onto the φ_0-plane must also be conformal. This means that φ_0 must be a regular, single-valued analytic function of z, and that $d\varphi_0/dz$ has no zeros in the region of the complex z-plane under consideration. Assuming $Q^2(z)$ to be regular or to have poles but no branch points in the region of the complex z-plane under consideration, we therefore realize from (2.2.17) that a point z where $Q^2(z)$ is regular and different from zero must correspond to a point φ_0 where $P^2(\varphi_0)$ is regular and different from zero, and that a zero or pole of $Q^2(z)$ of a certain order must correspond to a zero or pole, respectively, of $P^2(\varphi_0)$ of the same order. We now assume that in the actual region of the complex z-plane the function $Q^2(z)$ has r (≥ 0) zeros t_i of the orders m_i (where i assumes r different values) and s (≥ 0) poles t'_j of the orders p_j (where j assumes s different values, and p_j is equal to 1 or 2) but no branch points or essential singularities. The function $P^2(\varphi_0)$ is then chosen to be

$$P^2(\varphi_0) = c\frac{(\varphi_0 - \tau_1)^{m_1} \cdots (\varphi_0 - \tau_r)^{m_r}}{(\varphi_0 - \tau'_1)^{p_1} \cdots (\varphi_0 - \tau'_s)^{p_s}} = c\frac{\prod_i(\varphi_0 - \tau_i)^{m_i}}{\prod_j(\varphi_0 - \tau'_j)^{p_j}} \qquad (2.2.18)$$

where $\tau_1, \ldots, \tau_r$ correspond to $t_1, \ldots, t_r$, and $\tau'_1, \ldots, \tau'_s$ correspond to $t'_1, \ldots, t'_s$, and c is a constant factor. We obviously assume that no quantity τ_i is equal to any quantity τ'_j; thus, the numerator and the denominator in (2.2.18) have no common factor. We also remark that on the right-hand side of (2.2.18) it is understood that the numerator is unity when $r = 0$ and that the denominator is unity when $s = 0$.

The total number of parameters in the expression (2.2.18) is $r + s + 1$. We recall, however, the possibility, mentioned below (2.2.9c), of introducing from the beginning a new variable φ, which depends linearly on the original variable φ. In the limit $\lambda \to 0$ this corresponds to the introduction from the beginning of a new variable φ_0, which depends linearly on the original variable φ_0. This means that the expression on the right-hand side of (2.2.18) can be simplied from the beginning in a way which corresponds to the introduction of two conditions on the parameters in the right-hand member of (2.2.18), after which there remain $r + s - 1$ independent parameters. This procedure can be used to simplify the application of the comparison equation technique. The remaining $r + s - 1$ independent parameters are to be determined from the conditions imposed by the requirement that the mapping of the z-plane on the φ_0-plane be conformal. We shall now derive these conditions.

The complex z-plane is to be cut conveniently (or a convenient Riemann surface may be used for the variable z) so that $Q(z)$ becomes single-valued, and the complex φ_0-plane is to be cut correspondingly (or a convenient Riemann surface may be used for the variable φ_0) so that $P(\varphi_0)$ becomes single-valued. Furthermore, as already mentioned below (2.2.17), the phase of $P(\varphi_0)$ is chosen to correspond to that of $Q(z)$.

The *second-order poles* of $P^2(\varphi_0)$ must correspond to second-order poles of $Q^2(z)$ in an adequate way. In order that the mapping (2.2.17) be conformal in the neighborhood of a second-order pole t'_j of $Q^2(z)$, which corresponds to the second-order pole τ_j' of $P^2(\varphi_0)$, we require that

$$\lim_{\varphi_0 \to \tau_j'} (\varphi_0 - \tau_j') P(\varphi_0) = \lim_{z \to t_j'} (z - t_j') Q(z) \qquad (2.2.19)$$

for every relevant second-order pole. No sign ambiguity appears in (2.2.19), since the phases of $Q(z)$ and $P(\varphi_0)$ have already been assumed to be chosen to correspond to each other. We also remark that, alternatively, one can obtain the condition (2.2.19) from (2.2.6) by letting $\lambda \to 0$ and recalling that the phases of $Q(z)$ and $P(\varphi_0)$ correspond to each other.

From (2.2.17) it follows that

$$\int_{\varphi_0(z_0)}^{\varphi_0} P(\varphi_0) \, d\varphi_0 = \int_{z_0}^{z} Q(z) \, dz, \qquad (2.2.20)$$

where z_0 is an arbitrary fixed point in the complex z-plane, and $\varphi_0(z_0)$ is the corresponding fixed point in the complex φ_0-plane. We shall now consider more closely the relation (2.2.20), to be fulfilled by the conformal mapping of the complex z-plane onto the complex φ_0-plane, and the inferences which can be drawn from this relation.

Let us assume that there is at least one relevant *arbitrary-order* (m_ith-*order) zero* t_i or *first-order pole* t_j' of $Q^2(z)$. The condition that this zero or pole must correspond to the zero τ_i (of the same arbitrary order m_i) or pole τ_j' (of the first order) of $P^2(\varphi_0)$ is, according to (2.2.20),

$$\int_{\varphi_0(z_0)}^{\tau_i} P(\varphi_0) \, d\varphi_0 = \int_{z_0}^{t_i} Q(z) \, dz \qquad (2.2.21)$$

for every relevant arbitrary-order zero, or

$$\int_{\varphi_0(z_0)}^{\tau_j'} P(\varphi_0) \, d\varphi_0 = \int_{z_0}^{t_j'} Q(z) \, dz \qquad (2.2.22)$$

for every relevant first-order pole. Since the zeros t_i and τ_i are assumed to be of the same order m_i, it follows from (2.2.20) and (2.2.21) that φ_0 is a regular analytic function of z in the neighborhood of t_i. Similarly, since the poles t_j' and τ_j' are both assumed to be of the first order, it follows from (2.2.20) and (2.2.22) that φ_0 is a regular analytic function of z in the neighborhood of t_j'. Thus, in a region of the complex z-plane containing only regular points, including arbitrary-order zeros of $Q^2(z)$, and first-order poles of $Q^2(z)$, the mapping (2.2.20) is conformal, provided that the conditions (2.2.21) and (2.2.22) are fulfilled. If there are also second-order poles of $Q^2(z)$, the mapping (2.2.20) remains conformal, provided that the condition (2.2.19) is fulfilled for every one of these second-order poles.

If, as assumed above, there is at least one arbitrary-order zero or one first-order pole of $Q^2(z)$, then we can determine z_0 and $\varphi_0(z_0)$ such that either the condition (2.2.21) or the condition (2.2.22), is fulfilled. If the total number of arbitrary-order zeros and first-order poles of $Q^2(z)$, and hence also of $P^2(\varphi_0)$, is at least two, then according to (2.2.21) and (2.2.22) we obtain further conditions of the form

$$\int P(\varphi_0)\, d\varphi_0 = \int Q(z)\, dz, \qquad (2.2.23)$$

where the limits of integration in the right-hand member are any two of the arbitrary-order zeros and first-order poles of $Q^2(z)$, and the limits of integration in the left-hand member are the corresponding arbitrary-order zeros or first-order poles of $P^2(\varphi_0)$. The paths of integration in both members of (2.2.23), which should correspond to each other according to the conformal mapping of the z- and φ_0-planes, are assumed to avoid branch points as well as singularities that are too strong. They can, however, be deformed as we shall describe below.

Those of the conditions (2.2.23) that are associated with odd-order zeros or first-order poles can be written as

$$\frac{1}{2}\int_\Lambda P(\varphi_0)\, d\varphi_0 = \frac{1}{2}\int_\Lambda Q(z)\, dz. \qquad (2.2.24)$$

Alternatively, since $Y_0 = 1$ according to (2.1.5a), they can be written [in analogy to the higher-order conditions (2.2.59a,b) to be given in §2.2.5] as

$$\int_\Lambda P(\varphi_0)\, d\varphi_0 = \int_\Lambda Y_0 Q(z)\, dz, \qquad (2.2.24')$$

where Λ in either member is a closed contour encircling both points corresponding to the limits of integration on the right-hand or left-hand side of (2.2.23). The contours Λ in the z- and φ_0-planes correspond to each other in the sense that, after appropriate continuous deformation, they can be mapped conformally onto each other; therefore we do not distinguish these contours by different notations. The even-order zeros are regular points of $Q(z)$ and $P(\varphi_0)$; therefore the value of the integrals in (2.2.24) and (2.2.24') remains unchanged if Λ is deformed, so as to enclose any number of even-order zeros. This is also true if the deformation of the contours Λ in the z-plane and in the φ_0-plane is made differently so that these contours enclose different numbers of even-order zeros. Originally, the contour Λ in every condition (2.2.24) or (2.2.24') encloses precisely two points among the odd-order zeros and first-order poles but no second-order pole. However, since for a second-order pole the residue of $Q(z)$ is equal to that of $P(\varphi_0)$ according to (2.2.19), it does not matter for the validity of (2.2.24) and (2.2.24') if Λ (in the complex z-plane as well as in the complex φ_0-plane) is deformed so as to also enclose one or more second-order poles; provided that one ensures that the contours in the complex z-plane and in the complex φ_0-plane enclose *corresponding* second-order poles of $Q^2(z)$

and $P^2(\varphi_0)$. We have thus found that the contours Λ can be deformed so as to enclose any number of even-order zeros, which does not change the value of the integral in question, and to enclose any number of corresponding second-order poles in the complex z- and φ_0-planes, which changes the value of the integral in the z-plane by the same amount as the value of the integral in the φ_0-plane, so that the equality is preserved in (2.2.24) and (2.2.24'). Of course, one must not cross any cuts when one deforms the contours Λ. It also deserves to be remarked that the necessity of the condition (2.2.24) is an immediate consequence of (2.2.20) and the single-valuedness of $\varphi_0(z)$.

If, on the other hand, *at least one* of the limits of integration in (2.2.23) is an even-order zero, the integrals in question cannot be rewritten as integrals over closed contours as in (2.2.24). However, the paths of integration in the complex z-plane and φ_0-plane can be displaced so as to "sweep over" any number of even-order zeros (not necessarily corresponding ones in the z-plane and in the φ_0-plane) and any number of corresponding second-order poles in the complex z- and φ_0-planes. The displacement of the paths of integration over even-order zeros does not change the value of the integrals, while the simultaneous displacement over corresponding second-order poles in the z- and φ_0-planes for both integrals in (2.2.23) changes both members by the same amount, so that the relation (2.2.23) remains valid.

The number of the conditions (2.2.19) and (2.2.23) or (2.2.24) that are independent of each other is $r + s - 1$. This number is sufficient to determine $r + s - 1$ parameters in the right-hand member of (2.2.18); this is the number of independent parameters that remain after we have taken the option mentioned above of introducing, from the beginning, two convenient relations between the original $r + s + 1$ parameters.

When one has determined the parameters in the right-hand member of (2.2.18) as described above, one can easily transform the numerator and denominator there into polynomials in φ_0, and thus obtain the values of the coefficients $A_{n,0}$ and B_q in (2.2.16a–c). Of course, one can multiply all these coefficients by the same factor, since the function $P^2(\varphi_0)$ given by (2.2.16a–c) is then unchanged.

We summarize the results of the present subsection as follows. For the situation that exists when, in the region of the complex z-plane under consideration, the function $Q^2(z)$ has r zeros of the orders m_i and s poles of the orders p_j ($= 1$ or 2) that are transformed by the *conformal* mapping (2.2.20) into corresponding zeros and poles of the function $P^2(\varphi_0)$; we have derived conditions which determine the parameters in $P^2(\varphi_0)$. These conditions derive from the requirement that $d\varphi_0/dz$ be a regular, single-valued analytic function without zeros in the region of the complex z-plane under consideration.

2.2.3 Differential Equation for φ_{2N} when $N > 0$

We now return to (2.2.15), where we split off those terms in which one of the quantities β_i is equal to N (> 0). Using (2.2.16a–c), as well as (2.2.17), we obtain after some calculations

$$\frac{1}{P(\varphi_0)}\frac{d}{d\varphi_0}\left(P(\varphi_0)\varphi_{2N}\right) = I_{2N}, \qquad N \geq 1, \qquad (2.2.25)$$

that is,

$$\varphi_{2N} = \frac{1}{P(\varphi_0)}\int^{\varphi_0} I_{2N}P(\varphi_0)\,d\varphi_0, \qquad N \geq 1, \qquad (2.2.25')$$

where

$$
\begin{aligned}
I_{2N} = \frac{1}{2T(\varphi_0)}&\Bigg[\frac{1}{P^2(\varphi_0)(d\varphi_0/dz)^4}\sum_{q=0}^{\infty}B_q\sum_{\beta_1+\cdots+\beta_{q+2}=N-1}\varphi_{2\beta_1}\cdots\varphi_{2\beta_q}\\
&\times\left([R(z)-Q^2(z)]\frac{d\varphi_{2\beta_{q+1}}}{dz}\frac{d\varphi_{2\beta_{q+2}}}{dz}+\frac{3}{4}\frac{d^2\varphi_{2\beta_{q+1}}}{dz^2}\frac{d^2\varphi_{2\beta_{q+2}}}{dz^2}\right.\\
&\left.-\frac{1}{4}\frac{d\varphi_{2\beta_{q+1}}}{dz}\frac{d^3\varphi_{2\beta_{q+2}}}{dz^3}-\frac{1}{4}\frac{d^3\varphi_{2\beta_{q+1}}}{dz^3}\frac{d\varphi_{2\beta_{q+2}}}{dz}\right)\\
&+\frac{1}{(d\varphi_0/dz)^2}\sum_{q=0}^{\infty}B_q\sum_{\beta_1+\cdots+\beta_{q+2}=N}\varphi_{2\beta_1}\cdots\varphi_{2\beta_q}\frac{d\varphi_{2\beta_{q+1}}}{dz}\frac{d\varphi_{2\beta_{q+2}}}{dz}\\
&-\frac{1}{P^2(\varphi_0)}\sum_{n=0}^{\infty}\left(A_{n,2N}\varphi_0^n+\frac{1}{(d\varphi_0/dz)^4}\right.\\
&\left.\times\sum_{\beta_0+\cdots+\beta_{n+4}=N}A_{n,2\beta_0}\varphi_{2\beta_1}\cdots\varphi_{2\beta_n}\frac{d\varphi_{2\beta_{n+1}}}{dz}\cdots\frac{d\varphi_{2\beta_{n+4}}}{dz}\right)\Bigg],
\end{aligned}
$$

$$0 \leq \beta_i \leq N-1, \qquad N \geq 1. \qquad (2.2.26)$$

With the aid of the identity

$$
\begin{aligned}
\frac{3}{4}&\frac{d^2\varphi_{2\beta_{q+1}}}{dz^2}\frac{d^2\varphi_{2\beta_{q+2}}}{dz^2}-\frac{1}{4}\frac{d\varphi_{2\beta_{q+1}}}{dz}\frac{d^3\varphi_{2\beta_{q+2}}}{dz^3}-\frac{1}{4}\frac{d^3\varphi_{2\beta_{q+1}}}{dz^3}\frac{d\varphi_{2\beta_{q+2}}}{dz}\\
=&\left(\frac{d\varphi_0}{dz}\right)^4\left\{\left(\frac{d\varphi_0}{dz}\right)^{-3/2}\left[\frac{d^2}{dz^2}\left(\frac{d\varphi_0}{dz}\right)^{-1/2}\right]\frac{d\varphi_{2\beta_{q+1}}}{d\varphi_0}\frac{d\varphi_{2\beta_{q+2}}}{d\varphi_0}\right.\\
&+\frac{3}{4}\frac{d^2\varphi_{2\beta_{q+1}}}{d\varphi_0^2}\frac{d^2\varphi_{2\beta_{q+2}}}{d\varphi_0^2}-\frac{1}{4}\frac{d\varphi_{2\beta_{q+1}}}{d\varphi_0}\frac{d^3\varphi_{2\beta_{q+2}}}{d\varphi_0^3}\\
&\left.-\frac{1}{4}\frac{d^3\varphi_{2\beta_{q+1}}}{d\varphi_0^3}\frac{d\varphi_{2\beta_{q+2}}}{d\varphi_0}\right\}
\end{aligned}
$$

$$(2.2.27)$$

we can rewrite (2.2.26) into the form

$$
\begin{aligned}
I_{2N} = \frac{1}{2T(\varphi_0)} &\Bigg[\frac{1}{P^2(\varphi_0)} \sum_{q=0}^{\infty} B_q \sum_{\beta_1+\cdots+\beta_{q+2}=N-1} \varphi_{2\beta_1} \cdots \varphi_{2\beta_q} \\
&\times \left(g(\varphi_0) \frac{d\varphi_{2\beta_{q+1}}}{d\varphi_0} \frac{d\varphi_{2\beta_{q+2}}}{d\varphi_0} + \frac{3}{4} \frac{d^2\varphi_{2\beta_{q+1}}}{d\varphi_0^2} \frac{d^2\varphi_{2\beta_{q+2}}}{d\varphi_0^2} \right. \\
&\left. - \frac{1}{4} \frac{d\varphi_{2\beta_{q+1}}}{d\varphi_0} \frac{d^3\varphi_{2\beta_{q+2}}}{d\varphi_0^3} - \frac{1}{4} \frac{d^3\varphi_{2\beta_{q+1}}}{d\varphi_0^3} \frac{d\varphi_{2\beta_{q+2}}}{d\varphi_0} \right) \\
&+ \sum_{q=0}^{\infty} B_q \sum_{\beta_1+\cdots+\beta_{q+2}=N} \varphi_{2\beta_1} \cdots \varphi_{2\beta_q} \frac{d\varphi_{2\beta_{q+1}}}{d\varphi_0} \frac{d\varphi_{2\beta_{q+2}}}{d\varphi_0} \\
&- \frac{1}{P^2(\varphi_0)} \sum_{n=0}^{\infty} \left(A_{n,2N}\varphi_0^n + \sum_{\beta_0+\cdots+\beta_{n+4}=N} \Lambda_{n,2\beta_0} \right. \\
&\left. \times \varphi_{2\beta_1} \cdots \varphi_{2\beta_n} \frac{d\varphi_{2\beta_{n+1}}}{d\varphi_0} \cdots \frac{d\varphi_{2\beta_{n+4}}}{d\varphi_0} \right) \Bigg],
\end{aligned}
$$

$$
0 \le \beta_i \le N-1, \quad N \ge 1, \tag{2.2.28}
$$

where

$$
g(\varphi_0) = \frac{R(z) - Q^2(z)}{(d\varphi_0/dz)^2} + \left(\frac{d\varphi_0}{dz} \right)^{-3/2} \frac{d^2}{dz^2} \left(\frac{d\varphi_0}{dz} \right)^{-1/2} \tag{2.2.29}
$$

Since $\varphi_0(z)$ has been determined to be a regular analytic function of z, the derivative of which is different from zero in the whole region under consideration, then the last term on the right-hand side of (2.2.29) is a regular analytic function of z in that region. In many physically important situations it happens, however, that the first term on the right-hand side of (2.2.29) may be singular at particular points in the complex z-plane; this will be discussed in §2.2.4. Using (2.2.17), we can write (2.2.29) as follows

$$
g(\varphi_0) = P^2 \left(\varepsilon_0 - P^{-3/2} \frac{d^2 P^{-1/2}}{d\varphi_0^2} \right) \tag{2.2.30}
$$

where ε_0, defined by (2.1.6), is obviously independent of λ.

Using the definition (2.2.13b) with φ replaced by φ_0, i.e.,

$$
S_{2\beta}(\varphi_0) = \sum_{n=0}^{\infty} A_{n,2\beta}\varphi_0^n, \tag{2.2.31}
$$

we obtain from (2.2.28) with $N = 1$

$$
I_2 = \frac{1}{2} \left(\frac{g}{P^2} - \frac{S_2/T}{P^2} \right), \tag{2.2.32}
$$

which, with the aid of (2.2.30) and the expression (2.1.5b) for Y_2, can be written

$$I_2 = \frac{1}{2}(\varepsilon_0 - h) = Y_2 - \frac{1}{2}h, \tag{2.2.33}$$

where

$$h = P^{-3/2}\frac{d^2 P^{-1/2}}{d\varphi_0^2} + \frac{S_2/T}{P^2}. \tag{2.2.34}$$

Recalling the definitions (2.2.16c) and (2.2.31), we obtain from (2.2.28) with $N = 2$

$$I_4 = \frac{1}{2}\left[\frac{g}{P^2}\left(\frac{\varphi_2}{T}\frac{dT}{d\varphi_0} + 2\frac{d\varphi_2}{d\varphi_0}\right) - \frac{1}{2P^2}\frac{d^3\varphi_2}{d\varphi_0^3} - \frac{S_4/T}{P^2}\right.$$

$$- \frac{1}{P^2}\frac{\varphi_2}{T}\frac{dS_2}{d\varphi_0} - 4\frac{S_2/T}{P^2}\frac{d\varphi_2}{d\varphi_0} - \frac{\varphi_2^2}{2T}\left(\frac{1}{P^2}\frac{d^2 S_0}{d\varphi_0^2} - \frac{d^2 T}{d\varphi_0^2}\right) \tag{2.2.35}$$

$$\left. - 2\frac{\varphi_2}{T}\frac{d\varphi_2}{d\varphi_0}\left(\frac{2}{P^2}\frac{dS_0}{d\varphi_0} - \frac{dT}{d\varphi_0}\right) - \left(\frac{d\varphi_2}{d\varphi_0}\right)^2\left(6\frac{S_0/T}{P^2} - 1\right)\right].$$

Using (2.2.16a), we can rewrite the last three terms in (2.2.35), getting

$$I_4 = \frac{1}{2}\left[\frac{g}{P^2}\left(\frac{\varphi_2}{T}\frac{dT}{d\varphi_0} + 2\frac{d\varphi_2}{d\varphi_0}\right) - \frac{1}{2P^2}\frac{d^3\varphi_2}{d\varphi_0^3} - \frac{S_4/T}{P^2}\right.$$

$$- \frac{1}{P^2}\frac{\varphi_2}{T}\frac{dS_2}{d\varphi_0} - 4\frac{S_2/T}{P^2}\frac{d\varphi_2}{d\varphi_0} - \frac{\varphi_2^2}{P^2}\left(\frac{1}{2}\frac{d^2 P^2}{d\varphi_0^2} + \frac{dP^2}{d\varphi_0}\frac{1}{T}\frac{dT}{d\varphi_0}\right) \tag{2.2.36}$$

$$\left. - 2\varphi_2\frac{d\varphi_2}{d\varphi_0}\left(\frac{2}{P^2}\frac{dP^2}{d\varphi_0} + \frac{1}{T}\frac{dT}{d\varphi_0}\right) - 5\left(\frac{d\varphi_2}{d\varphi_0}\right)^2\right].$$

After a rather lengthy calculation, in which no additional equations are used, one can rewrite (2.2.36) in the form

$$I_4 = \left[\frac{1}{2}\left(\frac{g}{P^2} - \frac{S_2/T}{P^2}\right) - \frac{1}{P}\frac{d(P\varphi_2)}{d\varphi_0}\right]\left(\frac{1}{T}\frac{dT}{d\varphi_0}\varphi_2 + 2\frac{d\varphi_2}{d\varphi_0}\right)$$

$$- \frac{1}{2}\left[\frac{S_2/T}{P^2}\frac{1}{P}\frac{d(P\varphi_2)}{d\varphi_0} + \left(\frac{1}{P}\frac{d(P\varphi_2)}{d\varphi_0}\right)^2 + \frac{1}{2P^2}\frac{d^3\varphi_2}{d\varphi_0^3}\right. \tag{2.2.37}$$

$$\left. + \frac{S_4/T}{P^2} + \frac{1}{P}\frac{d}{d\varphi_0}\left(\frac{S_2/T}{P^2}P\varphi_2 + \frac{dP}{d\varphi_0}\varphi_2^2\right)\right].$$

We note that according to (2.2.32) and (2.2.25), the first bracket on the right-hand side of (2.2.37) is

$$\frac{1}{2}\left(\frac{g}{P^2} - \frac{S_2/T}{P^2}\right) - \frac{1}{P}\frac{d(P\varphi_2)}{d\varphi_0} = I_2 - \frac{1}{P}\frac{d(P\varphi_2)}{d\varphi_0} = 0. \tag{2.2.38}$$

Taking (2.2.38) into account and rewriting in (2.2.37) the two terms containing S_2 into a single term, we obtain the formula

$$I_4 = -\frac{1}{2}\left[\frac{1}{P^2\varphi_2}\frac{d}{d\varphi_0}\left(\frac{S_2\varphi_2^2}{T}\right) + \left(\frac{1}{P}\frac{d(P\varphi_2)}{d\varphi_0}\right)^2\right.$$

$$\left. + \frac{1}{2P^2}\frac{d^3\varphi_2}{d\varphi_0^3} + \frac{S_4/T}{P^2} + \frac{1}{P}\frac{d}{d\varphi_0}\left(\frac{dP}{d\varphi_0}\varphi_2^2\right)\right],$$

(2.2.39)

which will be used later. It is also useful to rewrite (2.2.37) in another way. To this purpose we introduce, instead of z, the λ-independent variable ζ defined by (2.1.7). Recalling (2.2.17) as well, we have

$$\zeta = \int^z Q(z)\,dz = \int^{\varphi_0} P(\varphi_0)\,d\varphi_0$$

(2.2.40)

and hence

$$\frac{d}{d\zeta} = \frac{1}{Q(z)}\frac{d}{dz} = \frac{1}{P(\varphi_0)}\frac{d}{d\varphi_0}.$$

(2.2.41)

Using (2.2.38), (2.2.41) and the identity

$$\frac{d^3\varphi_2}{d\varphi_0^3} = 2P^2\frac{d}{d\zeta}\left(P^{-3/2}\frac{d^2P^{-1/2}}{d\varphi_0^2}P\varphi_2\right)$$

$$+ 2P^{1/2}\frac{d^2P^{-1/2}}{d\varphi_0^2}\frac{d(P\varphi_2)}{d\zeta} + P^2\frac{d^3(P\varphi_2)}{d\zeta^3},$$

(2.2.42)

we can, with the aid of the definition (2.2.34), write (2.2.37) as

$$I_4 = -\frac{1}{2}\left[h\frac{d(P\varphi_2)}{d\zeta} + \left(\frac{d(P\varphi_2)}{d\zeta}\right)^2 + \frac{1}{2}\frac{d^3(P\varphi_2)}{d\zeta^3}\right.$$

$$\left. + \frac{S_4/T}{P^2} + \frac{d}{d\zeta}\left(h\,P\varphi_2 + \frac{dP}{d\varphi_0}\varphi_2^2\right)\right].$$

(2.2.43)

Rewriting (2.2.43), we obtain

$$I_4 = -\frac{1}{8}\left[\left(h + 2\frac{d(P\varphi_2)}{d\zeta}\right)^2 + \frac{d^2}{d\zeta^2}\left(h + 2\frac{d(P\varphi_2)}{d\zeta}\right)\right]$$

$$+ \frac{1}{8}\left(h^2 + \frac{d^2h}{d\zeta^2}\right) - \frac{S_4/T}{2P^2} - \frac{1}{2}\frac{d}{d\zeta}\left(h\,P\varphi_2 + \frac{dP}{d\varphi_0}\varphi_2^2\right).$$

(2.2.44)

From (2.2.41), (2.2.25), and (2.2.33) it follows that

$$h + 2\frac{d(P\varphi_2)}{d\zeta} = h + \frac{2}{P}\frac{d(P\varphi_2)}{d\varphi_0} = h + 2I_2 = \varepsilon_0,$$

(2.2.45)

and hence (2.2.44) can be written

$$I_4 = -\frac{1}{8}\left(\varepsilon_0^2 + \frac{d^2\varepsilon_0}{d\zeta^2}\right) + \frac{1}{8}\left(h^2 + \frac{d^2h}{d\zeta^2}\right) - \frac{S_4/T}{2P^2}$$
$$- \frac{d}{d\zeta}\left(\frac{1}{2}h\,P\varphi_2 + \frac{1}{2}\frac{dP}{d\varphi_0}\varphi_2^2\right),$$

(2.2.46)

where, according to (2.1.5c), the first term in the right-hand member is equal to Y_4, while the second term (except for the opposite sign) has the same analytical form, but with ε_0 replaced by h. It is easily seen that (2.2.46) can also be written as follows

$$I_4 = -\frac{1}{8}(\varepsilon_0 - h)^2 - \frac{1}{4}h(\varepsilon_0 - h) - \frac{S_4/T}{2P^2}$$
$$- \frac{d}{d\zeta}\left(\frac{1}{8}\frac{d(\varepsilon_0 - h)}{d\zeta} + \frac{1}{2}h\,P\varphi_2 + \frac{1}{2}\frac{dP}{d\varphi_0}\varphi_2^2\right).$$

(2.2.47)

Recalling (2.2.33), we can write (2.2.47) as follows

$$I_4 = -\left[\frac{1}{2}I_2^2 + \frac{1}{2}h\,I_2 + \frac{S_4/T}{2P^2}\right.$$
$$\left. + \frac{d}{d\zeta}\left(\frac{1}{4}\frac{dI_2}{d\zeta} + \frac{1}{2}h\,P\varphi_2 + \frac{1}{2}\frac{dP}{d\varphi_0}\varphi_2^2\right)\right].$$

(2.2.48)

This formula is particularly useful for calculating the limiting value of I_4 at a first-order pole of $Q^2(z)$; see §2.2.7.

2.2.4   Regularity Properties of I_{2N} and φ_{2N} when $N > 0$

We shall now study the regularity properties of I_{2N} and confirm the possibility of determining the functions φ_{2N} as regular functions of φ_0.

Let us first consider a point where $R(z)$ and $Q^2(z)$ are both regular and where $Q^2(z)$ is different from zero, which implies that $P^2(\varphi_0)$ is also regular and different from zero (see [2.2.17]). Since $d\varphi_0/dz$ is everywhere different from zero, it then follows from (2.2.29) that $g(\varphi_0)$ is regular. From (2.2.28) one then realizes that I_{2N} is regular, which implies that φ_{2N} can be determined as a regular analytic function of φ_0 from (2.2.25') for $N = 1, 2, \ldots$.

Next we shall consider the neighborhood of a second-order pole t'_j of $Q^2(z)$ in the complex z-plane, corresponding to a second-order pole of $\Pi(\varphi)$, situated at the point $\varphi = \tau'_j$, where τ'_j is a function of the λ-independent quantities B_q (cf. [2.2.9a–c]). Since $\tau'_j = \varphi(t'_j) = \varphi_0(t'_j) + \lambda^2\varphi_2(t'_j) + \ldots$, and τ'_j does not depend on λ, it can be seen that

$$\varphi_{2\beta}(t'_j) = 0 \qquad \text{for } \beta > 0. \tag{2.2.49}$$

Let us now consider the condition (2.2.6). Since $\varphi(t'_j) = \tau'_j$, we can, with the aid of (2.2.9a) and (2.2.13a), write the left-hand side of (2.2.6) as

$$\lim_{\varphi \to \tau'_j} (\varphi - \tau'_j)^2 \Pi(\varphi) = \sum_{\beta=0}^{\infty} \lim_{\varphi \to \tau'_j} (\varphi - \tau'_j)^2 \frac{S_{2\beta}(\varphi)\lambda^{2\beta}}{T(\varphi)}$$

$$= \sum_{\beta=0}^{\infty} \lim_{\varphi_0 \to \tau'_j} (\varphi_0 - \tau'_j)^2 \frac{S_{2\beta}(\varphi_0)\lambda^{2\beta}}{T(\varphi_0)},$$

(2.2.50)

where the last expression is obtained simply by renaming the variable. Equating the coefficients of the successive powers of λ^2 in this expression with the corresponding coefficients in the right-hand member of (2.2.6) and recalling (2.2.16a), we obtain

$$\lim_{\varphi_0 \to \tau'_j} (\varphi_0 - \tau'_j)^2 P^2(\varphi_0) = \lim_{z \to t'_j} (z - t'_j)^2 Q^2(z),$$

(2.2.51a)

$$\lim_{\varphi_0 \to \tau'_j} (\varphi_0 - \tau'_j)^2 \frac{S_2(\varphi_0)}{T(\varphi_0)} = \lim_{z \to t'_j} (z - t'_j)^2 [R(z) - Q^2(z)],$$

(2.2.51b)

$$\lim_{\varphi_0 \to \tau'_j} (\varphi_0 - \tau'_j)^2 \frac{S_{2\beta}(\varphi_0)}{T(\varphi_0)} = 0, \qquad 2\beta \geq 4.$$

(2.2.51c)

We would like to remark that (2.2.51a) can also be obtained from (2.2.19). Using the definition (2.2.29) of $g(\varphi_0)$ taking due regard of the fact that $d\varphi_0/dz$ is everywhere finite and different from zero, we can rewrite the right-hand member of (2.2.51b) as

$$\lim_{z \to t'_j} (z - t'_j)^2 [R(z) - Q^2(z)] = \lim_{z \to t'_j} (z - t'_j)^2 \left(\frac{d\varphi_0}{dz}\right)^2 g(\varphi_0)$$

$$= \lim_{z \to t'_j} \left(\frac{d\varphi_0/dz}{(\varphi_0 - \tau'_j)/(z - t'_j)}\right)^2 (\varphi_0 - \tau'_j)^2 g(\varphi_0)$$

(2.2.52)

$$= \lim_{\varphi_0 \to \tau'_j} (\varphi_0 - \tau'_j)^2 g(\varphi_0).$$

From (2.2.51b) and (2.2.52) we obtain

$$\lim_{\varphi_0 \to \tau'_j} (\varphi_0 - \tau'_j)^2 \left(g(\varphi_0) - \frac{S_2(\varphi_0)}{T(\varphi_0)}\right) = 0.$$

(2.2.51b')

From (2.2.32) and (2.2.51b') we obtain

$$\lim_{\varphi_0 \to \tau'_j} (\varphi_0 - \tau'_j)^2 P^2(\varphi_0) I_2 = 0.$$

(2.2.51b'')

Consider a situation when $Q^2(z)$ has a first- or second-order pole at $z = t_j'$ and, correspondingly, $P^2(\varphi_0)$ has a first- or second-order pole at $\varphi_0 = \tau_j'$; we conclude from (2.2.51b″) and (as regards a first-order pole) from calculations corresponding to those in the derivation of (7.3.14a) in §7.3 that the formula

$$I_{2N} = \begin{cases} O(\varphi_0 - \tau_j') \text{ as } \varphi_0 \to \tau_j', & \text{if } P^2(\varphi_0) \\ & \text{has a second-order pole at } \tau_j' \\ O(1) \text{ as } \varphi_0 \to \tau_j', & \text{if } P^2(\varphi_0) \\ & \text{has a first-order pole at } \tau_j' \end{cases}$$

(2.2.53)

is true for $N = 1$. From (2.2.25′) with $N = 1$ and (2.2.53) with $N = 1$ it is seen that, if $P^2(\varphi_0)$ has a second-order pole, then the function φ_2 is regular analytic at $\varphi_0 = \tau_j'$ independently of the choice of the integration constant in (2.2.25′); whereas if $P^2(\varphi_0)$ has a first-order pole at $\varphi_0 = \tau_j'$, then we must choose the lower limit of integration in (2.2.25′) to be the pole itself in order for φ_2 to be single-valued. Choosing the constant lower limit of integration in (2.2.25′) to be arbitrary when $P^2(\varphi_0)$ has a second-order pole at τ_j', but to be τ_j' when $P^2(\varphi_0)$ has a first-order pole at τ_j'; we find from (2.2.25′) in both cases that φ_2 behaves as $O(\varphi_0 - \tau_j')$ when φ_0 tends to τ_j'. The function S_2/T has, at the most, a second-order pole at $\varphi_0 = \tau_j'$, and S_4/T has, at the most, a first-order pole at $\varphi_0 = \tau_j'$; see (2.2.51c). Using the above results concerning the behavior of φ_2, S_2/T and S_4/T when φ_0 tends to τ_j'; we find from the expression (2.2.39) for I_4 that (2.2.53) is also true for $N = 2$. Our detailed examination of the regularity properties of I_{2N} in the neighborhood of a pole of $P^2(\varphi_0)$ has been restricted to $N = 1$ and $N = 2$, which should suffice in many practical applications. However, it is to be expected that analogous regularity properties prevail for $N > 2$. We therefore assume that (2.2.53) will be true for $N = 3, 4, \ldots$; that is, for every positive integer N. Accordingly, we will formulate our discussion below in general terms, not necessarily restricting ourselves to $N = 1$ or $N = 2$.

Continuing the treatment of the situation when $Q^2(z)$ has a first- or second-order pole at $z = t_j'$, and hence $P^2(\varphi_0)$ has a first- or second-order pole at $\varphi_0 = \tau_j'$; we obtain from (2.2.53) the formula

$$I_{2N} P(\varphi_0) = \begin{cases} O(1) & \text{as } \varphi_0 \to \tau_j', \text{ if } P^2(\varphi_0) \\ & \text{has a second-order pole at } \tau_j' \\ O(1/(\varphi_0 - \tau_j')^{1/2}) & \text{as } \varphi_0 \to \tau_j', \text{ if } P^2(\varphi_0) \\ & \text{has a first-order pole at } \tau_j'. \end{cases}$$

(2.2.54)

From (2.2.25′), where we choose the constant lower limit of integration to be arbitrary, when $P^2(\varphi_0)$ has a second-order pole at τ_j', but to be τ_j', when

$P^2(\varphi_0)$ has a first-order pole at τ_j'; we find, with the aid of (2.2.54), that

$$\varphi_{2N} = O(\varphi_0 - \tau_j') \text{ close to a first- or second-order pole of } P^2(\varphi_0).$$
$$(2.2.55)$$

We shall now consider the neighborhood of an *arbitrary-order zero* t_i of $Q^2(z)$, where $R(z)$ and hence $R(z) - Q^2(z)$ is regular analytic. The neighborhood of t_i corresponds to the neighborhood of a zero τ_i of $P^2(\varphi_0)$ that has the same multiplicity as the zero t_i of $Q^2(z)$ (see [2.2.17]). In the neighborhood of τ_i, the function $T(\varphi_0)$ is different from zero, and the functions $g(\varphi_0)$ and $S_{2\beta}(\varphi_0)/T(\varphi_0)$ are regular analytic (see [2.2.16a], [2.2.29] and [2.2.31]). Hence it can be seen from (2.2.32) and (2.2.39) that for $N = 1$ and $N = 2$ the function $I_{2N}P^2(\varphi_0)$ is also regular analytic in the neighborhood of τ_i. It is plausible to assume this assertion to be true also for $N = 3, 4, \ldots$ and hence for every positive integer N. As in (2.2.18) we denote the order of the zero τ_i by m_i. When $m_i = 1$ it is sufficient to choose the constant lower limit of integration in (2.2.25') to be τ_i in order to make φ_{2N} regular analytic at τ_i. However, when $m_i > 1$, one must also choose the coefficients $A_{n,2\beta}$ in the definition (2.2.31) of $S_{2\beta}(\varphi_0)$ so that, when the regular function $I_{2N}P^2(\varphi_0)$ is expanded in powers of $\varphi_0 - \tau_i$, the first $m_i - 1$ coefficients vanish, and $I_{2N}P^2(\varphi_0)$ thus becomes approximately proportional to $(\varphi_0 - \tau_i)^{m_i-1}$. Successively considering $N = 1, 2, 3, \ldots$, one thus obtains $m_i - 1$ conditions for the coefficients $A_{n,2N}$ for every value of N. When the constant lower limit of integration in (2.2.25') is then chosen to be τ_i, one obtains φ_{2N} as a regular analytic function of φ_0 in the neighborhood of τ_i.

2.2.5   Determination of the Coefficients $A_{n,2N}$ when $N > 0$

Assuming that for a given positive integer N the coefficients $A_{n,2\beta}$ with $\beta < N$, as well as the coefficients B_q, have been determined appropriately, and that the functions $\varphi_{2\beta}$ with $\beta < N$ have been obtained as *regular* (and hence single-valued) analytic functions of φ_0, we shall consider the determination of the coefficients $A_{n,2N}$ and the function φ_{2N}.

At a second-order pole τ_j' of $P^2(\varphi_0)$ the conditions (2.2.51a,b,c) must be fulfilled in order that the regular solution of (2.2.2), when expressed in terms of the regular solution of (2.2.5) according to (2.2.3), has the correct behavior at the pole. Furthermore, we have also found (see §2.2.4) that in the neighborhood of a second-order pole of $P^2(\varphi_0)$ a regular function φ_{2N} is automatically obtained from (2.2.25') for any choice of the integration constant; whereas in the neighborhood of a first-order pole or an arbitrary-order zero of $P^2(\varphi_0)$ one can obtain a regular, single-valued function φ_{2N} only with a convenient choice of the constant lower limit of integration in (2.2.25'). For a first-order pole or first-order zero of $P^2(\varphi_0)$, no further condition is required, but for a higher-order zero τ_i of $P^2(\varphi_0)$ with the multiplicity m_i, one must impose $m_i - 1$ conditions on the coefficients $A_{n,2N}$ (see the end of §2.2.4). These conditions make the integrand in (2.2.25') sufficiently small, close to τ_i, so that φ_{2N} will be regular when one chooses the constant lower limit of integration in (2.2.25′) to be τ_i. The value of this constant lower limit of integration must be the same whichever pole or zero of $P^2(\varphi_0)$ one considers, since there is only one integration constant in (2.2.25′). This requirement provides us with further conditions for determining the coefficients $A_{n,2N}$, if the total number of arbitrary-order zeros and first-order poles of $P^2(\varphi_0)$ is larger than 1. To elucidate this assertion, we assume that there is at least one arbitrary-order zero τ_1 or first-order pole τ_1' of $P^2(\varphi_0)$, and we choose the constant lower limit of integration in (2.2.25′) so as to make φ_{2N} regular and thus single-valued in the neighborhood of τ_1 or τ_1', respectively. Thus we get

$$\varphi_{2N} = \frac{1}{P(\varphi_0)} \int_{\tau_1}^{\varphi_0} I_{2N} \, P(\varphi_0) \, d\varphi_0 \qquad (2.2.56a)$$

or

$$\varphi_{2N} = \frac{1}{P(\varphi_0)} \int_{\tau_1'}^{\varphi_0} I_{2N} \, P(\varphi_0) \, d\varphi_0, \qquad (2.2.56b)$$

respectively. If there exist further arbitrary-order zeros or first-order poles of $P^2(\varphi_0)$ we require that the function φ_{2N}, given by (2.2.56a) or (2.2.56b), also be regular at those points, leading us to the conditions

$$\int_{\tau_i}^{\tau_i'} I_{2N} \, P(\varphi_0) \, d\varphi_0 = 0 \qquad (2.2.57a)$$

$$\int_{\tau_i}^{\tau_j'} I_{2N} \, P(\varphi_0) \, d\varphi_0 = 0 \qquad (2.2.57b)$$

$$\int_{\tau_j'}^{\tau_j'\,'} I_{2N} \, P(\varphi_0) \, d\varphi = 0, \qquad (2.2.57c)$$

which must be fulfilled for all possible relevant values of i, i', j and j'. The number of new independent conditions thus obtained on the coefficients $A_{n,2N}$ is equal to $r + s - 1$ minus the number of second-order poles in the relevant region; i.e., one less than the total number of first-order poles and arbitrary-order zeros. If the conditions (2.2.57a,b,c) are associated with first-order poles or odd-order zeros, then they may be replaced by conditions of the form

$$\int_\Lambda I_{2N} \, P(\varphi_0) \, d\varphi_0 = 0, \qquad (2.2.58)$$

where Λ is a contour encircling a first-order pole or odd-order zero as well as one (but only one) other first-order pole or odd-order zero. In the neighborhood of a second-order pole of $P^2(\varphi_0)$ the function $I_{2N} P(\varphi_0)$ is regular according to (2.2.54), and the integral $\int^{\varphi_0} I_{2N} P(\varphi_0) d\varphi_0$ is thus single-valued. Therefore, we can allow the contour Λ in (2.2.58) to enclose one or more second-order poles of $P^2(\varphi_0)$. It is often practical to do so in the applications. We would also like to remark that in the applications it is found that the number of the available coefficients $A_{n,2N}$ is sufficient to fulfill all the conditions described above. It also deserves to be remarked that the necessity of the condition (2.2.58) is an immediate consequence of (2.2.56a,b) and the single-valuedness of φ_{2N} as a function of φ_0 or z.

For $N = 1$ we can, with the aid of (2.2.33) and (2.2.17), write (2.2.58) as

$$\int_\Lambda \frac{1}{2} h\, P(\varphi_0)\, d\varphi_0 = \int_\Lambda Y_2\, Q(z)\, dz, \tag{2.2.59a}$$

and for $N = 2$ we can, with the aid of (2.2.46), (2.2.17) and (2.1.5c), write (2.2.58) as follows

$$\int_\Lambda \left(-\frac{1}{8} h^2 + \frac{S_4/T}{2P^2} \right) P(\varphi_0) d\varphi_0 = \int_\Lambda Y_4\, Q(z)\, dz. \tag{2.2.59b}$$

The conditions (2.2.59a) and (2.2.59b) for $N = 1$ and $N = 2$ are the third- and fifth-order counterparts of the first-order condition (2.2.24′).

2.2.6 Expressions for φ_2 and φ_4

When, for $N = 1$ or $N = 2$, the constant lower limit of integration in (2.2.25′) is a regular point of $P^2(\varphi_0)$ such that $P^2(\varphi_0) \neq 0$; the integral in (2.2.25′) can be written as a sum of convergent integrals over the separate terms in the expressions (2.2.33) for I_2 or (2.2.46) for I_4. However, this is no longer possible when the constant lower limit of integration in (2.2.25′) is a zero or first-order pole of $P^2(\varphi_0)$, since the *separate* integrals then no longer converge at the constant lower limit of integration. However, if, in particular, this lower limit of integration is an odd-order zero τ_i or first-order pole τ_j' of $P^2(\varphi_0)$, one realizes that, when φ_0 moves one turn around τ_i or τ_j', the integrand changes sign, since $P(\varphi_0)$ then changes sign, while I_{2N} remains unchanged. The convergent integral from τ_i or τ_j' to φ_0 in (2.2.25′) can therefore be written as half of the integral along a contour, called $\Gamma_{\tau_i}(\varphi_0)$ or $\Gamma_{\tau_j'}(\varphi_0)$, starting at the point corresponding to φ_0 on a Riemann sheet adjacent to that on which φ_0 lies, encircling the point τ_i or τ_j' one turn in the positive or negative sense, and ending at φ_0; the value of this contour integral is the same whether the encircling takes place in the positive or in the negative sense. Thus, the integral in (2.2.25′) can be written as a sum of convergent integrals corresponding to the separate terms in the integrand; and in particular those integrals, which correspond to total derivatives with respect to φ_0 in the integrand, can be evaluated. Since the mapping between z and φ_0 is conformal, the contour $\Gamma_{\tau_i}(\varphi_0)$ or $\Gamma_{\tau_j'}(\varphi_0)$ in the complex φ_0-plane corresponds to a similar contour $\Gamma_{t_i}(z)$ or $\Gamma_{t_j'}(z)$ in the complex z-plane. Thus, $\Gamma_{t_i}(z)$ or $\Gamma_{t_j'}(z)$ starts at the point corresponding to z on a Riemann sheet adjacent to that on which z

lies, encircles the point t_i or t'_j one turn in the positive or negative sense, and ends at z. Noting that $P(\varphi_0)d\varphi_0 = Q(z)dz$ according to (2.2.17) and introducing in an obvious way unspecified constants c_2 and c_4, which can be chosen conveniently; we therefore obtain from (2.2.25'), (2.2.33), (2.2.46), and (2.1.5c) the formulas

$$P(\varphi_0)\varphi_2 = \frac{1}{2}\int_{\Gamma_{t_i}(z)} (Y_2 - c_2)Q(z)\,dz$$

$$-\frac{1}{2}\int_{\Gamma_{\tau_i}(\varphi_0)} \left(\frac{1}{2}h - c_2\right) P(\varphi_0)\,d\varphi_0 \qquad (2.2.60\text{a})$$

$$P(\varphi_0)\varphi_4 = \frac{1}{2}\int_{\Gamma_{t_i}(z)} (Y_4 - c_4)Q(z)\,dz$$

$$-\frac{1}{2}\int_{\Gamma_{\tau_i}(\varphi_0)} \left[-\frac{1}{8}\left(h^2 + \frac{d^2h}{d\zeta^2}\right) + \frac{S_4/T}{2P^2} - c_4\right] P(\varphi_0)\,d\varphi_0$$

$$-\frac{1}{2}\left(h\,P\varphi_2 + \frac{dP}{d\varphi_0}\varphi_2^2\right) \qquad (2.2.60\text{b})$$

for the case of an odd-order zero. By replacing t_i by t'_j and τ_i by τ'_j in these formulas, one obtains the corresponding formulas for a first-order pole. Introducing short-hand notations for the integrals in (2.2.60a,b), as explained in connection with Eq. (4.3.3) in Chapter 4, we write (2.2.60a,b) as

$$P(\varphi_0)\varphi_2 = \int_{(t_i)}^z (Y_2 - c_2)Q(z)\,dz - \int_{(\tau_i)}^{\varphi_0} \left(\frac{1}{2}h - c_2\right) P(\varphi_0)\,d\varphi_0 \quad (2.2.60\text{a}')$$

$$P(\varphi_0)\varphi_4 = \int_{(t_i)}^z (Y_4 - c_4)Q(z)\,dz$$

$$-\int_{(\tau_i)}^{\varphi_0} \left[-\frac{1}{8}\left(h^2 + \frac{d^2h}{d\zeta^2}\right) + \frac{S_4/T}{2P^2} - c_4\right] P(\varphi_0)\,d\varphi_0$$

$$-\frac{1}{2}\left(h\,P\varphi_2 + \frac{dP}{d\varphi_0}\varphi_2^2\right). \qquad (2.2.60\text{b}')$$

2.2.7 Behavior of $\varphi_{2N}(z)$ in the Neighborhood of a First- or Second-Order Pole of $Q^2(z)$ when $N > 0$

According to (2.2.55) we have $\varphi_{2N} = O(\varphi_0 - \tau'_j) = O(z - t'_j)$ in the neighborhood of a first- or second-order pole t'_j of $Q^2(z)$. With the aim of obtaining formulas for calculating the limiting value of $\varphi_{2N}/(z - t'_j)$ as $z \to t'_j$, we shall now examine in detail the behavior of φ_{2N} in the neighborhood of a first- or second-order pole of $Q^2(z)$.

Case of a Second-Order Pole. Let us first assume that t'_j is a second-order pole of $Q^2(z)$ and hence that τ'_j is a second-order pole of $P^2(\varphi_0)$. If there is also a zero t_i of $Q^2(z)$ in the region of the complex z-plane under consideration, we have according to (2.2.20) with $z_0 = t_i$ and $\varphi_0(z_0) = \tau_i$

$$\int_{\tau_i}^{\varphi_0} P(\varphi_0)\, d\varphi_0 = \int_{t_i}^{z} Q(z)\, dz. \tag{2.2.61}$$

From (2.2.19) and (2.2.61) we obtain

$$\lim_{z \to t'_j} \frac{\varphi_0 - \tau'_j}{z - t'_j} = \exp\left[\lim_{z \to t'_j} \left(\frac{\int_{t_i}^{z} Q(z)\, dz}{\lim_{z \to t'_j}(z - t'_j)Q(z)} - \ln(z - t'_j) \right) \right.$$
$$\left. - \lim_{\varphi_0 \to \tau'_j} \left(\frac{\int_{\tau_i}^{\varphi_0} P(\varphi_0)\, d\varphi_0}{\lim_{\varphi_0 \to \tau'_j}(\varphi_0 - \tau'_j)P(\varphi_0)} - \ln(\varphi_0 - \tau'_j) \right) \right]. \tag{2.2.62}$$

According to (2.2.55) the function $\varphi_{2N}/(\varphi_0 - \tau'_j)$ and hence $\varphi_{2N}P(\varphi_0)$, in the present case of a second-order pole, is regular in the neighborhood of τ'_j. It follows from (2.2.19) that

$$\lim_{\varphi_0 \to \tau'_j} \frac{\varphi_{2N}}{\varphi_0 - \tau'_j} = \lim_{\varphi_0 \to \tau'_j} \frac{\varphi_{2N}P(\varphi_0)}{(\varphi_0 - \tau'_j)P(\varphi_0)}$$
$$= \frac{\lim_{\varphi_0 \to \tau'_j} \varphi_{2N}P(\varphi_0)}{\lim_{z \to t'_j}(z - t'_j)Q(z)}, \quad N > 0, \tag{2.2.63}$$

and

$$\lim_{\varphi_0 \to \tau'_j} P^{-3/2} \frac{d^2 P^{-1/2}}{d\varphi_0^2} = \lim_{z \to t'_j} Q^{-3/2}(z) \frac{d^2 Q^{-1/2}}{dz^2}, \tag{2.2.64}$$

where the limiting values are finite. From (2.2.51a,b,c) it follows that

$$\lim_{\varphi_0 \to \tau'_j} \frac{S_2/T}{P^2} = \lim_{\varphi_0 \to \tau'_j} \frac{(\varphi_0 - \tau'_j)^2 S_2/T}{(\varphi_0 - \tau'_j)^2 P^2} = \lim_{z \to t'_j} \frac{(z - t'_j)^2[R(z) - Q^2(z)]}{(z - t'_j)^2 Q^2(z)}$$
$$= \lim_{z \to t'_j} \frac{R(z) - Q^2(z)}{Q^2(z)}, \tag{2.2.65}$$

where the limiting value is finite, and

$$\lim_{\varphi_0 \to \tau'_j} \frac{S_{2\beta}/T}{P^2} = \lim_{\varphi_0 \to \tau'_j} \frac{(\varphi_0 - \tau'_j)^2 S_{2\beta}/T}{(\varphi_0 - \tau'_j)^2 P^2} = 0 \tag{2.2.66}$$

when $2\beta \geq 4$. Because of (2.2.64) and (2.2.65) we obtain from (2.1.6) and (2.2.34)

$$\lim_{\varphi_0 \to \tau'_j} h = \lim_{z \to t'_j} \varepsilon_0, \qquad (2.2.67)$$

where the limiting value is finite. This result is consistent with (2.2.33) and (2.2.53).

To obtain explicit formulas for the limiting values of $\varphi_2/(\varphi_0 - \tau'_j)$ and $\varphi_4/(\varphi_0 - \tau'_j)$ as $\varphi_0 \to \tau'_j$ from (2.2.63) and (2.2.60a',b'), we shall now choose the constants c_2 and c_4 in (2.2.60a',b') so that the integrals occurring there are convergent at $z = t'_j$ and $\varphi_0 = \tau'_j$, and we shall also obtain a convenient formula for the limiting value of the last term in (2.2.60b'). Recalling (2.1.5b), i.e., $Y_2 = \frac{1}{2}\varepsilon_0$, we can use (2.2.67) to define c_2 as follows:

$$c_2 = \lim_{z \to t'_j} Y_2 = \frac{1}{2} \lim_{z \to t'_j} \varepsilon_0 = \frac{1}{2} \lim_{\varphi_0 \to \tau'_j} h. \qquad (2.2.68a)$$

Recalling (2.1.5c), i.e., $Y_4 = -\frac{1}{8}\varepsilon_0^2 - \frac{1}{8}d^2\varepsilon_0/d\zeta^2$, we can use (2.2.41), (2.2.67), and (2.2.66) to define c_4 as follows:

$$c_4 = \lim_{z \to t'_j} Y_4 = -\frac{1}{8} \lim_{z \to t'_j} \varepsilon_0^2 = -\frac{1}{8} \lim_{\varphi_0 \to \tau'_j} h^2$$

$$= \lim_{\varphi_0 \to \tau'_j} \left[-\frac{1}{8}\left(h^2 + \frac{d^2 h}{d\zeta^2} \right) + \frac{S_4/T}{P^2} \right]. \qquad (2.2.68b)$$

Using (2.2.68a) and the fact that $P^2(\varphi_0)$ has a second-order pole at $\varphi_0 = \tau'_j$, we easily obtain the formula

$$\lim_{\varphi_0 \to \tau'_j} \frac{1}{2}\left(h\, P\varphi_2 + \frac{dP}{d\varphi_0}\varphi_2^2 \right) = \left(\lim_{\varphi_0 \to \tau'_j}(\varphi_0 - \tau'_j)\, P(\varphi_0) \right)$$

$$\times \left[c_2 \lim_{\varphi_0 \to \tau'_j} \frac{\varphi_2}{\varphi_0 - \tau'_j} - \frac{1}{2}\left(\lim_{\varphi_0 \to \tau'_j} \frac{\varphi_2}{\varphi_0 - \tau'_j} \right)^2 \right]. \qquad (2.2.69)$$

We can now write down the desired formulas for the limiting values of $\varphi_2/(\varphi_0 - \tau'_j)$ and $\varphi_4/(\varphi_0 - \tau'_j)$ as $\varphi_0 \to \tau'_j$. We obtain from (2.2.63) with $N = 1$, (2.2.60a') and (2.1.5b)

$$\lim_{\varphi_0 \to \tau'_j} \frac{\varphi_2}{\varphi_0 - \tau'_j} = \left(\lim_{z \to t'_j}(z - t'_j)Q(z) \right)^{-1}$$

$$\times \left\{ \int_{(t_i)}^{t'_j} \left(\frac{1}{2}\varepsilon_0 - c_2 \right) Q(z)\, dz \right.$$

$$- \int_{(\tau_i)}^{\tau'_j} \left(\frac{1}{2} h - c_2 \right) P(\varphi_0) \, d\varphi_0 \Bigg\} \tag{2.2.70a}$$

and from (2.2.63) with $N = 2$, (2.2.60b'), (2.1.5c), (2.2.41), (2.2.69), and (2.2.19), when we also note that those terms in the integrands which contain $d^2\varepsilon_0/d\zeta^2$ and $d^2h/d\zeta^2$ give no contributions to the integrals in (2.2.60b') when $z = t'_j$ and thus $\varphi_0 = \tau'_j$,

$$\lim_{\varphi_0 \to \tau'_j} \frac{\varphi_4}{\varphi_0 - \tau'_j} = \left(\lim_{z \to t'_j} (z_0 - t'_j) Q(z) \right)^{-1} \left[\int_{(t_i)}^{t'_j} \left(-\frac{1}{8} \varepsilon_0^2 - c_4 \right) Q(z) \, dz \right.$$

$$- \int_{(\tau_i)}^{\tau'_j} \left(-\frac{1}{8} h^2 + \frac{S_4/T}{2P^2} - c_4 \right) P(\varphi_0) \, d\varphi_0 \bigg]$$

$$- c_2 \lim_{\varphi_0 \to \tau'_j} \frac{\varphi_2}{\varphi_0 - \tau'_j} + \frac{1}{2} \left(\lim_{\varphi_0 \to \tau'_j} \frac{\varphi_2}{\varphi_0 - \tau'_j} \right)^2. \tag{2.2.70b}$$

With c_2 and c_4 given by (2.2.68a) and (2.2.68b), the formulas (2.2.70a) and (2.2.70b) are, in the present situation of a second-order pole of $Q^2(z)$, convenient expressions for obtaining the limiting behavior of φ_2 and φ_4.

Case of a First-Order Pole. Let us now assume that t'_j is a first-order pole of $Q^2(z)$ and that hence τ'_j is a first-order pole of $P^2(\varphi_0)$. According to (2.2.20) with $z_0 = t'_j$ and $\varphi_0(z_0) = \tau'_j$ we then have

$$\int_{\tau'_j}^{\varphi_0} P(\varphi_0) \, d\varphi_0 = \int_{t'_j}^{z} Q(z) \, dz, \tag{2.2.71}$$

and from this formula it follows that

$$\lim_{z \to t'_j} \frac{\varphi_0 - \tau'_j}{z - t'_j} = \frac{\lim_{z \to t'_j} (z - t'_j) Q^2(z)}{\lim_{\varphi_0 \to \tau'_j} (\varphi_0 - \tau'_j) P^2(\varphi_0)} = \frac{\operatorname{Res} Q^2(z)}{\operatorname{Res} P^2(\varphi_0)}, \tag{2.2.72}$$

where Res denotes the appropriate residue.

According to (2.2.53) the function I_{2N} is, in the present case of a first-order pole of $P^2(\varphi_0)$, regular in the neighborhood of τ'_j; hence it follows from (2.2.25'), with the constant lower limit of integration chosen to be τ'_j, that

$$\lim_{\varphi_0 \to \tau'_j} \frac{\varphi_{2N}}{\varphi_0 - \tau'_j} = 2 \lim_{\varphi_0 \to \tau'_j} I_{2N}, \qquad N > 0. \tag{2.2.73}$$

When z lies close to t'_j, and hence φ_0 lies close to τ'_j, one can use (2.2.71) to expand z in powers of $\varphi_0 - \tau'_j$. From (2.2.33), (2.1.6), and (2.2.34) one can then obtain I_2 as a power series in $\varphi_0 - \tau'_j$. The limiting values of I_2 and $dI_2/d\varphi_0$ as $\varphi_0 \to \tau'_j$ can thus be calculated.

We shall now describe how the limiting value of I_4 can be expressed in terms of the limiting values of I_2 and $dI_2/d\varphi_0$ as $\varphi_0 \to \tau'_j$. For this purpose it is convenient to use the formula (2.2.48) for I_4.

We first show that two of the terms in the right-hand member of (2.2.48) cancel out in the limit as $\varphi_0 \to \tau'_j$. Using (2.2.25) and (2.2.41), we rewrite the sum of the two terms as

$$\frac{1}{2} I_2^2 + \frac{d}{d\zeta}\left(\frac{1}{2}\frac{dP}{d\varphi_0}\varphi_2^2\right) = \frac{1}{2}\left(\frac{d\varphi_2}{d\varphi_0}\right)^2$$

$$+\frac{1}{P^2}\frac{dP^2}{d\varphi_0}\varphi_2\frac{d\varphi_2}{d\varphi_0} + \frac{1}{4P^2}\frac{d^2P^2}{d\varphi_0^2}\varphi_2^2. \qquad (2.2.74)$$

Since $P^2(\varphi_0)$ has a first-order pole at $\varphi_0 = \tau'_j$, and $\varphi_2 = O(\varphi_0 - \tau'_j)$ according to (2.2.55), the limiting value of the right-hand member of (2.2.74) is equal to

$$\lim_{\varphi_0\to\tau'_j}\left[\frac{1}{2}\left(\frac{d\varphi_2}{d\varphi_0}\right)^2 - \frac{\varphi_2}{\varphi_0 - \tau'_j}\frac{d\varphi_2}{d\varphi_0} + \frac{1}{2}\left(\frac{\varphi_2}{\varphi_0 - \tau'_j}\right)^2\right]$$

$$= \frac{1}{2}\lim_{\varphi_0\to\tau'_j}\left(\frac{d\varphi_2}{d\varphi_0} - \frac{\varphi_2}{\varphi_0 - \tau'_j}\right)^2 = 0,$$

and hence (2.2.74) yields

$$\lim_{\varphi_0\to\tau'_j}\left[\frac{1}{2}I_2^2 + \frac{d}{d\zeta}\left(\frac{1}{2}\frac{dP}{d\varphi_0}\varphi_2^2\right)\right] = 0. \qquad (2.2.75)$$

Using (2.2.25) and (2.2.41), we next rewrite the sum of two other terms in the right-hand member of (2.2.48) as

$$\frac{1}{2}h I_2 + \frac{d}{d\zeta}\left(\frac{1}{2}h P\varphi_2\right) = h I_2 + \frac{1}{2}\frac{dh}{d\varphi_0}\varphi_2. \qquad (2.2.76)$$

Since $P^2(\varphi_0)$ is assumed to have a first-order pole, h has, according to (2.2.34), a first-order pole (or is possibly regular) at $\varphi_0 = \tau'_j$. Using these facts and (2.2.25') with the constant lower limit of integration chosen to be τ'_j, we obtain from (2.2.76)

$$\lim_{\varphi_0\to\tau'_j}\left[\frac{1}{2}h I_2 + \frac{d}{d\zeta}\left(\frac{1}{2}h P\varphi_2\right)\right] = \left[\lim\left(h - \frac{\text{Res }h}{\varphi_0 - \tau'_j}\right)\right.$$

$$+ \frac{\text{Res }h}{3}\lim\left(\frac{P^2}{\text{Res }P^2} - \frac{1}{\varphi_0 - \tau'_j}\right)\left]\lim I_2 + \frac{2}{3}(\text{Res }h)\lim\frac{dI_2}{d\varphi_0}, \quad (2.2.77)\right.$$

where Res denotes the residue of the function in question at $\varphi_0 = \tau'_j$, and lim denotes the limit as $\varphi_0 \to \tau'_j$. Formula (2.2.77) is valid whether h has a first-order pole or is regular at $\varphi_0 = \tau'_j$. The exceptional situation that obtains when h is regular at $\varphi_0 = \tau'_j$ (i.e., Res $h = 0$) implies, according to (2.2.33) and the fact that lim I_2 is finite, that ε_0 is regular at $z = t'_j$; which, since $Q^2(z)$, like $P^2(\varphi_0)$, has a first-order pole at $z = t'_j$, can occur only when $R(z)$ has a second-order pole with the coefficient $3/16$ at $z = t'_j$; see (2.1.6).

Examining the term in (2.2.48) which involves $d^2I_2/d\zeta^2$, we obtain with the aid of (2.2.41)

$$\lim_{\varphi_0 \to \tau'_j} \frac{d}{d\zeta}\left(\frac{1}{4}\frac{dI_2}{d\zeta}\right) = \lim\left(\frac{1}{4P^2}\frac{d^2I_2}{d\varphi_0^2} - \frac{1}{8P^4}\frac{dP^2}{d\varphi_0}\frac{dI_2}{d\varphi_0}\right)$$

$$= \frac{1}{8 \text{ Res } P^2}\lim\frac{dI_2}{d\varphi_0}. \tag{2.2.78}$$

Recalling that $P^2(\varphi_0)$ has a first-order pole at τ'_j, one immediately realizes that the limiting value of the term in (2.2.48) which involves S_4/T is

$$\lim_{\varphi_0 \to \tau'_j} \frac{S_4/T}{2P^2} = \frac{\text{Res}(S_4/T)}{2 \text{ Res } P^2}. \tag{2.2.79}$$

Using (2.2.75), (2.2.77), (2.2.78), and (2.2.79), we obtain from (2.2.48)

$$\lim_{\varphi_0 \to \tau'_j} I_4 = -\left\{\frac{\text{Res}(S_4/T)}{2 \text{ Res } P^2} + \left[\lim\left(h - \frac{\text{Res } h}{\varphi_0 - \tau'_j}\right)\right.\right.$$

$$+ \frac{\text{Res } h}{3}\lim\left(\frac{P^2}{\text{Res } P^2} - \frac{1}{\varphi_0 - \tau'_j}\right)\right] \lim I_2$$

$$+ \left[\frac{1}{8 \text{ Res } P^2} + \frac{2}{3}\text{Res } h\right]\lim\frac{dI_2}{d\varphi_0}\right\}. \tag{2.2.80}$$

2.3 Derivation of the Arbitrary-Order Phase-Integral Approximation from the Comparison Equation Solution

To illuminate the connection between the comparison equation solution and the arbitrary-order phase-integral approximation generated from the unspecified base function $Q(z)$, we shall show how one can arrive at the phase-integral functions $f_1(z)$ and $f_2(z)$, defined by (2.1.2a,b), by particularizing the results obtained in §2.2.

We consider a region of the complex z-plane where the function $R(z)$ has neither zeros or poles. We choose $Q^2(z)$ to be a regular function closely similar to $R(z)$, and $\Pi(\varphi)$ to be equal to a constant $(\neq 0)$, which, according to the discussion below (2.2.9c), can be chosen without restriction to be

$$\Pi(\varphi) = 1. \tag{2.3.1}$$

Recalling (2.2.9a,b,c), we therefore set

$$A_n = \delta_{n,0} \tag{2.3.2a}$$

$$B_q = \delta_{q,0}, \tag{2.3.2b}$$

where the quantities on the right-hand sides are Kronecker symbols. From (2.2.12) and (2.3.2a) we obtain

$$A_{n,2\beta} = \delta_{n,0}\,\delta_{\beta,0}. \tag{2.3.3}$$

Inserting (2.3.2b) and (2.3.3) with $\beta = 0$ into (2.2.16a,b,c), and choosing the convenient sign for $P(\varphi_0)$, we obtain

$$P(\varphi_0) = 1. \tag{2.3.4}$$

Inserting (2.3.1) into (2.2.5), we obtain the differential equation

$$\frac{d^2\Phi}{d\varphi^2} + \frac{1}{\lambda^2}\,\Phi = 0, \tag{2.3.5}$$

which has the linearly independent solutions

$$\Phi = \exp(\pm i\varphi/\lambda). \tag{2.3.6}$$

According to (2.2.3), (2.3.6) and (2.1.7) the corresponding linearly independent solutions of the differential equation (2.2.2) are

$$\psi = \left(\frac{d\varphi}{dz}\right)^{-1/2}\exp(\pm i\varphi/\lambda) = \lambda^{-1/2}\left(\frac{d(\varphi/\lambda)}{dz}\right)^{-1/2}\exp\left(\pm i\int^z \frac{d(\varphi/\lambda)}{dz}\,dz\right)$$

$$= \lambda^{-1/2}\left(\frac{Q(z)}{\lambda}\frac{d\varphi}{d\zeta}\right)^{-1/2}\exp\left(\pm i\int^z \frac{d\varphi}{d\zeta}\frac{Q(z)}{\lambda}\,dz\right). \tag{2.3.7}$$

Inserting here the truncated series expansion (2.2.11′) of φ in powers of λ^2, we obtain

$$\psi = \lambda^{-1/2}\left(\frac{Q(z)}{\lambda}\sum_{n=0}^{N}Y_{2n}\lambda^{2n}\right)^{-1/2}\exp\left(\pm i\int^z \sum_{n=0}^{N}Y_{2n}\lambda^{2n}\frac{Q(z)}{\lambda}\,dz\right) \tag{2.3.8}$$

with Y_{2n} defined by

$$Y_{2n} = \frac{d\varphi_{2n}}{d\zeta}.$$ (2.3.9)

It follows from (2.3.9) with $n = 0$, (2.2.41), and (2.3.4) that

$$Y_0 = 1$$ (2.3.10a)

and from (2.2.16c), (2.2.25), (2.2.28), (2.2.30), (2.2.41), (2.3.2b), (2.3.3), (2.3.4), and (2.3.9) that

$$Y_{2N} = \frac{1}{2} \sum_{\beta_1+\beta_2=N-1} \left(\varepsilon_0 Y_{2\beta_1} Y_{2\beta_2} + \frac{3}{4} \frac{dY_{2\beta_1}}{d\zeta} \frac{dY_{2\beta_2}}{d\zeta} \right.$$

$$\left. - \frac{1}{4} Y_{2\beta_1} \frac{d^2 Y_{2\beta_2}}{d\zeta^2} - \frac{1}{4} \frac{d^2 Y_{2\beta_1}}{d\zeta^2} Y_{2\beta_2} \right)$$

$$+ \frac{1}{2} \sum_{\beta_1+\beta_2=N} Y_{2\beta_1} Y_{2\beta_2} - \frac{1}{2} \sum_{\beta_1+\beta_2+\beta_3+\beta_4=N} Y_{2\beta_1} Y_{2\beta_2} Y_{2\beta_3} Y_{2\beta_4},$$

$$0 \le \beta_i \le N-1, \quad N \ge 1.$$ (2.3.10b)

It is seen that (2.3.10b) agrees with the recurrence formula (1.3.13b) in Chapter 1, and that (2.3.10a) agrees with Eq. (1.3.13a). The quantities Y_{2n} in the present section are thus the same as the quantities Y_{2n} in Chapter 1 and hence the same as the quantities Y_{2n} in §2.1 and §2.2 (see [2.1.5a–c] with [2.1.6] and [2.1.7]). Setting $\lambda = 1$ in the approximate solution (2.3.8) of the differential equation (2.2.2), one therefore obtains the phase-integral approximation generated from the unspecified base function $Q(z)$ that pertains to the differential equation (2.2.1).

The above derivation illustrates the connection between the comparison equation solution and the phase-integral approximation, and it also illuminates the complexity of the derivation based on the comparison equation technique compared with the direct derivation of the phase-integral approximation given in Chapter 1.

2.4   Summary of the Procedure and the Results

The comparison equation solution of the auxiliary differential equation (2.2.2) is given by (2.2.3), where $\Phi(\varphi)$ is a known solution of the differential equation (2.2.5), and φ is a function of z and λ satisfying the differential equation (2.2.8). With $\Pi(\varphi)$ assumed to be the quotient of two power series (polynomials) according to (2.2.9a–c), the differential equation (2.2.8) takes the form (2.2.10), from which φ is to be determined as a function of z and λ. To achieve this goal, we expand the variable φ and the coefficients A_n in (2.2.10) in powers of λ according to (2.2.11) and

(2.2.12). From (2.2.10) we then obtain the formulas (2.2.20) and (2.2.25′) for φ_0 and φ_{2N} ($N \geq 1$), respectively. The coefficients in the expansions (2.2.9c) and (2.2.12) are determined from the requirement that the functions $\varphi_{2\beta}$ be regular (and hence single-valued) analytic functions. Thus, one obtains the conditions (2.2.19), (2.2.24′), and (2.2.59a,b) when $Q^2(z)$ has only odd-order zeros and first-order poles. The corresponding explicit expressions for φ_0, φ_2, and φ_4 are given by (2.2.20), with $z_0 = t_i$ and $\varphi_0(z_0) = \tau_i$, and (2.2.60a′,b′). The behavior of φ_0, φ_2, and φ_4 is given by (2.2.62) and (2.2.70a,b) with (2.2.68a,b) at a second-order pole of $Q^2(z)$ and by (2.2.72), (2.2.73), (2.2.33), and (2.2.80) at a first-order pole of $Q^2(z)$. When the expression for φ is inserted into the solution (2.2.3), one does not use the infinite series (2.2.11), but the truncated series (2.2.11′), where N is a conveniently chosen nonnegative integer (usually 0, 1 or 2). The comparison equation solution of the auxiliary differential equation (2.2.2) now described has been constructed so that when one performs an appropriate expansion in powers of λ and then sets $\lambda = 1$, one obtains for the original differential equation (2.2.1), except in the neighborhood of the zeros and poles of $Q^2(z)$, the arbitrary-order phase-integral approximation generated from the unspecified base function $Q(z)$, with the influence of the transition points duly accounted for. The result of the treatment in §2.3 confirms this assertion.

The comparison equation solution of the original differential equation (2.2.1) is thus obtained when one sets $\lambda = 1$ in the formulas mentioned above. This comparison equation solution, which contains the unspecified base function $Q(z)$, is valid as a formal solution of the differential equation for the entirety of a certain region of the complex z-plane containing unspecified numbers of arbitrary-order zeros and first- and second-order poles of $Q^2(z)$.

The whole treatment in §2.2 was of a quite formal nature, since we derive a mathematical solution without discussing the convergence of the series involved. Furthermore, the Stokes phenomenon, decisive for the behavior of asymptotic solutions, must be taken into account in tracing the solutions from one region of the complex z-plane to another. Thus, the formulas in §2.2, although also valid as formal solutions through zeros and poles of $Q^2(z)$, *cannot* be used for circumventing *the inherently one-directional nature of the usual connection formulas at a simple classical turning point* (see pp. 390–391 in [4]).

An important use of the formal comparison equation solution presented in §2.2 is the determination of the appropriate linear combination of the phase-integral functions, which yields the wave function far away from the clusters of zeros and poles of $Q^2(z)$, but with their influence on the solution taken into account, for physically important configurations of these zeros and poles; in particular for the case that they may lie close to each other. The determination of this phase-integral wave function is closely associated with the determination of the Stokes constants, with the aid of which one can trace a certain wave function in the complex z-plane. As a result of the analysis one obtains the supplementary quantities mentioned in §2.1, which can be used to construct new general phase-integral formulas, in particular new connection formulas, for the solution of typical model problems, where zeros and poles of the function $Q^2(z)$ may approach each other. As examples of such formulas we mention the connection formula for a barrier [10] containing a supplementary quantity [18], which is of decisive importance for energies near the top of the barrier; and the new phase-integral formulas given in Chapter 6, which are applicable when a first- or second-order pole and one or two zeros, respectively, of $Q^2(z)$ may lie close to each other.

References

[1] Cherry, T.M., Uniform asymptotic formulae for functions with transition points, *Trans. Am. Math. Soc.* **68** (1950), 224–257.

[2] Erdélyi, A., Asymptotic solutions of differential equations with transition points, *Proc. Int. Congr. Math. 1954* **III** (1956), 92–101.

[3] Erdélyi, A., Asymptotic solutions of differential equations with transition points or singularities, *J. Math. Phys.* **1** (1960), 16–26.

[4] Fröman, N., Detailed analysis of some properties of the JWKB-approximation, *Ark. Fys.* **31** (1966), 381–408.

[5] Fröman, N., Outline of a general theory for higher order approximations of the JWKB-type, *Ark. Fys.* **32** (1966), 541–548.

[6] Fröman, N., Connection formulas for certain higher order phase-integral approximations, *Ann. Phys. (N.Y.)* **61** (1970), 451–464.

[7] Fröman, N., A simple formula for calculating quantal expectation values without the use of wave functions, *Phys. Lett.* **48A** (1974), 137–139.

[8] Fröman, N., Phase-integral formulas for level densities, normalization factors, and quantal expectation values, not involving wave functions, *Phys. Rev.* **A17** (1978), 493–504.

[9] Fröman, N. and Fröman, P.O., *JWKB Approximation, Contributions to the Theory*, North-Holland Publishing Company, Amsterdam, 1965. (Russian translation: MIR, Moscow, 1967.)

[10] Fröman, N. and Fröman, P.O., Transmission through a real potential barrier treated by means of certain phase-integral approximations, *Nucl. Phys.* **A147** (1970), 606–626.

[11] Fröman, N. and Fröman, P.O., A direct method for modifying certain phase-integral approximations of arbitrary order, *Ann. Phys. (N.Y.)* **83** (1974), 103–107.

[12] Fröman, N. and Fröman, P.O., On modifications of phase integral approximations of arbitrary order, *Nuovo Cimento* **20B** (1974), 121–132.

[13] Fröman, N. and Fröman, P.O., Phase-integral calculation of quantal matrix elements without the use of wave functions, *J. Math. Phys.* **18** (1977), 903–906.

[14] Fröman, N. and Fröman, P.O., Exact formulas and phase-integral formulas, not involving wave functions, for expectation values pertaining to general potentials, *Ann. Phys. (N.Y.)* **163** (1985), 215–226.

[15] Fröman, N. and Fröman, P.O., Phase-integral approximation of arbitrary order generated from an unspecified base function. Review article in: *Forty More Years of Ramifications: Spectral Asymptotics and Its Applications*, edited by S.A. Fulling and F.J. Narcowich, Discourses in Mathematics and Its Applications, No. 1, Department of Mathematics, Texas A & M University, College Station, Texas 1991, pp. 121–159. After minor changes reprinted as Chapter 1 in the present monograph.

[16] Fröman, N., Fröman, P.O., and Karlsson, F., Phase-integral calculation of quantal matrix elements between unbound states, without the use of wave functions, *Molec. Phys.* **38** (1979), 749–767.

[17] Fröman, N., Fröman, P.O., and Linnaeus, S., Phase-integral formulas for the regular wave function when there are turning points close to a pole of the potential. This is Chapter 6 in this monograph.

[18] Fröman, N., Fröman, P.O., Myhrman, U., and Paulsson, R., On the quantal treatment of the double-well potential problem by means of certain phase-integral approximations, *Ann. Phys. (N.Y.)* **74** (1972), 314–323.

[19] Miller, Jr., S.C. and Good, Jr., R.H., A WKB-type approximation to the Schrödinger equation, *Phys. Rev.* **91** (1953), 174–179.

[20] Zauderer, E., A uniform asymptotic turning point theory for second order linear ordinary differential equations, *Proc. Am. Math. Soc.* **31** (1972), 489–494.

[21] Zauderer, E., Miller–Good method for approximating bound states, *J. Chem. Phys.* **56** (1972), 5198.

Adjoined Papers

3
Problem Involving One Transition Zero

Nanny Fröman and Per Olof Fröman

Abstract. The comparison equation technique developed in Chapter 2 in this monograph is used for solving an unspecified Schrödinger-like differential equation with one relevant transition zero. By expansion of the resulting formal solution in terms of a "small" bookkeeping parameter one obtains, at a distance sufficiently far away from the transition zero, the arbitrary-order phase-integral approximation generated from an unspecified base function $Q(z)$, the square of which has one relevant isolated zero.

3.1 Introduction

In this chapter the comparison equation technique developed in [1], i.e., Chapter 2 of this monograph, is used for solving an unspecified Schrödinger-like differential equation with one relevant transition zero. When the resulting solution is then expanded formally in terms of a "small" bookkeeping parameter, the arbitrary-order phase-integral approximation, described in Chapter 1, is obtained, provided that one is sufficiently far away from the transition zero. The equations from which we start our treatment are obtained by particularization of equations in Chapter 2; therefore, we shall in several places refer to equations stated there.

If the function $R(z)$ in the original differential equation (2.2.1) has only one first-order zero, but no pole in the region under consideration, then we choose the square of the base function (i.e., $Q^2(z)$) to be a regular function of z, similar to $R(z)$, with one zero in that region, and we choose the function $\Pi(\varphi)$ in the comparison equation (2.2.5) to be a linear function of φ. It follows from the discussion below Eq. (2.2.9c) that we can, without essential restriction, choose $\Pi(\varphi)$ to be equal to φ, which means that (apart from a sign) the comparison equation (2.2.5) becomes the differential equation for the Airy functions Ai and Bi. It is thus natural to use this differential equation as a comparison equation for a region of the complex z-plane, where there is a simple, well isolated transition zero; i.e., a zero of $Q^2(z)$. It turns out that asymptotically, to the right and to the left of the transition zero, our comparison equation solution then yields nothing but the right- and left-hand members of the arbitrary-order connection formulas derived by N. Fröman [2]. Thus, it is worthwhile in this case to use the comparison equation technique only if one is interested in knowing the wave function in the neighborhood of the transition zero; which is the case, for instance, if there is a boundary condition in the neighborhood of the transition zero.

The comparison equation solution does not display explicitly the one-directional nature of the connection formulas. This property must be considered separately. The fact that the comparison equation solution is valid *through* the transition zero does not imply that the one-directional nature of the connection formulas can be circumvented; this has clearly been demonstrated by N. Fröman [3]. The derivation of the connection formulas by N. Fröman [2] has the merit of displaying their one-directional nature and yielding upper bounds for the errors involved.

3.2   Comparison Equation Solution

The differential equation to be treated is written in the form of Eq. (2.2.1), that is,

$$\frac{d^2\psi}{dz^2} + R(z)\psi = 0, \tag{3.2.1}$$

where $R(z)$ is an analytic function of z, which is regular and has a simple zero in the region of the complex z-plane under consideration.

Instead of the original differential equation (3.2.1) we now consider the auxiliary differential equation (2.2.2), that is,

$$\frac{d^2\psi}{dz^2} + \left(\frac{Q^2(z)}{\lambda^2} + [R(z) - Q^2(z)]\right)\psi = 0, \tag{3.2.2}$$

where $Q^2(z)$ is an analytic function, so far unspecified, which can be conveniently chosen, and λ is a "small" bookkeeping parameter. The differential equation (3.2.2) will be solved by the comparison equation technique described in Chapter 2. The solution of the original differential equation thus obtained, when λ is set equal to unity, is very flexible, since it contains the unspecified function $Q(z)$, called the base function. Far away from the zeros and singularities of $Q^2(z)$, this solution yields the arbitrary-order phase-integral approximation generated from the unspecified base function (described in Chapter 1), as will be shown in §3.3. We assume that in the relevant region of the complex z-plane the analytic function $Q^2(z)$ has no branch points or singularities and has precisely one zero, which is denoted by t. All other zeros or singularities of $Q^2(z)$ are assumed to be situated well outside the region of the complex z-plane where the comparison equation technique is used.

To obtain the comparison equation solution of the auxiliary differential equation (3.2.2) we set (according to Eq. [2.2.3]),

$$\psi = \left(\frac{d\varphi}{dz}\right)^{-1/2} \Phi(\varphi), \tag{3.2.3}$$

where $\Phi(\varphi)$ is a function of φ and λ that will soon be specified, and φ is a so far unknown function of z and λ. The function $\Phi(\varphi)$ is assumed to be a known solution of the differential equation (2.2.5), that is,

$$\frac{d^2\Phi}{d\varphi^2} + \frac{\Pi(\varphi)}{\lambda^2} \Phi = 0, \tag{3.2.4}$$

where $\Pi(\varphi)$ is a function of φ and λ that, in the region of the complex φ-plane corresponding to the region of the complex z-plane under consideration, is chosen to be similar to the function $Q^2(z)$.

For the purpose of the present chapter the function $\Pi(\varphi)$ is chosen to be

$$\Pi(\varphi) = \varphi. \tag{3.2.5}$$

This formula is the particular case of Eqs. (2.2.9a–c) which one obtains by setting

$$A_n = \delta_{n,1} \tag{3.2.6}$$

$$B_q = \delta_{q,0}. \tag{3.2.7}$$

With the aid of Eq. (2.2.12), that is,

$$A_n = \sum_{\beta=0}^{\infty} A_{n,2\beta}\lambda^{2\beta}, \tag{3.2.8}$$

we obtain from (3.2.6)

$$A_{n,2\beta} = \delta_{n,1}\,\delta_{\beta,0}. \tag{3.2.9}$$

In order to determine φ as a function of z and λ we expand φ formally according to Eq. (2.2.11), that is,

$$\varphi = \sum_{\beta=0}^{\infty} \varphi_{2\beta}\,\lambda^{2\beta}, \tag{3.2.10}$$

where the quantities $\varphi_{2\beta}$ depend on z but are independent of λ.

Inserting (3.2.9) with $\beta = 0$ and (3.2.7) into Eqs. (2.2.16a–c), we obtain

$$P(\varphi_0) = \varphi_0^{1/2}, \tag{3.2.11}$$

where the sign of the square root should be chosen conveniently, so as to correspond to that of $Q(z)$. The function φ_0 must satisfy the integral relation (2.2.20), that is,

$$\int_0^{\varphi_0} P(\varphi_0)\,d\varphi_0 = \int_t^z Q(z)\,dz, \tag{3.2.12}$$

where the constant lower limits of integration have been chosen so that the comparison equation technique used for solving the differential equation (3.2.2) works satisfactorily in the neighborhood of $z = t$.

The general solution of the differential equation (3.2.4) with $\Pi(\varphi)$ given by (3.2.5) is

$$\Phi = \varphi^{1/2} Z_{1/3}(\chi/\lambda), \tag{3.2.13}$$

where

$$\chi = \frac{2}{3} \varphi^{3/2}, \tag{3.2.14}$$

and $Z_{1/3}(\chi/\lambda)$ denotes an unspecified linear combination of Bessel functions of the order $1/3$ and with the argument χ/λ. Inserting (3.2.13) into (3.2.3) and using (3.2.14), we obtain the comparison equation solution of (3.2.2) expressed as

$$\psi = \left(\frac{3}{2}\right)^{1/2} \left(\frac{d(\chi/\lambda)}{dz}\right)^{-1/2} (\chi/\lambda)^{1/2} Z_{1/3}(\chi/\lambda). \tag{3.2.15}$$

Introducing the definition (cf. Eq. [2.2.40])

$$\zeta = \int_t^z Q(z)\,dz, \tag{3.2.16}$$

which with the aid of (3.2.12) and (3.2.11) gives

$$\zeta = \frac{2}{3} \varphi_0^{3/2}, \tag{3.2.17}$$

that is,

$$\varphi_0 = (3\zeta/2)^{2/3}; \tag{3.2.18}$$

we can write (3.2.15) in the form

$$\psi = \left(\frac{3}{2}\right)^{1/2} Q^{-1/2}(z) \left(\frac{d(\chi/\lambda)}{d\zeta}\right)^{-1/2} (\chi/\lambda)^{1/2} Z_{1/3}(\chi/\lambda) \tag{3.2.19}$$

where, according to (3.2.14), (3.2.10), and (3.2.18),

$$\chi = \frac{2}{3}\left(\varphi_0 + \varphi_2\lambda^2 + \varphi_4\lambda^4 + \cdots\right)^{3/2} \tag{3.2.20}$$

$$= \zeta\left(1 + \frac{\varphi_2}{\varphi_0}\lambda^2 + \frac{\varphi_4}{\varphi_0}\lambda^4 + \cdots\right)^{3/2}. \tag{3.2.20'}$$

For $N > 0$ we have according to Eqs. (2.2.16c), (2.2.25), (2.2.28), (3.2.7), (3.2.9), (3.2.11), (3.2.16), and (3.2.18) the following equation for the determination of φ_{2N}

$$2\varphi_0^{1/2}\frac{d}{d\varphi_0}\left(\varphi_0^{1/2}\varphi_{2N}\right) = \sum_{\substack{\beta_1+\beta_2=N-1 \\ 0\leq\beta_i\leq N-1}} \left(g(\varphi_0)\frac{d\varphi_{2\beta_1}}{d\varphi_0}\frac{d\varphi_{2\beta_2}}{d\varphi_0} + \frac{3}{4}\frac{d^2\varphi_{2\beta_1}}{d\varphi_0^2}\frac{d^2\varphi_{2\beta_2}}{d\varphi_0^2}\right)$$

$$-\frac{1}{4}\frac{d\varphi_{2\beta_1}}{d\varphi_0}\frac{d^3\varphi_{2\beta_2}}{d\varphi_0^3}-\frac{1}{4}\frac{d^3\varphi_{2\beta_1}}{d\varphi_0^3}\frac{d\varphi_{2\beta_2}}{d\varphi_0}\Bigg)+\varphi_0\sum_{\substack{\beta_1+\beta_2=N\\0\le\beta_i\le N-1}}\frac{d\varphi_{2\beta_1}}{d\varphi_0}\frac{d\varphi_{2\beta_2}}{d\varphi_0}$$

$$-\sum_{\substack{\beta_1+\cdots+\beta_5=N\\0\le\beta_i\le N-1}}\varphi_{2\beta_1}\frac{d\varphi_{2\beta_2}}{d\varphi_0}\cdots\frac{d\varphi_{2\beta_5}}{d\varphi_0},\quad N\ge 1,\tag{3.2.21}$$

where according to Eq. (2.2.30)

$$g(\varphi_0)=P^2(\varphi_0)\left(\varepsilon_0-P^{-3/2}(\varphi_0)\frac{d^2P^{-1/2}(\varphi_0)}{d\varphi_0^2}\right)$$

$$=\varphi_0\left(\varepsilon_0-\frac{5}{16\varphi_0^3}\right)\tag{3.2.22}$$

with ε_0 defined by Eq. (2.1.6), that is,

$$\varepsilon_0=Q^{-3/2}(z)\frac{d^2Q^{-1/2}(z)}{dz^2}+\frac{R(z)-Q^2(z)}{Q^2(z)}.\tag{3.2.23}$$

We have here given (3.2.21) and (3.2.22) for future use, and we shall not dwell on these formulas in the present chapter.

Using Eqs. (2.2.60a',b') with $c_2=c_4=0$, (2.2.17) and (2.1.5b,c), we obtain

$$P(\varphi_0)\varphi_2=\int_{(t)}^z\frac{1}{2}(\varepsilon_0-h)Q(z)\,dz\tag{3.2.24a}$$

and

$$P(\varphi_0)\varphi_4=\int_{(t)}^z\left\{-\frac{1}{8}\left(\varepsilon_0^2+\frac{d^2\varepsilon_0}{d\zeta^2}\right)+\frac{1}{8}\left(h^2+\frac{d^2h}{d\zeta^2}\right)-\frac{S_4/T}{2P^2}\right\}Q(z)\,dz$$

$$-\frac{1}{2}\left(h\,P\varphi_2+\frac{dP}{d\varphi_0}\varphi_2^2\right)\tag{3.2.24b}$$

$$=\int_{(t)}^z\left\{-\frac{1}{8}(\varepsilon_0-h)^2-\frac{1}{4}h(\varepsilon_0-h)-\frac{S_4/T}{2P^2}\right\}Q(z)\,dz$$

$$-\frac{1}{8}\frac{d(\varepsilon_0-h)}{d\zeta}-\frac{1}{2}h(P\varphi_2)-\frac{dP/d\varphi_0}{2P^2}(P\varphi_2)^2.\tag{3.2.24b'}$$

For the definition of the short-hand notation for the integrals in (3.2.24a,b,b') we refer to the explanation in connection with Eq. (4.3.3) in Chapter 4.

Inserting (3.2.9) and (3.2.7) into Eqs. (2.2.13b) and (2.2.9c), with φ replaced by φ_0, we obtain

$$S_{2\beta}(\varphi_0)=\delta_{\beta,0}\,\varphi_0\tag{3.2.25a}$$

$$T(\varphi_0)=1.\tag{3.2.25b}$$

Inserting (3.2.11) and (3.2.25a,b) into the definition of the quantity h, given by Eq. (2.2.34), and using (3.2.18), we obtain

$$h = \frac{5}{36\,\zeta^2}. \tag{3.2.26}$$

Recalling (3.2.11), (3.2.18), (3.2.25a,b), and (3.2.26), we obtain from (3.2.24a, b,b') the formulas

$$\varphi_0^{1/2}\varphi_2 = \int_{(t)}^z \frac{1}{2}\,\varepsilon_0 Q(z)\,dz + \frac{5}{72\,\zeta} \tag{3.2.27a}$$

$$= \int_{(t)}^z \frac{1}{2}\left(\varepsilon_0 - \frac{5}{36\,\zeta^2}\right)Q(z)\,dz \tag{3.2.27a'}$$

and

$$\varphi_0^{1/2}\varphi_4 = \int_{(t)}^z \left(-\frac{1}{8}\varepsilon_0^2 - \frac{1}{8}\frac{d^2\varepsilon_0}{d\zeta^2}\right)Q(z)\,dz - \frac{1105}{31104\,\zeta^3}$$
$$- \frac{5}{72\,\zeta^2}(\varphi_0^{1/2}\varphi_2) - \frac{1}{6\,\zeta}(\varphi_0^{1/2}\varphi_2)^2 \tag{3.2.27b}$$

$$= \int_{(t)}^z \left[-\frac{1}{8}\left(\varepsilon_0 - \frac{5}{36\,\zeta^2}\right)^2 - \frac{5}{144\,\zeta^2}\left(\varepsilon_0 - \frac{5}{36\,\zeta^2}\right)\right]Q(z)\,dz$$
$$- \frac{1}{8}\frac{d}{d\zeta}\left(\varepsilon_0 - \frac{5}{36\,\zeta^2}\right) - \frac{5}{72\,\zeta^2}(\varphi_0^{1/2}\varphi_2) - \frac{1}{6\,\zeta}(\varphi_0^{1/2}\varphi_2)^2. \tag{3.2.27b'}$$

From the comparison equation solution obtained in the present section, given by (3.2.15) with (3.2.20), (3.2.27a,b), and (3.2.16), one can for $\zeta \neq 0$ and $\lambda \to 0$ derive the arbitrary-order phase-integral approximation generated from the unspecified base function $Q(z)$, as will be shown in §3.3.

3.3   Phase-Integral Approximation Obtained from the Comparison Equation Solution

Allowing $Z_{1/3}(\chi/\lambda)$ to be a Hankel function, and, for the sake of simplicity, assuming z to be real, we shall first consider the case when z lies in the classically allowed region (which means that φ is positive). Recalling the connection of the Hankel functions with the Bessel functions J_ν and Y_ν, then using Eqs. (2.8), (2.9a,b,c), (2.10a,b), (2.11), and (2.12) with $\nu = 1/3$ in [4], when χ/λ is positive and sufficiently large, we obtain the formal expansions

$$(\chi/\lambda)^{1/2}H_{1/3}^{(1)}(\chi/\lambda) = \left(\frac{2}{\pi}\right)^{1/2}\left(\frac{d}{d(\chi/\lambda)}\sum_{p=0}^\infty \frac{a_{2p}}{1-2p}(\chi/\lambda)^{1-2p}\right)^{-1/2}$$

$$\times \exp\left[+i\left(\sum_{p=0}^{\infty} \frac{a_{2p}}{1-2p}\,(\chi/\lambda)^{1-2p} - \frac{5\pi}{12}\right)\right] \tag{3.3.1a}$$

and

$$(\chi/\lambda)^{1/2} H_{1/3}^{(2)}(\chi/\lambda) = \left(\frac{2}{\pi}\right)^{1/2}\left(\frac{d}{d(\chi/\lambda)}\sum_{p=0}^{\infty}\frac{a_{2p}}{1-2p}\,(\chi/\lambda)^{1-2p}\right)^{-1/2}$$

$$\times \exp\left[-i\left(\sum_{p=0}^{\infty} \frac{a_{2p}}{1-2p}\,(\chi/\lambda)^{1-2p} - \frac{5\pi}{12}\right)\right], \tag{3.3.1b}$$

where the first few coefficients a_{2p} are

$$a_0 = 1 \tag{3.3.2a}$$

$$a_2 = \frac{5}{72} \tag{3.3.2b}$$

$$a_4 = -\frac{1105}{10368}. \tag{3.3.2c}$$

These formal expansions have a wider range of validity than indicated above (where we assumed χ/λ to be positive and sufficiently large), but for the sake of simplicity we will disregard their more general validity.

We shall now expand the quantity χ^{1-2p}, where χ is given by (3.2.20'), in powers of λ^2. The expansion thus obtained is valid when $z \neq t$ (i.e., when $\varphi_0 \neq 0$) but breaks down when $z = t$ (i.e., when $\varphi_0 = 0$). Assuming $\varphi_0 \neq 0$, we obtain from (3.2.20') and (3.2.18) the expansion

$$\chi^{1-2p} = \zeta^{1-2p} + \sum_{\mu=p+1}^{\infty}\left[\sum_{r=1}^{\mu-p}\left(\frac{2}{3}\right)^r\binom{\frac{3}{2}(1-2p)}{r}\zeta^{1-2p-r}\right.$$

$$\times \sum_{\substack{\beta_1+\cdots+\beta_r=\mu-p \\ \beta_i\geq 1}}(\varphi_0^{1/2}\varphi_{2\beta_1})\cdots(\varphi_0^{1/2}\varphi_{2\beta_r})\right]\lambda^{2\mu-2p}. \tag{3.3.3}$$

From (3.3.3) we obtain

$$\sum_{p=0}^{\infty}\frac{a_{2p}}{1-2p}\,(\chi/\lambda)^{1-2p} = a_0\zeta\lambda^{-1} + \sum_{\mu=1}^{\infty}\left[\frac{a_{2\mu}}{1-2\mu}\zeta^{1-2\mu}\right.$$

$$+ \sum_{p=0}^{\mu-1}\frac{a_{2p}}{1-2p}\sum_{r=1}^{\mu-p}\left(\frac{2}{3}\right)^r\binom{\frac{3}{2}(1-2p)}{r}\zeta^{1-2p-r} \tag{3.3.4}$$

$$\times \sum_{\substack{\beta_1+\cdots+\beta_r=\mu-p \\ \beta_i\geq 1}}(\varphi_0^{1/2}\varphi_{2\beta_1})\cdots(\varphi_0^{1/2}\varphi_{2\beta_r})\right]\lambda^{2\mu-1}.$$

Using (3.3.2a,b,c), we find that the first few terms in (3.3.4) are

$$\sum_{p=0}^{\infty} \frac{a_{2p}}{1-2p} (\chi/\lambda)^{1-2p} = \zeta \lambda^{-1} + \left(\varphi_0^{1/2} \varphi_2 - \frac{5}{72\,\zeta} \right) \lambda$$

$$+ \left(\varphi_0^{1/2} \varphi_4 + \frac{1}{6\,\zeta} (\varphi_0^{1/2} \varphi_2)^2 + \frac{5}{72\,\zeta^2} \varphi_0^{1/2} \varphi_2 + \frac{1105}{31104\,\zeta^3} \right) \lambda^3 + \cdots. \quad (3.3.5)$$

Inserting (3.2.27a,b) into (3.3.5), we obtain

$$\sum_{p=0}^{\infty} \frac{a_{2p}}{1-2p} (\chi/\lambda)^{1-2p} = \zeta \lambda^{-1} + \int_{(t)}^{z} \frac{1}{2} \varepsilon_0 Q(z)\, dz\, \lambda$$

$$+ \int_{(t)}^{z} \left(-\frac{1}{8} \varepsilon_0^2 - \frac{1}{8} \frac{d^2 \varepsilon_0}{d\zeta^2} \right) Q(z)\, dz\, \lambda^3 + \cdots. \quad (3.3.6)$$

Recalling (3.2.16) and the definition (4.3.3) of the short-hand notation for the integrals in (3.3.6), we obtain from (3.3.6)

$$\sum_{p=0}^{\infty} \frac{a_{2p}}{1-2p} (\chi/\lambda)^{1-2p} = \int_{(t)}^{z} q(z)\, dz, \quad (3.3.7)$$

where

$$q(z) = \frac{1}{\lambda} Q(z)(Y_0 + Y_2 \lambda^2 + Y_4 \lambda^4 + \cdots) \quad (3.3.8)$$

with

$$Y_0 = 1 \quad (3.3.9a)$$

$$Y_2 = \frac{1}{2} \varepsilon_0 \quad (3.3.9b)$$

$$Y_4 = -\frac{1}{8} (\varepsilon_0^2 + d^2 \varepsilon_0/d\zeta^2). \quad (3.3.9c)$$

Inserting (3.3.1a) and (3.3.1b) into (3.2.15) with $Z_{1/3}$ equal to $H_{1/3}^{(1)}$ and $H_{1/3}^{(2)}$, respectively, and using (3.3.7), we obtain

$$(\pi/3)^{1/2} \exp(\pm 5i\pi/12)\psi = q^{-1/2}(z) \exp \left(\pm i \int_{(t)}^{z} q(z)\, dz \right). \quad (3.3.10)$$

Recalling the exposition in Chapter 1, we realize that for $z \neq t$ the comparison equation solution (3.3.10) is the formal solution from which the arbitrary-order phase-integral approximation generated from the unspecified base function $Q(z)$ is obtained by truncation of the infinite series (3.3.8) with $\lambda = 1$.

If we next assume that z lies in the classically forbidden region (i.e., φ is negative) and consider the corresponding appropriate formal expansions of the functions $(\chi/\lambda)^{1/2}H_{1/3}^{(1)}(\chi/\lambda)$ and $(\chi/\lambda)^{1/2}H_{1/3}^{(2)}(\chi/\lambda)$, analogous to (3.3.1a,b), we then find the phase-integral approximation that in the classically forbidden region corresponds to the phase-integral approximation (3.3.10) in the classically allowed region. The result is consistent with the arbitrary-order phase-integral connection formulas [2]

$$|q^{-1/2}(z)| \exp\{\pm i[|w(z)| + \pi/4]\} \to |q^{-1/2}(z)| \exp[|w(z)|], \qquad (3.3.11)$$

where

$$w(z) = \int_{(t)}^{z} q(z)\, dz. \qquad (3.3.12)$$

The one-way nature of (3.3.11) is, however, not seen from the comparison equation solution, but requires special discussion. The connection formulas (3.3.11) are equivalent to the connection formula

$$|q^{-1/2}(z)| \cos[|w(z)| + \gamma - \pi/4] \to \sin\gamma |q^{-1/2}(z)| \exp[|w(z)|], \qquad (3.3.13)$$

where γ is a real constant subject only to the restriction of not being too close to a multiple of π.

Now let us choose the function Φ of (3.2.13) to be the Airy function $\mathrm{Ai}(-\varphi/\lambda^{2/3})$, which is related to the Hankel functions according to the formula [see formula 10.4.15 in [5] and (3.2.14)]

$$\mathrm{Ai}(-\varphi/\lambda^{2/3}) = 2^{-3/2}\lambda^{-1/3}\left(\frac{d(\chi/\lambda)}{d\varphi}\right)^{-1/2}(\chi/\lambda)^{1/2}\left[e^{i\pi/6}H_{1/3}^{(1)}(\chi/\lambda)\right.$$

$$\left. + e^{-i\pi/6}H_{1/3}^{(2)}(\chi/\lambda)\right]. \qquad (3.3.14)$$

Considering the formal expansions of $\mathrm{Ai}(-\varphi/\lambda^{2/3})$ when $\varphi/\lambda^{2/3}$ is sufficiently large (positive or negative), which can be obtained from the results in [4], analogously as above; we obtain from (3.2.3) expressions for the comparison equation solution that are consistent with the arbitrary-order phase-integral connection formula [2]

$$|q^{-1/2}(z)| \exp[-|w(z)|] \to 2|q^{-1/2}(z)| \cos[|w(z)| - \pi/4], \qquad (3.3.15)$$

but, as in the case of the connection formula (3.3.11), the understanding of its one-way nature requires special discussion.

We note that the phase of the cosine in the right-hand member of (3.3.15) is independent of the order of the phase-integral approximation used. Correspondingly, the Stokes constants, connecting solutions on three anti-Stokes lines, emerging in different directions (forming angles approximately equal to $2\pi/3$ with each other) from the region of the transition zero, are independent of the order of approximation. Thus, for any order of approximation, the connection matrices have the same simple form as for the first-order approximation.

The comparison equation solution obtained in §3.2 is valid in the region of the complex z-plane around the transition zero t; as shown in the present section, it yields for values of $|z - t|$ that are sufficiently large nothing but the arbitrary-order phase-integral approximation generated from an unspecified base function. Thus, the wave function can be traced around the transition zero with the aid of the known connection formulas, with due regard to their one-way nature, or, in more general situations, with the aid of the known Stokes constants. Therefore, there is no need for using comparison equation technique in order to obtain the wave function in the asymptotic region. The comparison equation technique, however, is very useful if one is really interested in knowing the wave function not only asymptotically, but *in the neighborhood of the transition zero*, a situation that occurs, for instance, when there is a boundary condition in the neighborhood of that point.

References

[1] Fröman, N. and Fröman, P.O., Technique of the comparison equation adapted to the phase-integral method. This is Chapter 2 in the present monograph.

[2] Fröman, N., *Ann. Phys. (N.Y.)* **61** (1970), 451–464.

[3] Fröman, N., *Ark. Fys.* **31** (1966), 381–408.

[4] Fröman, P.O., Karlsson, F., and Yngve, S., *J. Math. Phys.* **27** (1986), 2738–2747.

[5] Abramowitz, M. and Stegun, I.A. (Eds.), *Handbook of Mathematical Functions*, National Bureau of Standards, Applied Mathematics Series, 55. Fourth Printing, Washington, D.C., 1965.

4

Relations Between Different Nonoscillating Solutions of the q-Equation Close to a Transition Zero

Aleksander Dzieciol, Per Olof Fröman, and Nanny Fröman

Abstract. The numerical solution of a Schrödinger-like differential equation can be based on integration of the corresponding nonlinear q-equation. For the efficiency of the numerical integration, it is essential to select nonoscillating solutions of the q-equation. Analytical formulas that are useful for achieving this goal in a region of the complex z-plane, where there is one simple, well isolated transition zero, are derived.

From a simple, well isolated transition zero there emerge three first-order anti-Stokes lines; for each one of these lines there exists an almost uniquely determined solution of the q-equation that, in the neighborhood of the anti-Stokes line in question, is nonoscillating. The relations between these three particular solutions of the q-equation are determined in the neighborhood of the transition zero by means of comparison equation technique. Furthermore, for a given solution of the Schrödinger-like differential equation, from which the q-equation arises, the relations between the a-coefficients associated with the above-mentioned particular solutions of the q-equation are derived. A numerically exact condition for the determination of Regge pole positions, as well as the corresponding approximate condition obtained with the aid of comparison equation technique, is also derived.

4.1 Introduction

The present investigation has been performed in connection with the development of a method for obtaining a numerically exact solution of a Schrödinger-like differential equation by integration of the corresponding q-equation (1.3.3) in [1]. The q-equation is nonlinear, and there exist an infinite number of exact solutions, in terms of which a particular solution of the original Schrödinger-like differential equation can, in principle, be exactly expressed. However, in practice it is necessary to select nonoscillatory functions $q(z)$ in order that the numerical integration of the q-equation be efficient and sufficiently accurate. For this integration to be successful, one must therefore choose the starting values judiciously. This can be achieved by the use of the phase-integral approximation of a convenient order, when one is far away from the transition points, and by the use of formulas obtained by comparison equation technique, when one is in the neighborhood of a transition zero.

The integration procedure can be briefly described as follows. One starts by integrating the q-equation close to an appropriate anti-Stokes line, let us say the first-order anti-Stokes line A in Figure 4.1. For this integration the initial values of the function $q(z)$ and its first derivative, far away from the transition zero t in Figure 4.1, are chosen as the phase-integral function $q(z)$ of a conveniently high order; numerically one thus obtains a function $q_A(z)$ that is nonoscillatory in the neighborhood of the line A. In the neighborhood of the transition zero t the function $q_A(z)$ starts oscillating rapidly in sectors containing the first-order anti-Stokes lines B and C in Figure 4.1; therefore, it soon becomes numerically useless. To continue the numerical integration of the q-equation from the neighborhood of t in such a way that one obtains another solution $q_B(z)$, which is nonoscillating close to the anti-Stokes line B, one must know the initial values of the new q-function, $q_B(z)$, and its first derivative at the point in the neighborhood of t where the first integration of the q-equation is stopped and the second integration of that equation starts. These initial values for the second integration can be obtained from the values of $q_A(z)$ and $q_A'(z)$ at the end point of the first integration. In fact, in §4.2 and §4.3 we use comparison equation technique in an approximation of arbitrary order [2, 3] to determine the relations between $q_A(z)$ and $q_B(z)$ and the derivatives of these functions in the neighborhood of the transition zero t, which is assumed to be simple and well isolated. A corresponding treatment for the function $q_C(z)$, which is nonoscillating close to the first-order anti-Stokes line C in Figure 4.1, is also given. In §4.4 we simplify the formulas by means of which $q_B(z)$, $q_B'(z)$ and $q_C(z)$, $q_C'(z)$ can be expressed in terms of $q_A(z)$ and $q_A'(z)$ in the neighborhood of t by using only the first-order approximation of the comparison equation technique. It turns out that these first-order formulas are sufficient for the purpose of determining the starting values for the numerical integrations.

For any given solution of the Schrödinger-like differential equation we express in §4.5 the a-coefficients associated with $q_B(z)$ in terms of the a-coefficients associated with $q_A(z)$; we also describe the result of a similar treatment for the a-coefficients associated with $q_C(z)$ and $q_A(z)$.

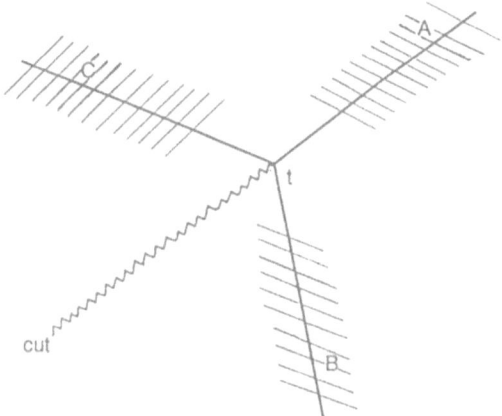

FIGURE 4.1. This figure shows the three first-order anti-Stokes lines A, B, and C which emerge from a simple transition zero t, i.e., a simple zero of the function $Q^2(z)$; see Chapter 1. To make the base function $Q(z)$ single-valued we have introduced a cut which also emerges from t and proceeds to infinity. The striped regions lie close to the anti-Stokes lines but far away from the transition zero.

In §4.6 we derive in a systematic way a numerically exact condition for the determination of Regge pole positions, and also the corresponding approximate condition that is obtained with the use of comparison equation technique in arbitrary-order approximation.

4.2 Comparison Equation Solutions

According to Eqs. (3.2.10), (3.2.14), and (3.2.15) the comparison equation solution, which pertains to the region of the complex z-plane around a simple transition zero, is

$$\psi = \left(\frac{3}{2}\right)^{1/2} \left(\frac{d(\chi/\lambda)}{dz}\right)^{-1/2} (\chi/\lambda)^{1/2} Z_{1/3}(\chi/\lambda), \qquad (4.2.1)$$

where $Z_{1/3}(\chi/\lambda)$ denotes an unspecified Bessel function of the order $1/3$ and

$$\chi = \frac{2}{3} \varphi^{3/2} \qquad (4.2.2)$$

with

$$\varphi = \sum_{n=0}^{N} \varphi_{2n} \lambda^{2n}. \qquad (4.2.3)$$

The series for φ in Eq. (3.2.10) has been truncated here so that the comparison equation solution is of the order $2N + 1$. The function φ_0 is given by Eqs. (3.2.18) with (3.2.16), that is,

$$\varphi_0 = (3\zeta/2)^{2/3} \tag{4.2.4}$$

with

$$\zeta = \int_t^z Q(z)\, dz, \tag{4.2.5}$$

where t is the transition zero. The phase of φ_0, which is not uniquely defined by (4.2.4), will be specified conveniently below. The phases of φ_2, φ_4, ... are then uniquely determined; see Chapter 3. To specify the phase of χ, we use (4.2.3) and (4.2.4) to rewrite (4.2.2) into the form

$$\chi = \zeta \left(\sum_{n=0}^{N} \frac{\varphi_{2n}}{\varphi_0} \lambda^{2n} \right)^{3/2}, \tag{4.2.6}$$

where we choose the phase of the factor to the right of ζ such that this factor is close to unity. For $n \geq 1$ we have according to Eqs. (2.1.6), (2.2.16c), (2.2.25), (2.2.28), (2.2.30), (3.2.7), (3.2.9), (3.2.11), (3.2.16), and (3.2.18) the following differential equation for φ_{2n}

$$2\varphi_0^{1/2} \frac{d}{d\varphi_0}\left(\varphi_0^{1/2}\varphi_{2n}\right) = \sum_{\substack{\beta_1+\beta_2=n-1 \\ 0\leq\beta_i\leq n-1}} \left(g(\varphi_0)\frac{d\varphi_{2\beta_1}}{d\varphi_0}\frac{d\varphi_{2\beta_2}}{d\varphi_0} + \frac{3}{4}\frac{d^2\varphi_{2\beta_1}}{d\varphi_0^2}\frac{d^2\varphi_{2\beta_2}}{d\varphi_0^2} \right.$$

$$\left. -\frac{1}{4}\frac{d\varphi_{2\beta_1}}{d\varphi_0}\frac{d^3\varphi_{2\beta_2}}{d\varphi_0^3} - \frac{1}{4}\frac{d^3\varphi_{2\beta_1}}{d\varphi_0^3}\frac{d\varphi_{2\beta_2}}{d\varphi_0} \right) + \varphi_0 \sum_{\substack{\beta_1+\beta_2=n \\ 0\leq\beta_i\leq n-1}} \frac{d\varphi_{2\beta_1}}{d\varphi_0}\frac{d\varphi_{2\beta_2}}{d\varphi_0}$$

$$- \sum_{\substack{\beta_1+\cdots+\beta_5=n \\ 0\leq\beta_i\leq n-1}} \varphi_{2\beta_1}\frac{d\varphi_{2\beta_2}}{d\varphi_0}\cdots\frac{d\varphi_{2\beta_5}}{d\varphi_0}, \quad n \geq 1, \tag{4.2.7}$$

where

$$g(\varphi_0) = \varphi_0\left(\varepsilon_0 - \frac{5}{16\,\varphi_0^3}\right) \tag{4.2.8}$$

with

$$\varepsilon_0 = Q^{-3/2}(z)\frac{d^2 Q^{-1/2}(z)}{dz^2} + \frac{R(z) - Q^2(z)}{Q^2(z)} \tag{4.2.9}$$

$$= \frac{1}{16\,Q^6(z)}\left[5\left(\frac{dQ^2(z)}{dz}\right)^2 - 4Q^2(z)\frac{d^2Q^2(z)}{dz^2}\right] + \frac{R(z) - Q^2(z)}{Q^2(z)}. \tag{4.2.9'}$$

Since the transition zero t is assumed to be a simple zero of $Q^2(z)$, we can expand this function as follows [4]

$$Q^2(z) = a(z-t)\left(1 + \sum_{n=1}^{\infty} b_n(z-t)^n\right), \quad a \neq 0, \tag{4.2.10}$$

and similarly we introduce the expansion

$$R(z) - Q^2(z) = \sum_{n=0}^{\infty} c_n (z - t)^n. \tag{4.2.11}$$

By inserting (4.2.10) and (4.2.11) into (4.2.9′), one can expand ε_0 in a series containing positive and some negative powers of $z-t$. From (4.2.10) it may be seen that $Q(z)$ can be written as $(z-t)^{1/2}$ times a series in powers of $z-t$. When this series is inserted into (4.2.5), one finds that ζ is equal to $(z-t)^{3/2}$ times a series in powers of $z-t$. Hence, $\zeta^{2/3}$ is a series in powers of $z-t$, which by inversion yields $z-t$ as a series in powers of $\zeta^{2/3}$; thus, according to (4.2.4), it yields $z-t$ as a series in powers of φ_0. When this series is inserted into the series for ε_0 in powers of $z-t$, one obtains ε_0 as a series containing positive and two negative powers of φ_0, the latter being cancelled in (4.2.8) so that one obtains for $g(\varphi_0)$ a series expansion of the form

$$g(\varphi_0) = \sum_{n=0}^{\infty} g_n \varphi_0^n, \tag{4.2.12}$$

where the coefficients g_n are expressed in terms of the coefficients a, b_n and c_n in (4.2.10) and (4.2.11). The determination of the coefficients g_n is tedious and is best performed numerically on a computer, since the analytical expressions for g_n soon become very complicated as the index n increases. Here we shall not worry about the determination of these coefficients but simply assume that one has a computer program for their numerical calculation. All series involved here have radii of convergence that are larger than zero and may be either finite or infinite. As we are interested in values of some quantities very close to the transition zero t, there is no need in this context to determine those radii explicitly. When the expansion (4.2.12) is known, one can obtain from (4.2.7) expressions for the functions φ_{2n} of the form

$$\varphi_{2n} = \sum_{\alpha=0}^{\infty} d_{2n,\alpha} \varphi_0^\alpha \tag{4.2.13}$$

by means of a computer program which has been written by Trygg (unpublished).

Inserting the Hankel functions $H_{1/3}^{(1)}(\chi/\lambda)$ and $H_{1/3}^{(2)}(\chi/\lambda)$ for $Z_{1/3}(\chi/\lambda)$ in (4.2.1), we get two linearly independent comparison equation solutions ψ_1 and ψ_2, respectively. The Wronskian of these solutions is

$$W = \psi_1 \frac{d\psi_2}{dz} - \psi_2 \frac{d\psi_1}{dz} = -\frac{6i}{\pi}. \tag{4.2.14}$$

If we set $\lambda = 1$, we obtain from (4.2.1) the formulas

$$\psi_1 \psi_2 = \frac{3\chi}{2\, d\chi/dz} H_{1/3}^{(1)}(\chi) H_{1/3}^{(2)}(\chi), \tag{4.2.15}$$

$$\frac{\psi_1}{\psi_2} = \frac{H_{1/3}^{(1)}(\chi)}{H_{1/3}^{(2)}(\chi)}. \tag{4.2.16}$$

With the aid of 9.1.3, 9.1.4 and 9.1.10 on pp. 358 and 360 in [5] and (4.2.2) in this chapter in addition to the reflection and recurrence formulas for the Gamma function, one can rewrite (4.2.15) and (4.2.16) as

$$\psi_1\psi_2 = \frac{2\chi}{d\chi/dz}\left\{[J_{-1/3}(\chi)]^2 - J_{-1/3}(\chi)J_{1/3}(\chi) + [J_{1/3}(\chi)]^2\right\}$$

$$= \frac{4}{[\Gamma(2/3)]^2}\frac{(\chi/2)^{1/3}}{d\chi/dz}\left[1 - \frac{3^{3/2}[\Gamma(2/3)]^2}{2\pi}(\chi/2)^{2/3}\right.$$

$$\left. + \text{ higher powers of } (\chi/2)^{2/3}\right]$$

$$= \frac{4}{3^{1/3}[\Gamma(2/3)]^2 d\varphi/dz}\left[1 - \frac{3^{5/6}[\Gamma(2/3)]^2}{2\pi}\varphi + \text{higher powers of } \varphi\right], \tag{4.2.17}$$

$$\frac{\psi_1}{\psi_2} = -\frac{1 - \exp(-i\pi/3)J(\varphi)}{1 - \exp(+i\pi/3)J(\varphi)}, \tag{4.2.18}$$

where

$$J(\varphi) = \frac{J_{1/3}(\chi)}{J_{-1/3}(\chi)} = \left(\frac{1}{3}\right)^{2/3}\frac{\sum_{k=0}^{\infty}\frac{(-1)^k}{k!3^{2k}\Gamma(k+4/3)}\varphi^{3k+1}}{\sum_{k=0}^{\infty}\frac{(-1)^k}{k!3^{2k}\Gamma(k+2/3)}\varphi^{3k}}. \tag{4.2.19}$$

4.3   Comparison Equation Expressions for Nonoscillating Solutions of the q-Equation

The exact, general solution of the q-equation is given by (1.5.14), that is,

$$\frac{1}{q_{\text{exact}}} = \alpha(\psi_1)^2 + \beta(\psi_2)^2 \pm 2\left(\alpha\beta - \frac{1}{W^2}\right)^{1/2}\psi_1\psi_2, \tag{4.3.1}$$

where α and β are arbitrary constants, and ψ_1 and ψ_2 are two linearly independent, but otherwise arbitrary, solutions of the differential equation (1.3.1) with the Wronskian W. We shall choose these solutions to be the comparison equation solutions ψ_1 and ψ_2 introduced in §4.1; therefore we insert for W the Wronskian (4.2.14) into (4.3.1), obtaining

$$\frac{1}{q_{\text{exact}}} = \alpha(\psi_1)^2 + \beta(\psi_2)^2 \pm (4\alpha\beta + \pi^2/9)^{1/2}\psi_1\psi_2. \tag{4.3.2}$$

For the phase-integral approximation we now define

$$\int_{(t)}^{z} q(z)\,dz = \frac{1}{2}\int_{\Gamma_t(z)} q(z)\,dz, \tag{4.3.3}$$

where $\Gamma_t(z)$ is a path that starts at the point corresponding to z on a Riemann sheet adjacent to the complex z-plane under consideration, encircles t in the positive or in the negative sense, and ends at z. In the first order of the phase-integral approximation $\int_{(t)}^z q(z)dz$ is thus the same as $\int_t^z q(z)dz$. The simplified notation in the left-hand member of (4.3.3) for the integral in the right-hand member of (4.3.3) was introduced by Fröman, Fröman and Lundborg [6], pp. 160–161. It makes it possible to use, for an arbitrary order of the phase-integral approximation, a similar simple notation and almost the same simple language (although in a generalized sense) as for the first order of the phase-integral approximation. This really simplifies the treatment of concrete problems when an arbitrary order of the phase-integral approximation is used.

From the transition zero t there emerge three first-order anti-Stokes lines A, B, and C, as shown in Figure 4.1. In the following treatment we shall separately consider the two possible cases which will be discussed in §4.3.1 and §4.3.2.

4.3.1 The Case when Re ζ Increases as z Moves Away from t in the Neighborhood of the Anti-Stokes Line A

Similarly to Re ζ, the real part of the phase-integral $\int_{(t)}^z q(z)dz$ increases as z moves away from t in the neighborhood of the anti-Stokes line A in Figure 4.1. We choose the phase of ζ to be equal to 0 on the anti-Stokes line A. According to (4.2.6) the phase of χ is then close to 0 on the anti-Stokes line A. According to (4.2.4) we can choose the phase of φ_0 to be equal to 0 on the anti-Stokes line A, which implies that the phase of φ is close to 0 in the neighborhood of the anti-Stokes line A. Then, generalizing Eq. (3.3.10), we conclude that in a region not too far away from the anti-Stokes line A, but sufficiently far away from the transition zero t (i.e., in the striped region pertaining to the anti-Stokes line A in Figure 4.1), the solutions ψ_1 and ψ_2 are very accurately given by the phase-integral formulas

$$\psi_1 = \left(\frac{3}{\pi}\right)^{1/2} \exp(-5i\pi/12)q^{-1/2}(z) \exp\left(+i\int_{(t)}^z q(z)\,dz\right) \qquad (4.3.4a)$$

$$\psi_2 = \left(\frac{3}{\pi}\right)^{1/2} \exp(+5i\pi/12)q^{-1/2}(z) \exp\left(-i\int_{(t)}^z q(z)dz\right), \qquad (4.3.4b)$$

which yield

$$\frac{1}{q} = \frac{\pi}{3}\,\psi_1\psi_2. \qquad (4.3.5)$$

When z does not lie too far away from the anti-Stokes line A, but lies sufficiently far away from the transition zero t, the exponentials in the phase-integral functions in (4.3.4a,b) vary rapidly with z; it is seen from

(4.3.2) that q_{exact} is an oscillating function of z, unless $\alpha = \beta = 0$. Introducing these particular values into (4.3.2), and choosing the $+$ sign in order that the resulting formula for q_{exact} be compatible with the phase-integral formula (4.3.5), we obtain

$$\frac{1}{q_A} = \frac{\pi}{3}\,\psi_1\psi_2, \qquad (4.3.6)$$

where we have written q_A instead of q_{exact} in order to indicate that q_A is the particular function q_{exact} which is nonoscillating in the neighborhood of the anti-Stokes line A. Inserting (4.3.4a,b) into (4.3.6), we obtain the approximate formula

$$q_A = q. \qquad (4.3.7)$$

We emphasize that (4.3.6), being a special case of (4.3.2), is an exact formula for q_A, valid in the whole region under consideration around the transition zero t; whereas (4.3.7) is an approximate formula for q_A, valid sufficiently far away from t, but not too far away from the anti-Stokes line A.

Using the matrix (7.5b) in [7], we can trace the solutions given by (4.3.4a,b) from the striped region in the neighborhood of the anti-Stokes line A to the corresponding region in the neighborhood of the anti-Stokes line B, which is also striped in Figure 4.1. The solutions ψ_1 and ψ_2 are given there by the phase-integral formulas

$$\psi_1 = \left(\frac{3}{\pi}\right)^{1/2} \exp(-5i\pi/12)q^{-1/2}(z)\left[\exp\left(+i\int_{(t)}^{z} q(z)\,dz\right)\right.$$

$$\left. - i\,\exp\left(-i\int_{(t)}^{z} q(z)\,dz\right)\right] \qquad (4.3.8a)$$

$$\psi_2 = \left(\frac{3}{\pi}\right)^{1/2} \exp(+5i\pi/12)q^{-1/2}(z)\,\exp\left(-i\int_{(t)}^{z} q(z)\,dz\right). \qquad (4.3.8b)$$

From (4.3.8a,b) it follows that

$$\frac{1}{q} = \frac{\pi}{3}\,\psi_1\psi_2[1 + \exp(-i\pi/3)\psi_2/\psi_1], \qquad (4.3.9)$$

which is thus an approximate formula, derived under the restrictive assumption that z lies not too far away from the anti-Stokes line B in Figure 4.1, but sufficiently far away from the transition zero t.

Inserting (4.3.8a,b) into (4.3.2), we obtain for q_{exact} the approximate formula

$$\frac{1}{q_{exact}} = \frac{3}{\pi q(z)}\left\{\left[\pm(4\alpha\beta + \pi^2/9)^{1/2} - 2i\,\alpha\,\exp(-5i\pi/6)\right]\right.$$

$$+ \alpha \exp(-5i\pi/6) \exp\left(+2i \int_{(t)}^{z} q(z)\,dz\right)$$

$$+ \left[\mp i(4\alpha\beta + \pi^2/9)^{1/2} - \alpha \exp(-5i\pi/6) \right.$$

$$\left. + \beta \exp(+5i\pi/6) \right] \exp\left(-2i \int_{(t)}^{z} q(z)\,dz\right) \Big\}. \qquad (4.3.10)$$

In the region under consideration in the neighborhood of the anti-Stokes line B, the exponentials in (4.3.10) are rapidly oscillating; it can be seen that q_{exact} is an oscillating function of z, unless the coefficients of these exponentials are equal to zero, in which case

$$\alpha = 0 \qquad (4.3.11a)$$

$$\beta = \pm\frac{\pi}{3} \exp(-i\pi/3). \qquad (4.3.11b)$$

In (4.3.11b) the upper and lower signs correspond to the upper and lower signs, respectively, in (4.3.2), when the square root in (4.3.2) is considered to be positive and therefore equal to $\pi/3$ for $\alpha = 0$. Taking notice of the fact that one has to choose the $+$ sign in (4.3.2) and (4.3.11b) in order for the resulting formula for q_{exact} to be compatible with (4.3.9); we obtain from (4.3.2), using the above values of α and β,

$$\frac{1}{q_B} = \frac{\pi}{3} \psi_1\psi_2[1 + \exp(-i\pi/3)\psi_2/\psi_1], \qquad (4.3.12)$$

where we have written q_B instead of q_{exact}, in order to indicate that q_B is the particular function q_{exact} that is nonoscillating when z lies in the neighborhood of the anti-Stokes line B in Figure 4.1. Inserting (4.3.8a,b) into (4.3.12), we obtain the approximate formula

$$q_B = q, \qquad (4.3.13)$$

which is valid sufficiently far away from t, but not too far away from the anti-Stokes line B. We emphasize that (4.3.12) is an exact formula for q_B, valid in the whole region under consideration around the transition zero t, whereas (4.3.13) is an approximate formula for q_B, valid in the region in the neighborhood of the anti-Stokes line B that is indicated by stripes in Figure 4.1.

From (4.3.6) and (4.3.12) we obtain the exact formula

$$q_B = \frac{q_A}{1 + \exp(-i\pi/3)\psi_2/\psi_1}, \qquad (4.3.14)$$

which with the aid of the formula (4.2.18), obtained from the comparison equation solutions, can be written as

$$q_B = \beta(\varphi)q_A, \qquad (4.3.15)$$

where

$$\beta(\varphi) = \frac{1 - \exp(-i\pi/3)J(\varphi)}{1 + J(\varphi)} \exp(-i\pi/3). \qquad (4.3.16)$$

The function $\beta(\varphi)$, defined by (4.3.16), must not be confused with the constant β in (4.3.2). From (4.3.6) and (4.3.12) we also obtain the formula

$$q_A - q_B = \frac{1}{\frac{\pi}{3}\psi_1\psi_2[1 + \exp(+i\pi/3)\psi_1/\psi_2]}, \qquad (4.3.17)$$

which will be used below.

From (4.3.14) we obtain by logarithmic differentiation and insertion of the value $-6i/\pi$ for the Wronskian (4.2.14)

$$\frac{d}{dz}\ln q_B = \frac{d}{dz}\ln q_A + \frac{2i}{\frac{\pi}{3}\psi_1\psi_2[1 + \exp(+i\pi/3)\psi_1/\psi_2]}, \qquad (4.3.18)$$

and hence, with the aid of (4.3.17),

$$\frac{d}{dz}\ln q_B = \frac{d}{dz}\ln q_A + 2i(q_A - q_B). \qquad (4.3.19)$$

Using the matrix (7.5a) in [7], one finds that the solutions ψ_1 and ψ_2, which are given by the phase-integral formulas (4.3.4a,b) in the neighborhood of the anti-Stokes line A, are in the neighborhood of the anti-Stokes line C in Figure 4.1 given by the phase-integral formulas

$$\psi_1 = \left(\frac{3}{\pi}\right)^{1/2} \exp(-5i\pi/12)q^{-1/2}(z) \exp\left(+i\int_{(t)}^{z} q(z)\,dz\right) \qquad (4.3.20a)$$

$$\psi_2 = \left(\frac{3}{\pi}\right)^{1/2} \exp(+5i\pi/12)q^{-1/2}(z)\left\{\exp\left(-i\int_{(t)}^{z} q(z)\,dz\right)\right.$$

$$\left. + i\exp\left(+i\int_{(t)}^{z} q(z)\,dz\right)\right\}, \qquad (4.3.20b)$$

which yield

$$\frac{1}{q} = \frac{\pi}{3}\psi_1\psi_2[1 + \exp(+i\pi/3)\psi_1/\psi_2]. \qquad (4.3.21)$$

Because this formula is a consequence of (4.3.20a,b), it is approximate. It has been derived under the restrictive assumption that z lies in the striped neighborhood of the anti-Stokes line C in Figure 4.1.

Inserting (4.3.20a,b) into (4.3.2), we obtain for q_{exact} the approximate formula

$$\frac{1}{q_{\text{exact}}} = \frac{3}{\pi q(z)}\left\{\left[\pm(4\alpha\beta + \pi^2/9)^{1/2} - 2\beta\exp(+i\pi/3)\right] + \left[\pm i(4\alpha\beta + \pi^2/9)^{1/2}\right.\right.$$

$$+\alpha \exp(-5i\pi/6) - \beta \exp(+5i\pi/6)\Big] \exp\left(+2i\int_{(t)}^{z} q(z)\,dz\right)$$

$$+ \beta \exp(+5i\pi/6) \exp\left(-2i\int_{(t)}^{z} q(z)\,dz\right)\Big\}. \tag{4.3.22}$$

In the region under consideration close to the anti-Stokes line C the exponentials in the right-hand member of (4.3.22) are rapidly oscillating, and therefore q_{exact} is an oscillating function of z unless the coefficients of those exponentials are equal to zero, yielding

$$\alpha = \pm\frac{\pi}{3}\exp(+i\pi/3), \tag{4.3.23a}$$

$$\beta = 0. \tag{4.3.23b}$$

In (4.3.23a) the upper and lower signs pertain to the upper and lower signs, respectively, in (4.3.2), when the square root in (4.3.2) is considered to be positive and therefore equal to $\pi/3$ for $\beta = 0$. With the values (4.3.23a,b) of α and β we obtain from (4.3.2), when one notices that one has to choose the $+$ sign in (4.3.2) and (4.3.23a) in order for the resulting formula for q_{exact} to be compatible with (4.3.21),

$$\frac{1}{q_C} = \frac{\pi}{3}\psi_1\psi_2[1 + \exp(+i\pi/3)\psi_1/\psi_2], \tag{4.3.24}$$

where we have written q_C instead of q_{exact} in order to indicate that we consider the particular function q_{exact}, which is nonoscillating when z lies in the neighborhood of the anti-Stokes line C in Figure 4.1. Inserting (4.3.20a,b) into (4.3.24), we obtain the approximate formula

$$q_C = q, \tag{4.3.25}$$

which is valid sufficiently far from t; however, it is not valid far from the anti-Stokes line C. We emphasize that (4.3.24) is an exact formula for q_C, valid in the whole region under consideration around the transition zero t, whereas (4.3.25) is an approximate formula for q_C, valid only in the restricted region just mentioned.

From (4.3.6) and (4.3.24) we obtain the formula

$$q_C = \frac{q_A}{1 + \exp(+i\pi/3)\psi_1/\psi_2}, \tag{4.3.26}$$

which with the aid of (4.2.18) can be written as

$$q_C = \gamma(\varphi)q_A, \tag{4.3.27}$$

where

$$\gamma(\varphi) = \frac{1 - \exp(+i\pi/3)J(\varphi)}{1 + J(\varphi)}\exp(+i\pi/3). \tag{4.3.28}$$

From (4.3.6) and (4.3.26) we obtain the formula

$$q_A - q_C = \frac{1}{\frac{\pi}{3}\psi_1\psi_2[1 + \exp(-i\pi/3)\psi_2/\psi_1]}, \tag{4.3.29}$$

and from (4.3.12) and (4.3.29) it may be seen that

$$q_B + q_C = q_A. \tag{4.3.30}$$

We obtain the following formula from (4.3.26) by logarithmic differentiation, also making use of (4.2.14):

$$\frac{d}{dz}\ln q_C = \frac{d}{dz}\ln q_A - \frac{2i}{\frac{\pi}{3}\psi_1\psi_2[1 + \exp(-i\pi/3)\psi_2/\psi_1]}, \tag{4.3.31}$$

and hence, with the aid of (4.3.29),

$$\frac{d}{dz}\ln q_C = \frac{d}{dz}\ln q_A - 2i(q_A - q_C). \tag{4.3.32}$$

4.3.2   The Case when Re ζ Decreases as z Moves Away from t in the Neighborhood of the Anti-Stokes Line A

As is the case with Re ζ, the real part of the phase-integral $\int_{(t)}^{z} q(z)dz$ decreases as z moves away from t in the neighborhood of the anti-Stokes line A shown in Figure 4.1. The final formulas pertinent to that case can be obtained from those pertinent to the case treated in §4.3.1 simply by the simultaneous replacement of q_A, q_B, and q_C by $-q_A$, $-q_B$, and $-q_C$. In this way one obtains from (4.3.6), (4.3.12), and (4.3.24)

$$\frac{1}{q_A} = -\frac{\pi}{3}\psi_1\psi_2, \tag{4.3.33}$$

$$\frac{1}{q_B} = -\frac{\pi}{3}\psi_1\psi_2[1 + \exp(-i\pi/3)\psi_2/\psi_1], \tag{4.3.34}$$

$$\frac{1}{q_C} = -\frac{\pi}{3}\psi_1\psi_2[1 + \exp(+i\pi/3)\psi_1/\psi_2], \tag{4.3.35}$$

from (4.3.15), (4.3.27), and (4.3.30)

$$q_B = \beta(\varphi)q_A, \tag{4.3.36}$$

$$q_C = \gamma(\varphi)q_A, \tag{4.3.37}$$

$$q_B + q_C = q_A, \tag{4.3.38}$$

and from (4.3.19) and (4.3.32)

$$\frac{d}{dz}\ln q_B = \frac{d}{dz}\ln q_A - 2i(q_A - q_B), \tag{4.3.39}$$

$$\frac{d}{dz}\ln q_C = \frac{d}{dz}\ln q_A + 2i(q_A - q_C). \tag{4.3.40}$$

The functions $\beta(\varphi)$ and $\gamma(\varphi)$ in (4.3.36) and (4.3.37) are the same as those in §4.3.1 and are thus given by (4.3.16) and (4.3.28) along with (4.2.19). The function φ_0 is calculated from (4.2.4) with the same requirement as in §4.3.1; i.e., the phase of φ_0 must be equal to 0 on the anti-Stokes line A.

4.3.3 Summary of the Results for the Two Cases in Sections 4.3.1 and 4.3.2

The particular, exact solutions q_A, q_B, and q_C of the q-equation, which are characterized by the property of being nonoscillating in the neighborhood of the anti-Stokes lines A, B, and C, respectively, are expressed in terms of two linearly independent solutions, ψ_1 and ψ_2, of the Schrödinger-like differential equation (1.3.1), according to the formulas (4.3.6), (4.3.33), (4.3.12), (4.3.34), (4.3.24), and (4.3.35); that is,

$$\frac{1}{q_A} = \pm\frac{\pi}{3}\psi_1\psi_2, \tag{4.3.41}$$

$$\frac{1}{q_B} = \pm\frac{\pi}{3}\psi_1\psi_2[1 + \exp(-i\pi/3)\psi_2/\psi_1], \tag{4.3.42}$$

$$\frac{1}{q_C} = \pm\frac{\pi}{3}\psi_1\psi_2[1 + \exp(+i\pi/3)\psi_1/\psi_2], \tag{4.3.43}$$

where the upper or the lower signs are appropriate depending on whether ζ, defined by (4.2.5), increases or decreases, respectively, as z moves away from the transition zero t along the anti-Stokes line A shown in Figure 4.1.

When q_A is known, and when ψ_1/ψ_2 is expressed in terms of comparison equation solutions, then according to (4.3.15), (4.3.36), (4.3.27), and (4.3.37), one can obtain q_B and q_C from the formulas

$$q_B = \beta(\varphi)q_A, \tag{4.3.44}$$

$$q_C = \gamma(\varphi)q_A, \tag{4.3.45}$$

where according to (4.3.16) and (4.3.28)

$$\beta(\varphi) = \frac{1 - \exp(-i\pi/3)J(\varphi)}{1 + J(\varphi)}\exp(-i\pi/3), \tag{4.3.46}$$

$$\gamma(\varphi) = \frac{1 - \exp(+i\pi/3)J(\varphi)}{1 + J(\varphi)}\exp(+i\pi/3), \tag{4.3.47}$$

with $J(\varphi)$ given by (4.2.19), that is,

$$J(\varphi) = \frac{J_{1/3}(\chi)}{J_{-1/3}(\chi)} = \left(\frac{1}{3}\right)^{2/3}\frac{\sum_{k=0}^{\infty}\frac{(-1)^k}{k!3^{2k}\Gamma(k+4/3)}\varphi^{3k+1}}{\sum_{k=0}^{\infty}\frac{(-1)^k}{k!3^{2k}\Gamma(k+2/3)}\varphi^{3k}}. \tag{4.3.48}$$

From (4.3.46) and (4.3.47) it follows that

$$\beta(\varphi) + \gamma(\varphi) = 1, \tag{4.3.49}$$

and

$$\beta(\varphi)\gamma(\varphi) = \frac{1 - J(\varphi) + [J(\varphi)]^2}{[1 + J(\varphi)]^2}. \tag{4.3.50}$$

Using (4.3.49), we obtain from (4.3.44) and (4.3.45)

$$q_B + q_C = q_A, \tag{4.3.51}$$

$$\frac{q_B - q_A}{q_A} = -\gamma(\varphi), \tag{4.3.52}$$

$$\frac{q_C - q_A}{q_A} = -\beta(\varphi). \tag{4.3.53}$$

From (4.3.46) and (4.3.47) along with (4.3.48) it follows that

$$\beta(\varphi e^{-2i\pi/3}) = \frac{1}{\gamma(\varphi)}, \tag{4.3.54}$$

$$\gamma(\varphi e^{+2i\pi/3}) = \frac{1}{\beta(\varphi)}. \tag{4.3.55}$$

The derivative dq_B/dz can be expressed in terms of q_A, dq_A/dz, and q_B by means of either (4.3.19) or (4.3.39); the derivative dq_C/dz can be expressed in terms of q_A, dq_A/dz, and q_C by means of either (4.3.32) or and (4.3.40). We may write these formulas as

$$\frac{d}{dz} \ln q_B = \frac{d}{dz} \ln q_A \pm 2i(q_A - q_B), \tag{4.3.56}$$

$$\frac{d}{dz} \ln q_C = \frac{d}{dz} \ln q_A \mp 2i(q_A - q_C), \tag{4.3.57}$$

where the upper signs are to be used if ζ increases as z moves away from t on the anti-Stokes line A in Figure 4.1, whereas the lower signs are to be used if ζ decreases as z moves away from t on the anti-Stokes line A. With the aid of (4.3.44) and (4.3.52) we can rewrite (4.3.56), obtaining

$$\frac{dq_B/dz}{dq_A/dz} = \beta(\varphi) \pm 2i\beta(\varphi)\gamma(\varphi) \frac{q_A^2}{dq_A/dz}. \tag{4.3.56'}$$

With the aid of (4.3.45) and (4.3.53) we can rewrite (4.3.57), obtaining

$$\frac{dq_C/dz}{dq_A/dz} = \gamma(\varphi) \mp 2i\beta(\varphi)\gamma(\varphi) \frac{q_A^2}{dq_A/dz}. \tag{4.3.57'}$$

To calculate the function φ one may start by calculating the function ζ given by (4.2.5); that is,

$$\zeta = \int_{(t)}^{z} Q(z)\, dz, \tag{4.3.58}$$

and then one can obtain the function φ_0 from (4.2.4); that is,

$$\varphi_0 = (3\zeta/2)^{2/3}, \tag{4.3.59}$$

where one chooses the branch of this function that fulfills the condition that φ_0 be real and positive on the anti-Stokes line A. Then one calculates $\varphi_2, \varphi_4, \ldots$ and obtains φ from (4.2.3), that is,

$$\varphi = \sum_{n=0}^{N} \varphi_{2n}. \tag{4.3.60}$$

It should be emphasized that in the formulas in the present section only the function φ, but not its derivative $d\varphi/dz$, appears. It should also be noticed that in (4.3.51), (4.3.56), and (4.3.57) the function φ does not appear *explicitly* at all.

4.3.4 Application Illustrating the Consistency of the Formulas Obtained

Consider the first-order anti-Stokes lines 1 and 2 shown in Figure 4.2. The double arrows on these anti-Stokes lines indicate the directions in which ζ, defined by (4.2.5), is assumed to increase on these lines.

Let us first identify the anti-Stokes line 2 in Figure 4.2 with the anti-Stokes line A in Figure 4.1 and the anti-Stokes line 1 in Figure 4.2 with the anti-Stokes line C in Figure 4.1. From (4.3.45) and (4.3.57) we obtain

$$q_1 = \gamma(\varphi)q_2. \tag{4.3.61a}$$

$$\frac{d}{dz} \ln q_1 = \frac{d}{dz} \ln q_2 - 2i(q_2 - q_1). \tag{4.3.61b}$$

The phase of the function φ in (4.3.61a) is close to 0 on the anti-Stokes line 2 and hence close to $2\pi/3$ on the anti-Stokes line 1.

Let us then identify the anti-Stokes line 1 in Figure 4.2 with the anti-Stokes line A in Figure 4.1 and the anti-Stokes line 2 in Figure 4.2 with the anti-Stokes line B in Figure 4.1. From (4.3.44) and (4.3.56) we then obtain

$$q_2 = \beta(\bar{\varphi})q_1, \tag{4.3.62a}$$

$$\frac{d}{dz} \ln q_2 = \frac{d}{dz} \ln q_1 - 2i(q_1 - q_2), \tag{4.3.62b}$$

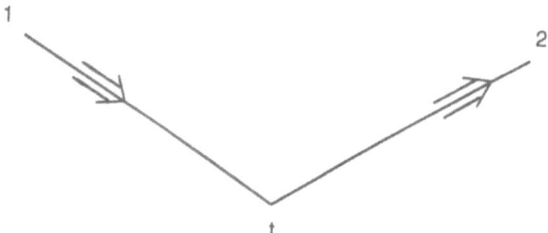

FIGURE 4.2. This figure shows two of the three first-order anti-Stokes lines that emerge from the simple transition zero t. These two lines, which are called 1 and 2, can correspond to lines in Figure 4.1 in two different ways: the lines 1 and 2 can correspond either to C and A, respectively, or to A and B, respectively. The double arrows on these anti-Stokes lines indicate the directions in which ζ, defined by (4.2.5), is assumed to increase on these lines.

where we have denoted the φ-function by $\bar{\varphi}$ in order to distinguish it from the function φ in (4.3.61a). The phase of the function $\bar{\varphi}$ in (4.3.62a) is close to 0 on the anti-Stokes line 1; therefore

$$\arg \bar{\varphi} - \arg \varphi \approx 0 - 2\pi/3 = -2\pi/3, \qquad (4.3.63)$$

and hence

$$\bar{\varphi} \approx \exp(-2i\pi/3)\varphi. \qquad (4.3.64)$$

From (4.3.54) and (4.3.64) it follows that

$$\beta(\bar{\varphi}) \approx \frac{1}{\gamma(\varphi)} \, ; \qquad (4.3.65)$$

hence (4.3.61a) is approximately the same as (4.3.62a). Furthermore, rewriting (4.3.61b), we may see that this formula is the same as (4.3.62b).

4.4   Simple First-Order Formulas

Considering the neighborhood of the transition zero t, we shall use the first-order approximation of the comparison equation solution and thus obtain simple formulas for expressing q_B, dq_B/dz and q_C, dq_C/dz in terms of q_A and dq_A/dz. From (4.2.5) and (4.2.10) we obtain

$$\zeta = \frac{2}{3} a^{1/2}(z-t)^{3/2} \left(1 + \frac{3b_1}{10}(z-t) + \cdots \right). \qquad (4.4.1)$$

From (4.2.4) and (4.4.1) we obtain

$$\varphi_0 = a^{1/3}(z-t)\left(1 + \frac{b_1}{5}(z-t) + \cdots\right);\qquad(4.4.2)$$

hence

$$\frac{d\varphi_0}{dz} = a^{1/3}\left(1 + \frac{2b_1}{5}(z-t) + \cdots\right).\qquad(4.4.3)$$

In the first-order approximation we replace φ by φ_0 in (4.2.17), use (4.4.2) and (4.4.3), and introduce the definition

$$\bar{b}_1 = \frac{4\pi b_1 a^{-1/3}}{5\cdot 3^{5/6}[\Gamma'(2/3)]^2},\qquad(4.4.4)$$

obtaining

$$\psi_1\psi_2 = \frac{4a^{-1/3}}{3^{1/3}[\Gamma(2/3)]^2} - \frac{2\sqrt{3}}{\pi}(1+\bar{b}_1)(z-t) + \cdots;\qquad(4.4.5)$$

hence

$$\frac{d(\psi_1\psi_2)}{dz} = -\frac{2\sqrt{3}}{\pi}(1+\bar{b}_1) + O(z-t).\qquad(4.4.6)$$

From (4.3.41) one obtains

$$\frac{dq_A/dz}{q_A^2} = \mp\frac{\pi}{3}\frac{d(\psi_1\psi_2)}{dz};\qquad(4.4.7)$$

hence, with the use of (4.4.6),

$$\frac{dq_A/dz}{q_A^2} = \pm\frac{2}{\sqrt{3}}(1+\bar{b}_1) + O(z-t),\qquad(4.4.8)$$

where the upper or lower signs are to be used depending on whether ζ increases or decreases, respectively, as z moves away from t on the anti-Stokes line A.

As we are now restricting ourselves to using the first-order approximation, we obtain from (4.3.48) and (4.4.2)

$$J(\varphi) = O(z-t);\qquad(4.4.9)$$

hence according to (4.3.46) and (4.3.47) we have

$$\beta(\varphi) = \exp(-i\pi/3)[1 + O(z-t)]\qquad(4.4.10)$$

and

$$\gamma(\varphi) = \exp(+i\pi/3)[1 + O(z-t)].\qquad(4.4.11)$$

From (4.3.44) along with (4.4.10) and (4.3.45) along with (4.4.11) we obtain
the first-order formulas

$$\frac{q_B}{q_A} = \exp(-i\pi/3)[1 + O(z - t)] \tag{4.4.12a}$$

and

$$\frac{q_C}{q_A} = \exp(+i\pi/3)[1 + O(z - t)]. \tag{4.4.13a}$$

Inserting (4.4.8), (4.4.10), and (4.4.11) into (4.3.56') and (4.3.57'), we obtain the first-order formulas

$$\frac{dq_B/dz}{dq_A/dz} = \exp(+i\pi/3) - \frac{i\bar{b}_1\sqrt{3}}{1 + \bar{b}_1} + O(z - t) \tag{4.4.12b}$$

and

$$\frac{dq_C/dz}{dq_A/dz} = \exp(-i\pi/3) + \frac{i\bar{b}_1\sqrt{3}}{1 + \bar{b}_1} + O(z - t), \tag{4.4.13b}$$

where $\bar{b}_1$ is the quantity defined in (4.4.4).

4.5   Relations Between the a-Coefficients Associated with Different q-Functions, in Terms of which a Given Solution $\psi(z)$ is Expressed

In the procedure for numerical integration of the q-equation, described in §4.1, one has to change from one solution (let us say q_A) to another (q_B or q_C) in order to be able to deal with numerically convenient (i.e., nonoscillatory) functions. For a given solution $\psi(z)$ of the original Schrödinger-like differential equation, the change from one function q, in terms of which $\psi(z)$ is expressed, to another implies a corresponding change of the associated a-coefficients. Therefore, formulas relating the a-coefficients, pertaining to different functions q, to each other are of practical importance for the numerical method mentioned in §4.1; they may be useful also in other contexts.

Recalling that q_A, q_B, and q_C are exact solutions of the q-equation, we define

$$f_1(A; z) = q_A^{-1/2}(z) \exp[+iw_A(z)], \tag{4.5.1a}$$

$$f_2(A; z) = q_A^{-1/2}(z) \exp[-iw_A(z)], \tag{4.5.1b}$$

with

$$w_A(z) = \int^z q_A(z)\, dz. \tag{4.5.2}$$

We can then write a given exact solution $\psi(z)$ and its derivative $\psi'(z)$ in matrix form as

$$\begin{pmatrix} \psi(z) \\ \psi'(z) \end{pmatrix} = \mathbf{W}_f(A; z) \begin{pmatrix} a_1(A) \\ a_2(A) \end{pmatrix}, \qquad (4.5.3)$$

where $a_1(A)$ and $a_2(A)$ are constant coefficients and

$$\mathbf{W}_f(A; z) = \begin{pmatrix} f_1(A; z) & f_2(A; z) \\ f_1'(A; z) & f_2'(A; z) \end{pmatrix}. \qquad (4.5.4)$$

The prime mark denotes differentiation with respect to z. Defining similarly the functions $f_1(B; z)$, $f_2(B; z)$, and $w_B(z)$, the matrix $\mathbf{W}_f(B; z)$ and the constant coefficients $a_1(B)$ and $a_2(B)$; we can write the same exact solution and its derivative in matrix form as

$$\begin{pmatrix} \psi(z) \\ \psi'(z) \end{pmatrix} = \mathbf{W}_f(B; z) \begin{pmatrix} a_1(B) \\ a_2(B) \end{pmatrix}. \qquad (4.5.5)$$

From (4.5.3) and (4.5.5) we obtain

$$\begin{pmatrix} a_1(B) \\ a_2(B) \end{pmatrix} = \mathbf{T}(B, A; z) \begin{pmatrix} a_1(A) \\ a_2(A) \end{pmatrix}, \qquad (4.5.6)$$

where

$$\mathbf{T}(B, A; z) = \mathbf{W}_f^{-1}(B; z)\mathbf{W}_f(A; z). \qquad (4.5.7)$$

Since the determinant of $\mathbf{W}_f(B; z)$ is equal to $-2i$, the inverse of the matrix $\mathbf{W}_f(B; z)$ exists; it is given by the formula

$$\mathbf{W}_f^{-1}(B; z) = \frac{i}{2} \begin{pmatrix} f_2'(B; z) & -f_2(B; z) \\ -f_1'(B; z) & f_1(B; z) \end{pmatrix}. \qquad (4.5.8)$$

Using (4.5.4) and (4.5.8), we obtain for the elements of the matrix $\mathbf{T}(B, A; z)$, defined by (4.5.7), the formulas

$$T_{11}(B, A; z) = \frac{i}{2} f_2(B; z) f_1(A; z) \frac{d}{dz} \ln \frac{f_2(B; z)}{f_1(A; z)}, \qquad (4.5.9a)$$

$$T_{12}(B, A; z) = \frac{i}{2} f_2(B; z) f_2(A; z) \frac{d}{dz} \ln \frac{f_2(B; z)}{f_2(A; z)}, \qquad (4.5.9b)$$

$$T_{21}(B, A; z) = \frac{i}{2} f_1(B; z) f_1(A; z) \frac{d}{dz} \ln \frac{f_1(A; z)}{f_1(B; z)}, \qquad (4.5.9c)$$

$$T_{22}(B, A; z) = \frac{i}{2} f_1(B; z) f_2(A; z) \frac{d}{dz} \ln \frac{f_2(A; z)}{f_1(B; z)}. \qquad (4.5.9d)$$

From (4.5.1a,b) and (4.5.2) we obtain

$$\frac{d}{dz} \ln f_1(A; z) = iq_A(z) - \frac{1}{2} \frac{d}{dz} \ln q_A(z), \qquad (4.5.10a)$$

$$\frac{d}{dz} \ln f_2(A; z) = -iq_A(z) - \frac{1}{2} \frac{d}{dz} \ln q_A(z). \tag{4.5.10b}$$

With the aid of (4.5.10a,b) and the corresponding formulas that have A replaced by B, one obtains

$$\frac{d}{dz} \ln \frac{f_2(B; z)}{f_1(A; z)} = -i[q_A(z) + q_B(z)] + \frac{1}{2} \frac{d}{dz} \ln \frac{q_A(z)}{q_B(z)}, \tag{4.5.11a}$$

$$\frac{d}{dz} \ln \frac{f_2(B; z)}{f_2(A; z)} = i[q_A(z) - q_B(z)] + \frac{1}{2} \frac{d}{dz} \ln \frac{q_A(z)}{q_B(z)}, \tag{4.5.11b}$$

$$\frac{d}{dz} \ln \frac{f_1(A; z)}{f_1(B; z)} = i[q_A(z) - q_B(z)] + \frac{1}{2} \frac{d}{dz} \ln \frac{q_B(z)}{q_A(z)}, \tag{4.5.11c}$$

$$\frac{d}{dz} \ln \frac{f_2(A; z)}{f_1(B; z)} = -i[q_A(z) + q_B(z)] + \frac{1}{2} \frac{d}{dz} \ln \frac{q_B(z)}{q_A(z)}. \tag{4.5.11d}$$

Inserting (4.5.11a–d) into (4.5.9a–d), we obtain

$$T_{11}(B, A; z) = \frac{i}{2} f_2(B; z) f_1(A; z) \left(\frac{1}{2} \frac{d}{dz} \ln \frac{q_A(z)}{q_B(z)} - i[q_A(z) + q_B(z)] \right),$$
$$\tag{4.5.12a}$$

$$T_{12}(B, A; z) = \frac{i}{2} f_2(B; z) f_2(A; z) \left(\frac{1}{2} \frac{d}{dz} \ln \frac{q_A(z)}{q_B(z)} + i[q_A(z) - q_B(z)] \right),$$
$$\tag{4.5.12b}$$

$$T_{21}(B, A; z) = \frac{i}{2} f_1(B; z) f_1(A; z) \left(\frac{1}{2} \frac{d}{dz} \ln \frac{q_B(z)}{q_A(z)} + i[q_A(z) - q_B(z)] \right),$$
$$\tag{4.5.12c}$$

$$T_{22}(B, A; z) = \frac{i}{2} f_1(B; z) f_2(A; z) \left(\frac{1}{2} \frac{d}{dz} \ln \frac{q_B(z)}{q_A(z)} - i[q_A(z) + q_B(z)] \right).$$
$$\tag{4.5.12d}$$

In the derivation of (4.5.12a–d) we have not introduced any approximations; therefore, these formulas give exact expressions for the elements of the matrix $\mathbf{T}(B, A; z)$. To obtain corresponding expressions for the elements of the matrix $\mathbf{T}(C, A; z)$ one simply replaces B by C in (4.5.12a–d).

4.6 Condition for Determination of Regge Pole Positions

Regular solutions of the radial Schrödinger equation which behave as purely outgoing waves at $z = +\infty$ correspond to poles, called Regge poles, of the one-by-one scattering matrix in the complex angular momentum plane. The determination of the positions of the Regge poles is of central importance in Regge pole theory. We shall here derive a condition suitable for exact determination of these poles with the aid of the numerical method mentioned in this chapter. From this exact condition we derive an approximate comparison equation formula, which is then further approximated into a very simple formula.

Figure 4.3 shows the two transition zeros t_1 and t_2, which are of significant importance in the integration procedure, and the paths of integration used. The origin is assumed to lie in or on the border of the sector delimited by the anti-Stokes line 1 and the cut emerging from t_1. The solution which is regular at the origin corresponds to a wave propagating away from t_1 on the anti-Stokes line 1. The purely outgoing wave at infinity corresponds to a wave propagating away from t_2 on the anti-Stokes line 3.

We start the numerical integration of the q-equation (1.3.3) far to the right of the transition zero t_2 in Figure 4.3, and perform this integration along the line 3 which lies close to the first-order anti-Stokes line emerging from t_2 and proceeds in the direction towards $+\infty$. Thus we obtain the function $q_3(z)$ and its derivative $q_3'(z)$ numerically at a point z_2 that lies close to the transition zero t_2. At the point z_2 we use (4.3.44) and (4.3.56) with the upper sign to obtain

$$q_2(z_2) = \beta(\varphi_2(z_2))q_3(z_2), \tag{4.6.1a}$$

$$\frac{q_2'(z_2)}{q_2(z_2)} = \frac{q_3'(z_2)}{q_3(z_2)} + 2i[q_3(z_2) - q_2(z_2)], \tag{4.6.1b}$$

where $\varphi_2(z_2)$ is the value at the point z_2 of the function $\varphi(z)$, which is associated with the transition zero t_2. Starting from the initial values (4.6.1a,b), we then integrate the q-equation along the line 2 that lies close to the first-order anti-Stokes line emerging from t_2 towards the neighborhood of the other transition zero t_1 (see Figure 4.3). The result of the numerical integration of the q-equation is that we obtain values of the function $q_2(z)$ and its derivative $q_2'(z)$ at the point z_1 (defined below). We emphasize that we use the comparison equation formulas (4.6.1a,b) only to obtain initial values at the point z_2 for a function $q_2(z)$ that is nonoscillating on line 2. According to Hökback (personal communication) it is, as mentioned in §4.1, sufficient in general to use (4.6.1a,b) in the first order of the comparison equation technique.

Let us start another numerical integration of the q-equation far away from the transition zero t_1 and perform this integration along line 1, which lies close to a first-order anti-Stokes line emerging from t_1 and intersects line 2 at the point z_1 (see Figure 4.3). In this way we may obtain the values of the function $q_1(z)$ and its derivative $q_1'(z)$ at the point z_1.

If we require that far away from t_2 on line 3 there must be a purely outgoing wave (i.e., a wave travelling in the direction away from t_2), and if we assume that (4.5.1a) with A replaced by 3 represents such a wave (which determines the directions of the double arrows in Figure 4.3), we then obtain from (4.5.6), (4.5.12a,c), and (4.5.1a,b) with A replaced by 3

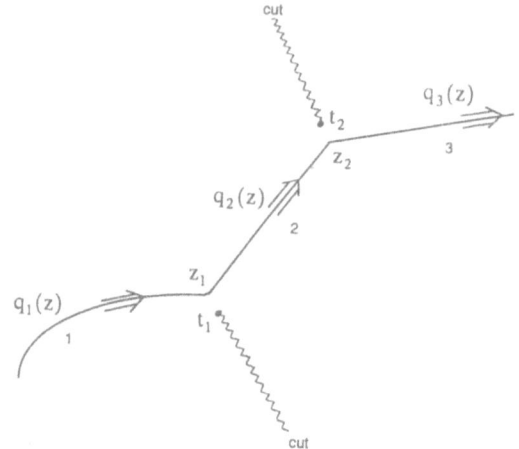

FIGURE 4.3. This figure is used in connection with the derivation of the numerically exact condition (4.6.6). The points t_1 and t_2 are transition zeros from each of which there emerges a cut towards infinity. Lines 1, 2, and 3, which lie close to first-order anti-Stokes lines emerging from t_1 or t_2, are the paths along which one performs the numerical integration of the q-equation in order to obtain the functions $q_1(z)$, $q_2(z)$, and $q_3(z)$, respectively. The double arrows on lines 1, 2, and 3 indicate the directions in which the real part of ζ, defined by (4.2.5), is assumed to increase along these lines.

and B replaced by 2

$$\frac{a_2(2)}{a_1(2)} = \frac{2i[q_2(z_2) - q_3(z_2)] - \left(\frac{q_2'(z_2)}{q_2(z_2)} - \frac{q_3'(z_2)}{q_3(z_2)}\right)}{2i[q_2(z_2) + q_3(z_2)] + \left(\frac{q_2'(z_2)}{q_2(z_2)} - \frac{q_3'(z_2)}{q_3(z_2)}\right)} \; \exp[2iw_2(z_2)], \quad (4.6.2)$$

where $a_1(2)$, $a_2(2)$, and $w_2(z)$ are the constant a-coefficients and the w-integral, associated with line 2. Using (4.6.1b), we obtain from (4.6.2) the formula

$$\frac{a_2(2)}{a_1(2)} = \frac{q_2(z_2) - q_3(z_2)}{q_3(z_2)} \; \exp[2iw_2(z_2)], \quad (4.6.3)$$

which with the aid of (4.6.1a) and (4.3.49) can be written as

$$\frac{a_2(2)}{a_1(2)} = -\gamma(\varphi_2(z_2)) \; \exp[2iw_2(z_2)]. \quad (4.6.4)$$

The use of the first-order approximation in the comparison equation formulas (4.6.1a,b) to simplify (4.6.2) into (4.6.4) does not affect the accuracy of the condition to be derived, since (4.6.1a,b) are used only to obtain a function $q_2(z)$, which is slowly varying on line 2.

If we require that far away from t_1 on line 1 there must be a purely outgoing wave (as seen from the transition zero t_1); i.e., a wave travelling in the direction away from t_1; we obtain from (4.5.6), (4.5.12b,d), and

(4.5.1a,b) with A replaced by 1 and B replaced by 2

$$\frac{a_2(2)}{a_1(2)} = -\frac{2i[q_1(z_1) + q_2(z_1)] + \frac{q_1'(z_1)}{q_1(z_1)} - \frac{q_2'(z_1)}{q_2(z_1)}}{2i[q_1(z_1) - q_2(z_1)] + \frac{q_1'(z_1)}{q_1(z_1)} - \frac{q_2'(z_1)}{q_2(z_1)}} \exp[2iw_2(z_1)]. \qquad (4.6.5)$$

From (4.6.4), (4.6.5), and (4.5.2) we obtain the numerically exact condition

$$\exp\left(2i \int_{z_1}^{z_2} q_2(z)\, dz\right) = \frac{1}{\gamma(\varphi_2(z_2))} \frac{2i[q_1(z_1) + q_2(z_1)] + \frac{q_1'(z_1)}{q_1(z_1)} - \frac{q_2'(z_1)}{q_2(z_1)}}{2i[q_1(z_1) - q_2(z_1)] + \frac{q_1'(z_1)}{q_1(z_1)} - \frac{q_2'(z_1)}{q_2(z_1)}}, \qquad (4.6.6)$$

from which the positions of the Regge poles can be determined. This condition, which requires the integration of the q-equation along the three lines 1, 2, and 3 in Figure 4.3, may be used to obtain numerically exact results. It is, however, of interest to see how, with the aid of comparison equation technique, one can transform (4.6.6) into a very simple approximate condition.

For this purpose we note that we can obtain from (4.3.56) with the lower sign the approximate comparison equation formula

$$\frac{q_1'(z_1)}{q_1(z_1)} - \frac{q_2'(z_1)}{q_2(z_1)} = 2i[q_1(z_1) - q_2(z_1)]. \qquad (4.6.7)$$

Inserting (4.6.7) into the exact condition (4.6.6), we obtain the approximate condition

$$\exp\left(2i \int_{z_1}^{z_2} q_2(z)\, dz\right) = \frac{q_1(z_1)}{\gamma(\varphi_2(z_2))[q_1(z_1) - q_2(z_1)]}. \qquad (4.6.8)$$

Using (4.3.52), we obtain from (4.6.8) the approximate comparison equation condition

$$\exp\left(2i \int_{z_1}^{z_2} q_2(z)\, dz\right) = \frac{1}{\gamma(\varphi_1(z_1))\gamma(\varphi_2(z_2))}, \qquad (4.6.9)$$

where $\varphi_1(z_1)$ is the value at the point z_1 of the function $\varphi(z)$, which is associated with the transition zero t_1; the quantity $\varphi_2(z_2)$, as we have mentioned, is the value at the point z_2 of the function $\varphi(z)$ which is associated with the transition zero t_2. When one uses the approximate comparison equation quantization condition (4.6.9), the q-equation is to be integrated only along line 2. According to (4.4.11) we have for the first order of the phase-integral approximation

$$\gamma(\varphi_1(t_1)) = \gamma(\varphi_2(t_2)) = \exp(i\pi/3). \qquad (4.6.10)$$

Letting $z_1 = t_1$ and $z_2 = t_2$ in (4.6.9), we therefore obtain the simple approximate condition

$$\int_{t_1}^{t_2} q_2(z)\, dz = (s + 2/3)\pi, \qquad (4.6.11)$$

where s is a nonnegative integer. To determine the function $q_2(z)$ in (4.6.11) one starts the numerical integration of the q-equation somewhere in the middle of a line joining t_1 and t_2, where one can use the phase-integral expression $q(z)$ of a conveniently high order to obtain initial values for $q_2(z)$ and $q_2'(z)$. We note that in the neighborhood of t_1 and t_2 the function $q_2'(z)$ in (4.6.11) is a quite different function than the phase-integral function $q(z)$ that occurs in the analogous phase-integral condition

$$\int_{(t_1)}^{(t_2)} q(z)\, dz = (s + 1/2)\pi, \qquad (4.6.12)$$

where s is still a nonnegative integer.

References

[1] Fröman, N. and Fröman, P.O., Phase-integral approximation of arbitrary order generated from an unspecified base function. This is Chapter 1 in the present monograph.

[2] Fröman, N. and Fröman, P.O., Technique of the comparison equation adapted to the phase-integral method. This is Chapter 2 in the present monograph.

[3] Fröman, N. and Fröman, P.O., Problem involving one transition zero. This is Chapter 3 in the present monograph.

[4] Imai, I., Contributions from the Department of Physics, Faculty of Science, University of Tokyo, No. 341. [Reprinted from *IRE Transactions on Antennas and Propagation*, **AP-4** (1956), 233–239.]

[5] Abramowitz, M. and Stegun, I.A., (eds.), *Handbook of Mathematical Functions*, National Bureau of Standards, Applied Mathematics Series, 55. Fourth printing, Washington, 1965.

[6] Fröman, N., Fröman, P.O., and Lundborg, B., *Math. Proc. Camb. Phil. Soc.* **104** (1988), 153–179.

[7] Fröman, N. and Fröman, P.O., *JWKB Approximation, Contributions to the Theory*, North-Holland Publishing Co., Amsterdam, 1965. Russian translation: *MIR*, Moscow, 1967.

5
Cluster of Two Simple Transition Zeros

Nanny Fröman, Per Olof Fröman, and Bengt Lundborg

Abstract. The comparison equation technique described in Chapter 2 is used for solving differential equations of the Schrödinger type for a general situation involving two complex transition zeros. The connection problem is analyzed by combination of general results concerning a cluster of two simple transition zeros and results obtained by means of comparison equation technique. Thus, one obtains the Stokes constants associated with solutions on anti-Stokes lines that recede away from the cluster in four different directions. When an exact solution is given far away from the cluster on such an anti-Stokes line, one therefore has the means of constructing the corresponding phase-integral solution on anti-Stokes lines receding away from the cluster in all four directions mentioned. From such an anti-Stokes line the solution in question can then be further traced in the complex plane along directions in which the dominating term in the linear combination of the two phase-integral functions increases. This result has important applications; for example, in the theory of transmission through a complex potential barrier, in the theory of bound states and its generalization to localized perturbations in plasma physics, in the theory of complex angular momentum, and in the theory of quasi-bound electron states at crystal surfaces.

5.1 Introduction

In the present chapter we shall address the connection problem for a cluster of two simple transition zeros. To this purpose we use the F-matrix technique [13] combined with an appropriate comparison equation technique [22], adapted to the phase-integral method in which one uses the phase-integral approximation generated from an unspecified base function $Q(z)$ [21]. For further background of the present chapter we refer the reader to §2 of [26].

As pointed out in [23], there is no need to use comparison equation technique in the treatment of the problem involving one transition zero, unless one is interested in the wave function in the neighborhood of the transition zero. For the problem involving two transition zeros; i.e., for a cluster of two simple zeros of $Q^2(z)$, which may lie close together; comparison equation technique is useful for deriving analytic expressions for the Stokes constants associated with the four sectors, into which anti-Stokes lines, emerging from the cluster, divide the asymptotic region of the complex z-plane around the cluster. The Stokes constants in question are *order-dependent*; i.e., their analytic expressions depend on the order $2N + 1$ of the phase-integral approximation used, and they contain closed-loop integrals of various terms in the expression for the function $q(z)$; see Chapter 1.

In a previous paper by Fröman, Fröman and Lundborg [26] concerning the Stokes constants for a cluster consisting of an unspecified number of transition points, it has been shown that for a cluster of two transition zeros one can express three of the Stokes constants in terms of the fourth, which is not initially known. In the present chapter we shall determine the remaining Stokes constant by means of the comparison equation technique described in Chapter 2, while choosing the "comparison model potential" to be parabolic.

The case of a real potential barrier has already been treated in a previous paper by Fröman and Fröman [14]. In that case the comparison equation solution yields the same absolute value of the transmission coefficient as was obtained in [14] without comparison equation technique. However, in barrier transmission problems the phase of the wave function is often extremely important; with the aid of the comparison equation solution we can determine the supplementary quantities that allow us to construct arbitrary-order connection formulas that pertain to a complex barrier and that are valid uniformly for all energies, even for those in the neighborhood of the top of the barrier. For a real barrier, expressions for these supplementary quantities (up to the fifth-order approximation) were published without derivation in [27]. In §5.4 of [11] a first-order connection formula for such a barrier was given and discussed. For the general case of a complex barrier with a parabolic top, the supplementary quantities are derived in the present chapter. Closely related quantities appear in Regge pole theory [20].

If we are concerned with a bound-state problem, the comparison equation solution, with a parabolic well as the "comparison model potential," yields the same arbitrary-order quantization conditions as can be more easily obtained with the aid of one of the phase-integral connection formulas in [7] or by tracing the phase-integral solution in the complex z-plane [11, 17]. For the bound-state problem, the merit of the comparison equation solution is that the wave function is yielded in the neighborhood of the turning points. Such information is, however, not needed as often as it might initially seem; in particular, it is not generally needed for the calculation of expectation values and matrix elements; see N. Fröman [8, 9], Fröman and Fröman

[16, 19], and N. Fröman et al. [24]. For the theory of barrier transmission and for the theory of complex angular momentum, comparison equation technique is more important. While the Regge pole positions are obtained from a "quantization condition" that can be derived without the use of comparison equation technique, the derivation of an accurate formula for the residues is based on the use of comparison equation technique [20].

Treatments of real or complex barriers in what corresponds to the first-order phase-integral approximation with $Q^2(z) = R(z)$ (where $R(z)$ is the coefficient function appearing in the differential equation of the Schrödinger type) can be found in early papers by Ford et al. [6], Moriguchi [34], and Heading [28].

Heading [28] considers complex potentials and gives a formula for the reflection amplitude that at first looks inconsistent with the corresponding formula obtained from the results presented in this chapter. The reason for this is that Heading uses a convention opposite to ours with regard to the direction of propagation of a wave; thus $\exp(ikx)$ with $k > 0$ is interpreted by him as a left-going wave. As a consequence, for the same physical problem, the transition points in his picture are the complex conjugates of those in our picture, and therefore the resulting formulas look different.

5.2 Wave Equation and Phase-Integral Approximation

The wave equation for the problem under consideration can be written in the form of Eq. (2.2.1), that is,

$$\frac{d^2\psi}{dz^2} + R(z)\psi = 0, \qquad (5.2.1)$$

where $R(z)$ is assumed to be an analytic function of z. In the case of the one-dimensional Schrödinger equation, possibly resulting from separation of variables; we have, using obvious notation,

$$R(z) = \frac{2m}{\hbar^2}[E - V(z)]. \qquad (5.2.2)$$

In the case of the wave equation for one of the characteristic modes of an electromagnetic wave, the function $R(z)$ is proportional to the square of the effective refractive index [32, 33]. Various analytic forms of $R(z)$ appear in the common equations of mathematical physics, e.g., the Mathieu equation, the Bessel equation and the Legendre equation. In the present chapter $R(z)$ is assumed to be a regular analytic function in the whole region of the complex z-plane under consideration in the comparison equation treatment.

We now introduce a "small," real, and positive expansion parameter λ by considering instead of the original wave equation (5.2.1) the auxiliary differential equation (2.2.2), that is,

$$\frac{d^2\psi}{dz^2} + \{Q^2(z)/\lambda^2 + [R(z) - Q^2(z)]\}\psi = 0, \qquad (5.2.3)$$

where $Q^2(z)$ is an analytic function to be further described presently. The auxiliary differential equation (5.2.3) will be solved by the comparison equation technique described in Chapter 2. At the end λ is set equal to unity, so that the original differential equation (5.2.1) is restored. The solution of (5.2.1) thus obtained is constructed so that, except in the neighborhood of the transition points (i.e., in the present case, the two zeros of $Q^2(z)$), it can be transformed into the phase-integral approximation of arbitrary order pertinent to the differential equation (5.2.1).

In the phase-integral method one uses approximate solutions of the original differential equation (5.2.1) that are generated from the base function $Q(z)$. This function, which is determined by the requirement that the first-order approximation be reasonably good in the region of the complex z-plane, where the phase-integral approximation is to be used, can in general be chosen so that $Q^2(z)$ is either equal to $R(z)$, or approximately equal to $R(z)$, except in the neighborhood of certain critical points where the phase-integral approximation would fail for $Q^2(z) = R(z)$. In this chapter we assume that $R(z)$ has two simple zeros but no pole in the region under consideration; in addition, we choose $Q^2(z)$ to be a regular function of z, similar to $R(z)$, having two simple zeros (transition zeros) t_1 and t_2 in the actual region of the complex z-plane. We emphasize that it is the zeros of $Q^2(z)$, and not those of $R(z)$, that appear as transition zeros in the phase-integral method. There may be other transition points as well, but they are assumed to lie far enough from t_1 and t_2 for the cluster to be considered to be well isolated. If the parameters in the base function $Q(z)$ are changed continuously, the zeros of $Q^2(z)$ correspondingly alter their positions in the complex z-plane. The first-order Stokes and anti-Stokes lines emerging from t_1 and t_2 are shown in Figure 5.1 for the case in which $Q^2(z) = (z - t_1)(z - t_2)$, $t_2 = -t_1$; and in which the argument of $t_1 - t_2$ is changed in steps of $\pi/8$ through the angle π, while the absolute value of $t_1 - t_2$ is kept constant.

In order to make the phase-integral functions (to be defined below) single-valued in the complex z-plane, we introduce a cut between the two zeros of $Q^2(z)$; from this cut we introduce another cut which proceeds away from the cluster in one of the four directions in which anti-Stokes lines recede from the cluster. We have found it convenient to introduce the last mentioned cut as shown in Figures 5.1 and 5.2, where the transition zero t_2 lies on the right-hand lip of the cut. If one starts from the left-hand lip of the cut proceeding away from the cluster, and if one moves in the negative sense around the cluster, the transition zero t_1 appears first and then t_2. If the parameters in the base function $Q(z)$ are changed continuously, the transition zeros, the anti-Stokes lines emerging from the transition zeros,

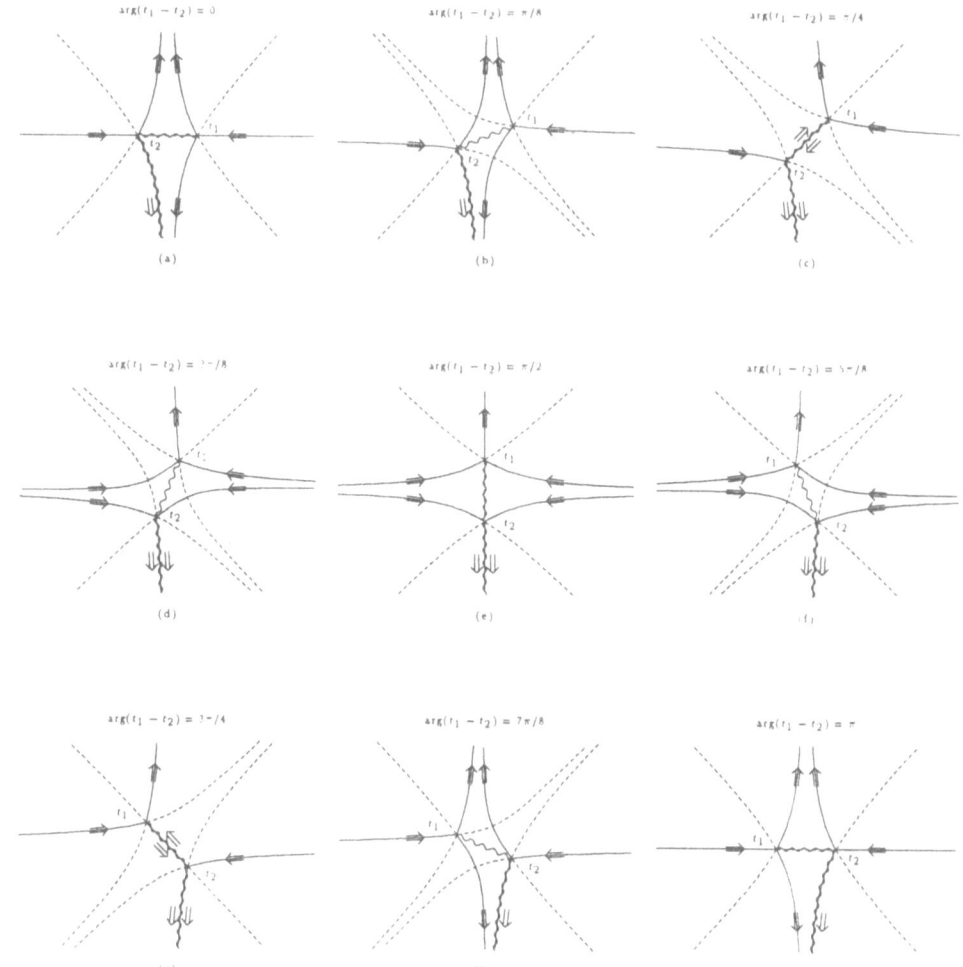

FIGURE 5.1. Patterns of the first-order Stokes (dashed) and anti-Stokes (full-drawn) lines emerging from the simple zeros t_1 and t_2 of $Q^2(z)$ are shown for the case when $Q^2(z) = (z - t_1)(z - t_2)$. The different sub-figures (a)–(i) indicate how these Stokes and anti-Stokes lines change as the argument of $t_1 - t_2$ changes, while $|t_1 - t_2|$ is kept constant. The argument of $t_1 - t_2$ is indicated in each sub-figure. The patterns of corresponding higher-order Stokes and anti-Stokes lines show the same general behavior, except in small regions close to each zero of $Q^2(z)$. The wavy lines are cuts in the complex z-plane. The double arrows associated with the anti-Stokes lines indicate the directions in which Re $w(z)$ increases along these lines. The transition zero t_2 lies on the right-hand lip of the cut.

and the cuts in Figures 5.1 and 5.2, simultaneously change in a continuous way.

In the phase-integral method the approximate solutions of the auxiliary differential equation (5.2.3) are linear combinations of the two linearly independent phase-integral functions

$$f_1(z) = q^{-\frac{1}{2}}(z) \exp[+iw(z)] \tag{5.2.4a}$$

$$f_2(z) = q^{-\frac{1}{2}}(z) \exp[-iw(z)], \tag{5.2.4b}$$

where

$$w(z) = \int_{(t_1)}^{z} q(z) \, dz \tag{5.2.5}$$

$$q(z) = \sum_{n=0}^{N} q^{(2n+1)}(z) \lambda^{2n-1} \tag{5.2.6}$$

with

$$q^{(2n+1)}(z) = Q(z) Y_{2n} \tag{5.2.7}$$

$$Y_0 = 1 \tag{5.2.8a}$$

$$Y_2 = \frac{1}{2}\varepsilon_0 \tag{5.2.8b}$$

$$Y_4 = -\frac{1}{8}\varepsilon_0^2 - \frac{1}{8}\frac{d^2\varepsilon_0}{d\zeta^2} \tag{5.2.8c}$$

$$\frac{d}{d\zeta} = \frac{1}{Q}\frac{d}{dz} \tag{5.2.9}$$

$$\varepsilon_0 = Q^{-\frac{3}{2}}\frac{d^2}{dz^2}Q^{-\frac{1}{2}} + \frac{R - Q^2}{Q^2}; \tag{5.2.10}$$

see §2.1 in [26]. In particular we refer to Eqs. (2.7) and (2.8) in [26], or to (4.3.3) in the present monograph, for the explanation of the short-hand notation in (5.2.5) above with a limit of integration within parentheses. According to the definition of this notation, the integral in (5.2.5) is performed on a Riemann surface for $q(z)$ along a path, which starts at the point corresponding to z but lies on a Riemann sheet adjacent to the complex z-plane of Figure 5.1 or Figure 5.2, encircles t_1 in the positive or negative sense, and ends at z. For the first-order approximation this integral agrees with an ordinary integral from t_1 to z.

As we have mentioned, we assume all other transition points to lie sufficiently far away from t_1 and t_2, so that one can draw a large circle around t_1 and t_2 along which $\mu \ll 1$; cf. Eq. (2.35) in [26]. Thus, we have an isolated second-degree cluster, and can use results from [26]. The first-order anti-Stokes lines emerging from the transition zeros t_1 and t_2 are shown in Figure 5.2(a) for the usual case in which six anti-Stokes lines recede away from the cluster; in Figure 5.2(b) the lines are illustrated for the exceptional case in which an anti-Stokes line joins t_1 and t_2, so that only four anti-Stokes lines recede from the cluster.

The F-matrices, pertaining to the four sectors into which the complex z-plane is divided by the emerging anti-Stokes lines, are also written in Figure 5.2, each matrix being characterized by one unknown element, the Stokes constant for the actual sector. With the aid of the appropriate F-matrix one can trace the phase-integral expression for a solution from one of the delimiting anti-Stokes lines of the sector to the other one; see §2.2 and §5 in [26]. Our aim in this chapter is to derive approximate, but generally quite accurate, expressions for the Stokes constants of a cluster of two transition zeros.

We assume that the phase of $Q(z)$, as well as the direction of the cut proceeding away from the cluster, has been chosen as in the beginning of §5 in [26], which means that Re $w(z)$ increases as z moves away from the cluster along the anti-Stokes lines proceeding in the same direction as the cut (see Figures 5.1 and 5.2). As a consequence of this assumption, Re $w(z)$ increases along the anti-Stokes lines running away from the cluster in the opposite direction, but decreases along the anti-Stokes lines receding away from the cluster in the other two directions. We would like to remark that this choice of the phase of $Q(z)$ and direction of the cut proceeding away from the cluster is consistent with the treatment of a real potential barrier in [14], if z_1 and z_{-1} in our Figure 5.2(a) correspond to the points $-\infty$ and $+\infty$, respectively, on the real z-axis in [14].

The appropriate linear combinations of the phase-integral functions at typical points, denoted in Figure 5.2 by z, z_{-1}, z_0, z_1, and z' (of which z and z' lie on opposite lips of the cut), are obtained by means of the matrices

$$\mathbf{F}(z_{-1}, z) = \begin{pmatrix} 1 & a_{-1} \\ 0 & 1 \end{pmatrix}, \tag{5.2.11a}$$

$$\mathbf{F}(z_0, z_{-1}) = \begin{pmatrix} 1 & 0 \\ b_0 & 1 \end{pmatrix}, \tag{5.2.11b}$$

$$\mathbf{F}(z_1, z_0) = \begin{pmatrix} 1 & a_1 \\ 0 & 1 \end{pmatrix}, \tag{5.2.11c}$$

$$\mathbf{F}(z', z_1) = \begin{pmatrix} 1 & 0 \\ b_2 & 1 \end{pmatrix}. \tag{5.2.11d}$$

For the cluster of two transition zeros now under consideration, the four Stokes constants, a_{-1}, b_0, a_1, b_2, are linked by three algebraic relations given by Eqs. (5.19$_{1,2,3}$) in [26], viz.

$$b_2 = \exp(+i\Omega)b_0, \tag{5.2.12a}$$

$$a_1 = \exp(-i\Omega)a_{-1}, \tag{5.2.12b}$$

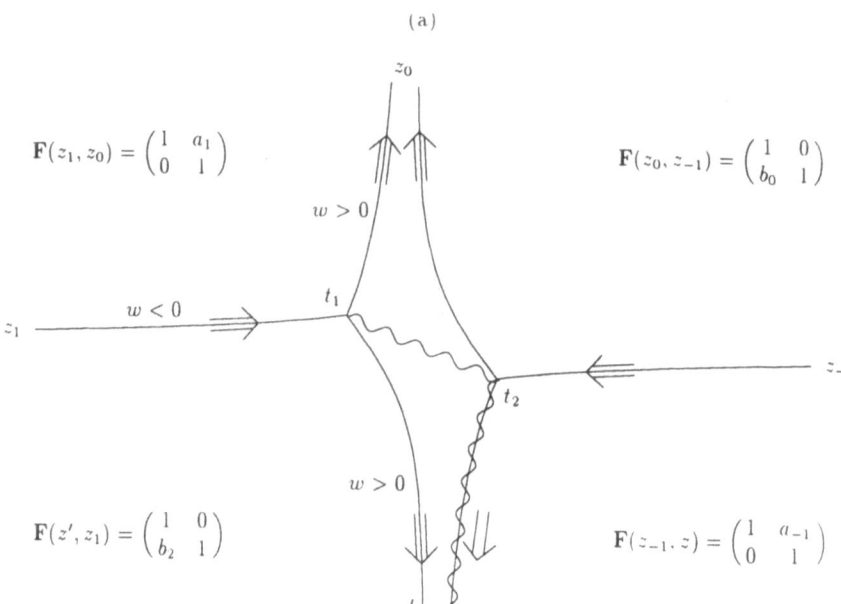

(a)

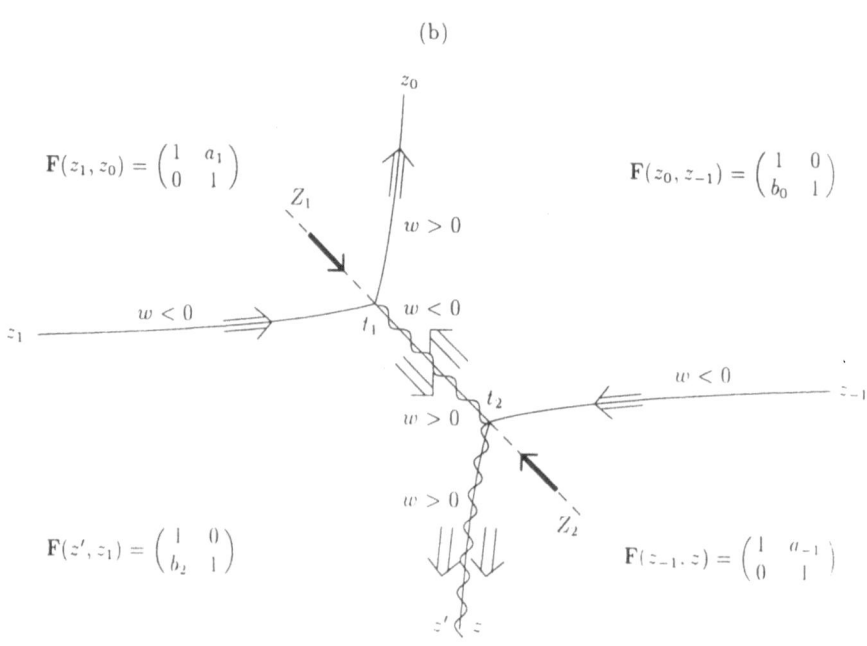

(b)

Caption to Figure 5.2. Cuts (wavy lines), first-order anti-Stokes lines (full drawn) emerging from the two simple zeros t_1 and t_2 of $Q^2(z)$, and connection matrices associated with these anti-Stokes lines are shown, in (a) for the usual case of a complex potential barrier, and in (b) for the exceptional case of a complex potential well which may support a localized state, i.e., for the particular case when an anti-Stokes line joins t_1 and t_2, which is the case when Ω, defined by (5.2.13), is real, i.e., according to (5.4.26'), when K is purely imaginary; see the discussion in the beginning of §5.7. The transition zero t_2 lies on the right-hand lip of the cut. The particular case in (b) occurs for instance when the Regge pole condition is fulfilled. In (b) two particular Stokes lines (dashed), $t_1 Z_1$ and $t_2 Z_2$, are also shown.

According to (5.2.5) the function $w(z)$ is real on the anti-Stokes lines emerging from t_1 but complex on those emerging from t_2 except in the particular case shown in (b), i.e., when an anti-Stokes line joins t_1 and t_2. The choice of phase for $Q(z)$ in the complex z-plane is reflected by the sign of $w(z)$ on any one of the anti-Stokes lines emerging from t_1. The double arrows associated with the anti-Stokes lines indicate the directions in which $\mathrm{Re}\ w(z)$ increases, and the simple arrows on the Stokes lines indicate the directions in which $|\exp[iw(z)]|$ increases.

Unless z lies too close to t_1 or t_2, the pattern of the anti-Stokes lines emerging from t_1 and t_2 is qualitatively the same as in this figure, also when a higher order of the phase-integral approximation is used. When one is well away from t_1 and t_2, the effect of using a higher-order approximation is only a slight distortion of the anti-Stokes lines; see Figure 1 in [7] or Figure 2 in [11].

$$a_{-1}b_0 = -[1 + \exp(+i\Omega)], \qquad (5.2.12c)$$

with the quantity Ω defined by (see pp. 170–171 in [26])

$$\Omega = \int_\Lambda q(z)\, dz, \qquad (5.2.13)$$

where the contour Λ is a closed loop encircling, in the positive sense, both t_1 and t_2 but no other transition points that may exist. For the first-order approximation the closed-contour integral in (5.2.13) can be replaced by twice the integral between t_1 and t_2. Because of the relations (5.2.12a–c), three of the Stokes constants can be expressed in terms of the fourth. Hence, there is only one unknown Stokes constant that remains to be determined by comparison equation technique.

Using (5.2.11b,c) and (5.2.12b,c), we obtain the formula for the F-matrix connecting z_{-1} to z_1 by a path running above the cluster in the complex z-plane:

$$\mathbf{F}(z_1, z_{-1}) = \mathbf{F}(z_1, z_0)\mathbf{F}(z_0, z_{-1}) = \begin{pmatrix} 1 + a_1 b_0 & a_1 \\ b_0 & 1 \end{pmatrix}$$

$$= \begin{pmatrix} -\exp(-i\Omega) & a_1 \\ b_0 & 1 \end{pmatrix}. \qquad (5.2.14)$$

In the exceptional case when $a_{-1}b_0 = 0$ we obtain from (5.2.12c)

$$\Omega = 2\pi \left(\nu + \frac{1}{2}\right), \qquad \nu \text{ integer,} \qquad (5.2.15)$$

which corresponds to the quantization condition for bound states. From (5.2.13) and Figure 5.2(b), with the double-arrows indicating the directions in which Re $w(z)$ increases; it may be seen that $\Omega > 0$ and hence $\nu \geq 0$ in (5.2.15). With the phase of $Q(z)$ and the cuts chosen according to Figure 5.2(b) it turns out that a_{-1} cannot be equal to zero (see the discussion at the end of §5.6). Therefore, $a_{-1}b_0$ is equal to zero only when b_0 is equal to zero. When $b_0 = 0$, there is a particular solution, which on all the anti-Stokes lines emerging from the cluster in Figure 5.2(b) is represented by $f_1(z)$, except for an arbitrary constant factor; this solution is dominant in the two sectors (z_{-1}, z_0) and (z_1, z'), but subdominant in the other two sectors. Here the particular solution is thus proportional to $f_1(z)$, whereas in [15] it was proportional to $f_2(z)$, owing to a different choice of phase for $Q(z)$.

We shall now consider the "quantization condition" $b_0 = 0$ and reformulate it as a condition on the matrix $\mathbf{F}(Z_1, Z_2)$, where Z_1 and Z_2 are points, far away from the cluster, on particular Stokes lines (i.e., lines along which Re $w(z)$ is constant) emerging from t_1 and t_2, respectively, as shown in Figure 5.2(b). According to (5.2.11b) the condition $b_0 = 0$ implies that $\mathbf{F}(z_0, z_{-1})$ is a unit matrix. Therefore, when we move from Z_2 to Z_1 above the cluster, we have

$$\mathbf{F}(Z_1, Z_2) = \mathbf{F}(Z_1, z_0)\mathbf{F}(z_0, z_{-1})\mathbf{F}(z_{-1}, Z_2) = \mathbf{F}(Z_1, z_0)\mathbf{F}(z_{-1}, Z_2),$$
$$(5.2.16)$$

and hence

$$F_{21}(Z_1, Z_2) = F_{21}(Z_1, z_0)F_{11}(z_{-1}, Z_2) + F_{22}(Z_1, z_0)F_{21}(z_{-1}, Z_2). \quad (5.2.17)$$

With the aid of the basic estimates (4.3a,c) in [13], the inversion formula (3.20) in [13] and our definition (5.2.5) of $w(z)$, we find that $F_{11}(z_{-1}, Z_2)$ and $F_{22}(Z_1, z_0)$ are approximately equal to unity, whereas $F_{21}(Z_1, z_0)$ and $F_{21}(z_{-1}, Z_2)$ are negligibly small. Thus, we obtain from (5.2.17) the approximate "quantization condition"

$$F_{21}(Z_1, Z_2) = 0, \qquad (5.2.18)$$

which corresponds to the quantization condition $F_{12}(-\infty, +\infty) = 0$ for a quantal particle in a real potential well, as given by Eq. (10.13) in [13]. In (5.2.18) the matrix element F_{21} appears, while in Eq. (10.13) in [13] the matrix element F_{12} appears. This difference is due to the different choices of phase for $Q(z)$. The parts of the real axis to the left and to the right of the classically allowed region in [13] correspond to the two Stokes lines on which Z_1 and Z_2 are located in our discussion above.

5.3 Comparison Equation

To solve the auxiliary differential equation (5.2.3) by the comparison equation technique described in Chapter 2, we set, following Eq. (2.2.3),

$$\psi = \left(\frac{d\varphi}{dz}\right)^{-\frac{1}{2}} \Phi(\varphi), \qquad (5.3.1)$$

where φ is a so far unknown function of z and λ (to be determined from the differential equation (2.2.8)). The function $\Phi(\varphi)$ is a known solution of the differential equation (2.2.5); that is,

$$\frac{d^2\Phi}{d\varphi^2} + \frac{\Pi(\varphi)}{\lambda^2}\Phi = 0, \qquad (5.3.2)$$

where $\Pi(\varphi)$ is a function of φ and λ that, in the region of the complex φ-plane corresponding to the region of the complex z-plane under consideration, has zeros in analogy to the function $Q^2(z) + [R(z) - Q^2(z)]\lambda^2$ (see [5.2.3]).

For the purpose of this chapter, in which we deal with a cluster of two simple zeros of $Q^2(z)$, the function $\Pi(\varphi)$ is chosen to be

$$\Pi(\varphi) = A_0 + A_1\varphi + A_2\varphi^2, \qquad A_2 \neq 0, \qquad (5.3.3)$$

where the coefficients A_0, A_1, and A_2 are independent of φ, but may depend on λ. Our formula (5.3.3) is the particular case of Eqs. (2.2.9a–c) which one obtains by setting

$$A_n = 0 \quad \text{for} \quad n > 2 \qquad (5.3.4)$$

and

$$B_q = \delta_{q,0}. \qquad (5.3.5)$$

Therefore, we can avoid lengthy calculations, which have been performed in a general way once and for all in Chapter 2, by particularizing according to (5.3.4) and (5.3.5) formulas given in Chapter 2.

By introducing in (5.3.2) and (5.3.3) a new independent variable φ, which is a linear function of the original independent variable φ, one realizes that it is no restriction to set A_1 equal to zero and A_2 equal to 1/4 (see the discussion below Eqs. [2.2.9a–c]). Using then instead of A_0 the notation $-\overline{K}$, we have

$$A_0 = -\overline{K} \qquad (5.3.6a)$$

$$A_1 = 0 \qquad (5.3.6b)$$

$$A_2 = 1/4 \qquad (5.3.6c)$$

$$A_n = 0 \quad \text{for} \quad n > 2, \qquad (5.3.6d)$$

and hence (5.3.3) gives

$$\Pi(\varphi) = \varphi^2/4 - \overline{K}. \qquad (5.3.7)$$

The comparison equation that we treat in this chapter is thus (5.3.2), using the particular expression (5.3.7) for $\Pi(\varphi)$. This differential equation can be written in the form

$$\frac{d^2\Phi}{d(\varphi/\sqrt{\lambda})^2} + \left(\frac{(\varphi/\sqrt{\lambda})^2}{4} - \overline{K}/\lambda\right)\Phi = 0. \tag{5.3.8}$$

The solutions of this differential equation are the parabolic cylinder functions, also called the Weber functions, of the complex variable $\varphi/\sqrt{\lambda}$ and the complex parameter $\overline{K}/\lambda$.

5.4   Comparison Equation Solution

The solution of (5.3.8) which for large positive values of $\varphi/\sqrt{\lambda}$ represents a single outgoing wave (i.e., a single wave which has a phase that increases with φ), is proportional to the parabolic cylinder function $E(\overline{K}/\lambda, \varphi/\sqrt{\lambda})$; cf. 19.17.9 in [1], which is valid for complex a (i.e., $\overline{K}/\lambda$ in our notation), if, instead of 19.17.10 in [1], we set

$$\phi_2 = \frac{1}{2i}\ln\frac{\Gamma(\frac{1}{2} + i\overline{K}/\lambda)}{\Gamma(\frac{1}{2} - i\overline{K}/\lambda)}, \tag{5.4.1}$$

with the branch of the logarithm function that is zero when $\overline{K}/\lambda = 0$. The corresponding solution (5.3.1) of the auxiliary differential equation (5.2.3) is

$$\psi = \mathcal{A}\left(\frac{d(\varphi/\sqrt{\lambda})}{dz}\right)^{-\frac{1}{2}} E(\overline{K}/\lambda, \varphi/\sqrt{\lambda}), \tag{5.4.2}$$

where $\mathcal{A}$ is a normalization factor which is independent of φ, but may depend on λ.

We shall determine φ as a function of z and λ. For this purpose we expand φ formally according to Eq. (2.2.11); that is,

$$\varphi = \sum_{\beta=0}^{\infty} \varphi_{2\beta}\lambda^{2\beta}, \tag{5.4.3}$$

where the quantities $\varphi_{2\beta}$ depend on z but are independent of λ. Similarly, $\overline{K}$, which may depend on λ, is assumed to be expanded in powers of λ^2:

$$\overline{K} = \sum_{\beta=0}^{\infty} \overline{K}_{2\beta}\lambda^{2\beta}, \tag{5.4.4}$$

where the coefficients $\overline{K}_{2\beta}$ are independent of λ. Recalling (5.3.6a–d) and (5.4.4), we expand the quantities A_n according to the formal series (2.2.12);

that is,

$$A_n = \sum_{\beta=0}^{\infty} A_{n,2\beta}\lambda^{2\beta}, \tag{5.4.5}$$

where in the case we are currently considering

$$A_{0,2\beta} = -\overline{K}_{2\beta} \tag{5.4.6a}$$

$$A_{1,2\beta} = 0 \tag{5.4.6b}$$

$$A_{2,2\beta} = \frac{1}{4}\delta_{\beta,0} \tag{5.4.6c}$$

$$A_{n,2\beta} = 0 \quad \text{for} \quad n > 2. \tag{5.4.6d}$$

From the definitions (2.2.31) and (2.2.16c), that is,

$$S_{2\beta}(\varphi_0) = \sum_{n=0}^{\infty} A_{n,2\beta}\varphi_0^n \tag{5.4.7}$$

and

$$T(\varphi_0) = \sum_{q=0}^{\infty} B_q\varphi_0^q, \tag{5.4.8}$$

we obtain with the aid of (5.3.5) and (5.4.6a–d)

$$\frac{S_{2\beta}(\varphi_0)}{T(\varphi_0)} = \frac{1}{4}\delta_{\beta,0}\varphi_0^2 - \overline{K}_{2\beta}. \tag{5.4.9}$$

Recalling the definition (2.2.16a), that is,

$$P^2(\varphi_0) = \frac{S_0(\varphi_0)}{T(\varphi_0)}, \tag{5.4.10}$$

we obtain from (5.4.9)

$$P(\varphi_0) = (\varphi_0^2/4 - \overline{K}_0)^{\frac{1}{2}}, \tag{5.4.11}$$

where the phase of the square root must be chosen appropriately, as will be discussed below.

5.4.1 Determination of $\varphi_0(z)$ and $\overline{K}_0$

In order for the comparison equation technique for solving the auxiliary differential equation (5.2.3) to work satisfactorily, the mapping of the z-plane onto the φ_0-plane must be conformal according to Chapter 2. This means that φ_0 must be a regular analytic function of z, the derivative of which has no zero in the region of the complex z-plane under consideration. As a consequence of the conformal mapping, it follows that the functions

$Q^2(z)$ and $P^2(\varphi_0)$ must have the same kind of zeros at corresponding points in the complex z- and φ_0-planes. The cluster in the z-plane must thus correspond to a cluster of the same kind in the φ_0-plane.

Pursuing the analogy with the situation in the z-plane, we introduce cuts in the φ_0-plane that correspond to those in the z-plane in the sense that the function $\varphi_0(z)$ maps the cuts in the z-plane onto those in the φ_0-plane; we choose the phase of the square root in (5.4.11) so that

$$P(\varphi_0) = (\varphi_0^2/4 - \overline{K}_0)^{\frac{1}{2}} = -\frac{1}{2}\varphi_0 \left(1 - \frac{4\overline{K}_0}{\varphi_0^2}\right)^{\frac{1}{2}}, \qquad (5.4.12)$$

where the square root in the last member is defined so as to tend to unity as $\varphi_0 \to \infty$. This implies that the phases of $P(\varphi_0)$ and $Q(z)$, as well as the cuts in the complex φ_0- and z-planes, correspond to one another in relevant corresponding regions of the two planes (see Figures 5.1 and 5.2). According to (5.4.11), the function $P^2(\varphi_0)$ has two simple zeros in the φ_0-plane, which we denote by τ_1 and τ_2, and which in the z-plane correspond to t_1 and t_2, respectively, so that $\tau_1 = \varphi_0(t_1)$ and $\tau_2 = \varphi_0(t_2)$. Thus, since when we move, from the cut tending to infinity, in the negative sense around the cluster, τ_1 appears first; we obtain from (5.4.12)

$$\tau_1 = 2\overline{K}_0^{\frac{1}{2}} \exp(i\pi), \qquad (5.4.13a)$$

$$\tau_2 = 2\overline{K}_0^{\frac{1}{2}}, \qquad (5.4.13b)$$

where the branch of the square root is to be chosen appropriately, so that the phases of τ_1 and τ_2 conform to the actual situation relevant to the cuts introduced in the φ_0-plane. For the case corresponding to Figure 5.1(i) (i.e., for sub-barrier penetration through a real potential barrier with t_1 lying to the left of t_2, and hence τ_1 lying to the left of τ_2) the quantity $\overline{K}_0^{\frac{1}{2}}$ in (5.4.13a,b) is chosen to be positive. This is the reason it is natural to introduce the factor $\exp(+i\pi)$ in (5.4.13a) instead of the factor $\exp(-i\pi)$ in (5.4.13b).

The function $\varphi_0(z)$ must satisfy the differential equation (2.2.17); that is,

$$\frac{d\varphi_0}{dz} = \frac{Q(z)}{P(\varphi_0)}. \qquad (5.4.14)$$

From (5.4.14) and from the requirement that the zero t_1 of $Q^2(z)$ must be mapped on the zero τ_1 of $P^2(\varphi_0)$ we obtain the relation

$$\int_{\tau_1}^{\varphi_0} P(\varphi_0)d\varphi_0 = \int_{t_1}^{z} Q(z)\, dz, \qquad (5.4.15a)$$

which determines φ_0 as a function of z.

The function $\varphi_0(z)$ must be regular analytic, and hence single-valued, in the whole region containing the zeros of $Q^2(z)$. To determine the parameter

$\overline{K}_0$ we therefore impose the condition that $\varphi_0(z)$ must be single-valued on a closed curve that encloses the two transition zeros t_1 and t_2. Recalling that $Y_0 = 1$ according to (5.2.8a), we obtain from (5.4.15a) the equation (2.2.24'); that is,

$$\int_\Lambda P(\varphi_0)d\varphi_0 = \int_\Lambda Y_0 Q(z)\, dz, \qquad (5.4.16a)$$

where Λ is a closed contour of integration enclosing both t_1 and t_2 in the complex z-plane; it was already introduced in (5.2.13). The corresponding curve enclosing the points τ_1 and τ_2 in the complex φ_0-plane is also denoted by Λ.

For large values of φ_0 it follows from (5.4.12) that

$$P(\varphi_0) = -\frac{1}{2}\varphi_0 + \frac{\overline{K}_0}{\varphi_0} + O(\varphi_0^{-3}), \qquad \varphi_0 \to \infty. \qquad (5.4.17a)$$

Therefore, one can evaluate the integral in the left-hand member of (5.4.16a) by deforming the contour Λ into a very large circle and using residue calculus. Thus, one obtains from (5.4.16a) and (5.4.17a) the formula

$$\overline{K}_0 = -\frac{i}{2\pi} \int_\Lambda Y_0 Q(z)\, dz, \qquad (5.4.18a)$$

which, according to (5.4.13a) and (5.4.13b), determines the position of the zeros τ_1 and τ_2 of $P^2(\varphi_0)$ in the φ_0-plane. If the parameters in the base function $Q(z)$ are varied, it is obvious from (5.4.18a) and (5.4.13a,b) that the positions of τ_1 and τ_2 vary correspondingly. The cuts in the z- and φ_0-planes proceed in such a way that the correspondence between the zeros t_1, t_2 and τ_1, τ_2 is preserved during such variation of the parameters.

5.4.2 Determination of $\varphi_{2\beta}$ and $\overline{K}_{2\beta}$ for $\beta > 0$

The functions φ_2 and φ_4 are given by Eqs. (2.2.60a',b'). Setting $c_2 = c_4 = 0$ and recalling (2.2.41), (5.4.11), and (5.4.9) with $\beta = 2$, we thus have

$$(\varphi_0^2/4 - \overline{K}_0)^{\frac{1}{2}}\varphi_2 = \int_{(t_1)}^z Y_2 Q(z)\, dz - \int_{(\tau_1)}^{\varphi_0} \frac{1}{2}h(\varphi_0^2/4 - \overline{K}_0)^{\frac{1}{2}}d\varphi_0 \quad (5.4.15b)$$

and

$$(\varphi_0^2/4 - \overline{K}_0)^{\frac{1}{2}}\varphi_4 = \int_{(t_1)}^z Y_4 Q(z)\, dz$$

$$+ \int_{(\tau_1)}^{\varphi_0} \left[\frac{1}{8}h^2 + \frac{1}{8(\varphi_0^2/4 - \overline{K}_0)^{\frac{1}{2}}}\frac{d}{d\varphi_0}\left(\frac{1}{(\varphi_0^2/4 - \overline{K}_0)^{\frac{1}{2}}}\frac{dh}{d\varphi_0} \right) \right.$$

$$\left. + \frac{\overline{K}_4}{2(\varphi_0^2/4 - \overline{K}_0)} \right] (\varphi_0^2/4 - \overline{K}_0)^{\frac{1}{2}}d\varphi_0$$

$$-\frac{1}{2}\left(h(\varphi_0^2/4 - \overline{K}_0)^{\frac{1}{2}}\varphi_2 + \frac{\varphi_0}{4(\varphi_0^2/4 - \overline{K}_0)^{\frac{1}{2}}}\varphi_2^2\right),\tag{5.4.15c}$$

where the choice of the constant lower limits of integration ensures that the functions φ_2 and φ_4 will be regular analytic at τ_1. According to (2.2.34), (5.4.9) with $\beta = 1$, and (5.4.11) the quantity h in (5.4.15b,c) is given by

$$h = h_0(\varphi_0, \overline{K}_0) - \frac{\overline{K}_2}{\varphi_0^2/4 - \overline{K}_0},\tag{5.4.19}$$

where, by definition,

$$h_0(\varphi_0, \overline{K}_0) = (\varphi_0^2/4 - \overline{K}_0)^{-\frac{3}{4}}\frac{d^2}{d\varphi_0^2}(\varphi_0^2/4 - \overline{K}_0)^{-\frac{1}{4}}.\tag{5.4.20}$$

One can determine the quantities $\overline{K}_2$ and $\overline{K}_4$ from (5.4.15b,c) by requiring that φ_2 and φ_4 be regular analytic in the region under consideration and hence single-valued on the closed curve Λ. Thus, one can obtain

$$\int_\Lambda \frac{1}{2}h(\varphi_0^2/4 - \overline{K}_0)^{\frac{1}{2}}d\varphi_0 = \int_\Lambda Y_2 Q(z)\,dz\tag{5.4.16b}$$

and

$$\int_\Lambda \left(-\frac{1}{8}h^2 - \frac{\overline{K}_4}{2(\varphi_0^2/4 - \overline{K}_0)}\right)(\varphi_0^2/4 - \overline{K}_0)^{\frac{1}{2}}d\varphi_0 = \int_\Lambda Y_4 Q(z)\,dz,\tag{5.4.16c}$$

where Λ is the closed contour of integration encircling the two relevant transition zeros (t_1 and t_2 or τ_1 and τ_2) mentioned above.

The integrals in the left-hand members of (5.4.16b,c) are easily evaluated when the contour Λ is deformed into a very large circle and when use is made of the formulas

$$\frac{1}{2}h(\varphi_0^2/4 - \overline{K}_0)^{\frac{1}{2}} = \frac{\overline{K}_2}{\varphi_0} + O(\varphi_0^{-3}), \qquad \varphi_0 \to \infty,\tag{5.4.17b}$$

and

$$\left(-\frac{1}{8}h^2 - \frac{\overline{K}_4}{2(\varphi_0^2/4 - \overline{K}_0)}\right)(\varphi_0^2/4 - \overline{K}_0)^{\frac{1}{2}} = \frac{\overline{K}_4}{\varphi_0} + O(\varphi_0^{-3}), \quad \varphi_0 \to \infty,$$
$$\tag{5.4.17c}$$

which are obtained with the use of (5.4.12), (5.4.19), and (5.4.20). Thus, from (5.4.16b,c) and (5.4.17b,c) one obtains

$$\overline{K}_2 = -\frac{i}{2\pi}\int_\Lambda Y_2 Q(z)\,dz\tag{5.4.18b}$$

$$\overline{K}_4 = -\frac{i}{2\pi}\int_\Lambda Y_4 Q(z)\,dz.\tag{5.4.18c}$$

As a consequence of the requirement that the functions $\varphi_{2\beta}$ are regular analytic in the region under consideration, it is apparent that we have derived explicit formulas for $\overline{K}_{2\beta}$ when $\beta = 0$, 1, and 2. Analogous formulas should be valid for all possible values of β:

$$\overline{K}_{2\beta} = -\frac{i}{2\pi} \int_\Lambda Y_{2\beta} Q(z)\, dz, \qquad \beta = 0, 1, 2, \dots. \tag{5.4.21}$$

We can obtain the comparison equation solution of the order $2N + 1$ by using in (5.4.2) for φ and $\overline{K}$ expressions which one obtains by truncating the series in the right-hand members of (5.4.3) and (5.4.4) at $\beta = N$; that is,

$$\varphi = \sum_{\beta=0}^{N} \varphi_{2\beta} \lambda^{2\beta} \tag{5.4.22}$$

and

$$\overline{K} = \sum_{\beta=0}^{N} \overline{K}_{2\beta} \lambda^{2\beta}. \tag{5.4.23}$$

Combining (5.4.21) and (5.4.23), we obtain

$$\overline{K} = -\frac{i}{2\pi} \sum_{\beta=0}^{N} \lambda^{2\beta} \int_\Lambda Y_{2\beta} Q(z)\, dz, \tag{5.4.24}$$

or, taking into account (5.2.6) and (5.2.7),

$$\overline{K}/\lambda = -\frac{i}{2\pi} \int_\Lambda q(z)\, dz, \tag{5.4.25}$$

and hence, according to (5.2.13),

$$\Omega = 2\pi i \overline{K}/\lambda. \tag{5.4.26}$$

For future purpose (see [5.7.2]) we also introduce the quantity

$$K = \pi \overline{K}/\lambda = -\frac{i}{2} \int_\Lambda q(z)\, dz. \tag{5.4.27}$$

Using this quantity, (5.4.26) can be written as

$$\Omega = 2iK. \tag{5.4.26'}$$

5.5 Phase-Integral Solution Obtained from the Comparison Equation Solution

The derivation of the phase-integral solution from the comparison equation solution in the treatment of the problem concerning one transition zero in

Chapter 3 was based on a direct and straightforward expansion of the comparison equation solution. However, the occurring integrals, which in that particular case were easy to evaluate, are in more general problems, involving more than one transition point, rather complicated. Therefore, in the problem concerning two transition zeros, we shall use an approach that is more subtle, but which has the merit of leading to the desired formulas in a general way and without complicated calculations. For the sake of completeness we present in the Appendix an alternative procedure involving direct, straightforward calculations.

When $\varphi/\sqrt{\lambda} \to \infty$, while the parameter $\overline{K}/\lambda$ remains fixed, we have the asymptotic formulas

$$E(\overline{K}/\lambda, \varphi/\sqrt{\lambda}) \sim \left(\frac{2}{\varphi/\sqrt{\lambda}}\right)^{\frac{1}{2}}$$

$$\times \exp\left[+i\left((\varphi/\sqrt{\lambda})^2/4 - (\overline{K}/\lambda)\ln(\varphi/\sqrt{\lambda})\right.\right.$$

$$\left.\left. + \pi/4 + \frac{1}{4i}\ln\frac{\Gamma(\frac{1}{2}+i\overline{K}/\lambda)}{\Gamma(\frac{1}{2}-i\overline{K}/\lambda)}\right)\right],$$

$\varphi/\sqrt{\lambda} \to \infty$ with fixed $\arg\varphi$ in the interval $-\pi/2 < \arg\varphi < \pi$, (5.5.1a)

and

$$E(\overline{K}/\lambda, \varphi/\sqrt{\lambda}) \sim \left(\frac{2}{\varphi/\sqrt{\lambda}}\right)^{\frac{1}{2}}$$

$$\times \exp\left[+i\left((\varphi/\sqrt{\lambda})^2/4 - (\overline{K}/\lambda)\ln(\varphi/\sqrt{\lambda})\right.\right.$$

$$\left.\left. + \pi/4 + \frac{1}{4i}\ln\frac{\Gamma(\frac{1}{2}+i\overline{K}/\lambda)}{\Gamma(\frac{1}{2}-i\overline{K}/\lambda)}\right)\right]$$

$$- \exp(\pi\overline{K}/\lambda)[1 + \exp(2\pi\overline{K}/\lambda)]^{\frac{1}{2}}\left(\frac{2}{\varphi/\sqrt{\lambda}}\right)^{\frac{1}{2}}$$

$$\times \exp\left[-i\left((\varphi/\sqrt{\lambda})^2/4 - (\overline{K}/\lambda)\ln(\varphi/\sqrt{\lambda})\right.\right.$$

$$\left.\left. + \pi/4 + \frac{1}{4i}\ln\frac{\Gamma(\frac{1}{2}+i\overline{K}/\lambda)}{\Gamma(\frac{1}{2}-i\overline{K}/\lambda)}\right)\right],$$

$\varphi/\sqrt{\lambda} \to \infty$ with fixed $\arg\varphi$ in the interval $\pi/2 < \arg\varphi < 3\pi/2$,

(5.5.1b)

where, as in (5.4.1), the logarithm of the quotient of the two Γ-functions is equal to zero for $\overline{K}/\lambda = 0$. The expressions (5.5.1a,b) cease to be valid when arg φ approaches the boundaries of the sectors defined by the conditions on arg φ in (5.5.1a,b).

To obtain phase-integral functions associated with the comparison equation (5.3.8), we choose the square of the base function to be equal to the coefficient function in that differential equation; hence, the base function is

$$Q_c(\varphi/\sqrt{\lambda}) = \left[(\varphi/\sqrt{\lambda})^2/4 - \overline{K}/\lambda\right]^{\frac{1}{2}} = -\frac{\varphi/\sqrt{\lambda}}{2}\left[1 - \frac{4\overline{K}/\lambda}{(\varphi/\sqrt{\lambda})^2}\right]^{\frac{1}{2}}, \quad (5.5.2)$$

where the square root in the last member should approach unity as $\varphi/\sqrt{\lambda}$ approaches infinity, in order that the phase of $Q_c(\varphi/\sqrt{\lambda})$ approaches zero as $\varphi/\sqrt{\lambda}$ approaches $-\infty$ (like the phase of $P(\varphi_0)$ as φ_0 approaches $-\infty$; see [5.4.12]). By an index c on a quantity we indicate, in (5.5.2) and in the following, that the quantity in question refers to the comparison equation (5.3.8). With an appropriate choice of phase, we obtain from (5.5.2)

$$Q_c^{-\frac{1}{2}}(\varphi/\sqrt{\lambda}) = i\left(\frac{2}{\varphi/\sqrt{\lambda}}\right)^{\frac{1}{2}}\left[1 - \frac{4\overline{K}/\lambda}{(\varphi/\sqrt{\lambda})^2}\right]^{-\frac{1}{4}}, \quad (5.5.3)$$

where the phase of $\varphi^{\frac{1}{2}}$ is zero when φ is real and positive, and where the last factor approaches unity as $\varphi/\sqrt{\lambda}$ approaches infinity. The $(2N+1)$th-order phase-integral approximation for the solution of the comparison equation (5.3.8) is a linear combination of the phase-integral functions

$$q_c^{-\frac{1}{2}}(\varphi/\sqrt{\lambda})\exp\left(\pm i\int_{(\vartheta_1)}^{\varphi/\sqrt{\lambda}} q_c(\varphi/\sqrt{\lambda})d(\varphi/\sqrt{\lambda})\right), \quad (5.5.4)$$

where ϑ_1 is a so-far unspecified "lower limit of integration" and

$$q_c(\varphi/\sqrt{\lambda}) = \sum_{n=0}^{N} q_c^{(2n+1)}(\varphi/\sqrt{\lambda}). \quad (5.5.5)$$

The expressions for the first few functions in the right-hand member of (5.5.5) are [cf. (5.2.7), (5.2.8a–c), (5.2.9) and (5.2.10)]

$$q_c^{(1)}(\varphi/\sqrt{\lambda}) = Q_c(\varphi/\sqrt{\lambda}) \quad (5.5.6a)$$

$$q_c^{(3)}(\varphi/\sqrt{\lambda}) = Q_c(\varphi/\sqrt{\lambda})\frac{1}{2}\varepsilon_{0c}(\varphi/\sqrt{\lambda}) \quad (5.5.6b)$$

$$q_c^{(5)}(\varphi/\sqrt{\lambda}) = Q_c(\varphi/\sqrt{\lambda})\left[-\frac{1}{8}\varepsilon_{0c}^2(\varphi/\sqrt{\lambda})\right.$$

$$\left. -\frac{1}{8Q_c(\varphi/\sqrt{\lambda})}\frac{d}{d(\varphi/\sqrt{\lambda})}\left(\frac{1}{Q_c(\varphi/\sqrt{\lambda})}\frac{d}{d(\varphi/\sqrt{\lambda})}\varepsilon_{0c}(\varphi/\sqrt{\lambda})\right)\right] \quad (5.5.6c)$$

with

$$\varepsilon_{0c} = Q_c^{-\frac{3}{2}}(\varphi/\sqrt{\lambda}) \frac{d^2}{d(\varphi/\sqrt{\lambda})^2} Q_c^{-\frac{1}{2}}(\varphi/\sqrt{\lambda}). \tag{5.5.7}$$

Recalling (5.5.3), one realizes that $\varepsilon_{0c} = O((\varphi/\sqrt{\lambda})^{-4})$ when $\varphi/\sqrt{\lambda} \to \infty$. Furthermore,

$$Q_c(\varphi/\sqrt{\lambda}) = -\left\{ \frac{\varphi/\sqrt{\lambda}}{2} - \frac{\overline{K}/\lambda}{\varphi/\sqrt{\lambda}} + O\left[(\varphi/\sqrt{\lambda})^{-3}\right] \right\}$$

according to (5.5.2); hence,

$$\int^{\varphi/\sqrt{\lambda}} Q_c(\varphi/\sqrt{\lambda})d(\varphi/\sqrt{\lambda}) \to -\left[(\varphi/\sqrt{\lambda})^2/4 - (\overline{K}/\lambda)\ln(\varphi/\sqrt{\lambda}) + \text{const}\right]$$

as $\varphi/\sqrt{\lambda} \to \infty$; thus, the phase-integral functions (5.5.4) approach [cf. (5.5.3)]

$$i\left(\frac{2}{\varphi/\sqrt{\lambda}}\right)^{\frac{1}{2}} \exp\left\{ \mp i\left[(\varphi/\sqrt{\lambda})^2/4 - (\overline{K}/\lambda)\ln(\varphi/\sqrt{\lambda}) + \delta_c\right] \right\} \text{ as } \varphi/\sqrt{\lambda} \to \infty,$$
$$\tag{5.5.8}$$

where

$$\delta_c = -\lim_{\varphi/\sqrt{\lambda}\to\infty} \left(\int_{(\vartheta_1)}^{\varphi/\sqrt{\lambda}} q_c(\varphi/\sqrt{\lambda})d(\varphi/\sqrt{\lambda}) \right.$$

$$\left. + (\varphi/\sqrt{\lambda})^2/4 - (\overline{K}/\lambda)\ln(\varphi/\sqrt{\lambda}) \right). \tag{5.5.9}$$

For sufficiently large values of $\varphi/\sqrt{\lambda}$ one can therefore replace the expressions

$$\left(\frac{2}{\varphi/\sqrt{\lambda}}\right)^{\frac{1}{2}} \exp\left\{ \mp i\left[(\varphi/\sqrt{\lambda})^2/4 - (\overline{K}/\lambda)\ln(\varphi/\sqrt{\lambda})\right] \right\}$$

in (5.5.1a,b) by the expressions (5.5.4) multiplied by $-i\exp(\pm i\delta_c)$. Thus, one can obtain from (5.5.1a) and (5.5.1b) the phase-integral formulas

$$E(\overline{K}/\lambda, \varphi/\sqrt{\lambda}) = -iq_c^{-\frac{1}{2}}(\varphi/\sqrt{\lambda}) \exp\left[-i\left(\int_{(\vartheta_1)}^{\varphi/\sqrt{\lambda}} q_c(\varphi/\sqrt{\lambda})d(\varphi/\sqrt{\lambda}) \right.\right.$$

$$\left.\left. + \delta_c - \pi/4 - \frac{1}{4i}\ln\frac{\Gamma(\frac{1}{2}+i\overline{K}/\lambda)}{\Gamma(\frac{1}{2}-i\overline{K}/\lambda)} \right) \right], \tag{5.5.10a}$$

and

$$E(\overline{K}/\lambda, \varphi/\sqrt{\lambda}) = -iq_c^{-\frac{1}{2}}(\varphi/\sqrt{\lambda}) \exp\left[-i\left(\int_{(\vartheta_1)}^{\varphi/\sqrt{\lambda}} q_c(\varphi/\sqrt{\lambda})d(\varphi/\sqrt{\lambda}) \right.\right.$$

$$+ \delta_c - \pi/4 - \frac{1}{4i} \ln \frac{\Gamma(\frac{1}{2} + i\overline{K}/\lambda)}{\Gamma(\frac{1}{2} - i\overline{K}/\lambda)} \Bigg) \Bigg]$$

$$+ i \exp(\pi\overline{K}/\lambda)[1 + \exp(2\pi\overline{K}/\lambda)]^{\frac{1}{2}} q_c^{-\frac{1}{2}}(\varphi/\sqrt{\lambda})$$

$$\times \exp\left[+i \left(\int_{(\vartheta_1)}^{\varphi/\sqrt{\lambda}} q_c(\varphi/\sqrt{\lambda}) d(\varphi/\sqrt{\lambda}) + \delta_c - \pi/4 - \frac{1}{4i} \ln \frac{\Gamma(\frac{1}{2} + i\overline{K}/\lambda)}{\Gamma(\frac{1}{2} - i\overline{K}/\lambda)} \right) \right],$$

$$(5.5.10b)$$

respectively. The asymptotic formulas (5.5.1a,b) are accurate only for large values of $\varphi/\sqrt{\lambda}$, whereas the phase-integral formulas (5.5.10a,b) are good approximations when $\varphi/\sqrt{\lambda}$ does not lie too close to one or both of the two transition zeros $\varphi/\sqrt{\lambda} = \pm 2(\overline{K}/\lambda)^{\frac{1}{2}}$ (see [5.5.2]). Furthermore, for the validity of (5.5.1a,b) and (5.5.10a,b), $\varphi/\sqrt{\lambda}$ must lie well inside the regions delimited by anti-Stokes lines, which for large values of $\varphi/\sqrt{\lambda}$ are determined by the inequalities $-\pi/2 < \arg\varphi < \pi$ and $\pi/2 < \arg\varphi < 3\pi/2$ in (5.5.1a) and (5.5.1b), respectively.

Inserting for $E(\overline{K}/\lambda, \varphi/\sqrt{\lambda})$ the phase-integral expressions (5.5.10a) and (5.5.10b) along with (5.5.5) into the comparison equation solution (5.4.2) of the auxiliary differential equation (5.2.3), we obtain

$$\psi = -i\mathcal{A} \left(\sum_{n=0}^{N} q_c^{(2n+1)}(\varphi/\sqrt{\lambda}) \frac{d(\varphi/\sqrt{\lambda})}{dz} \right)^{-\frac{1}{2}}$$

$$\times \exp\left\{ -i \left[\sum_{n=0}^{N} \left(\int_{(\vartheta_1)}^{\varphi/\sqrt{\lambda}} q_c^{(2n+1)}(\varphi/\sqrt{\lambda}) d(\varphi/\sqrt{\lambda}) + \delta_c^{(2n+1)} \right) \right. \right.$$

$$\left. \left. - \pi/4 - \frac{1}{4i} \ln \frac{\Gamma(\frac{1}{2} + i\overline{K}/\lambda)}{\Gamma(\frac{1}{2} - i\overline{K}/\lambda)} \right] \right\}, \qquad (5.5.11a)$$

valid in particular when z is the point z_{-1} in Figure 5.2, and

$$\psi = -i\mathcal{A} \left(\sum_{n=0}^{N} q_c^{(2n+1)}(\varphi/\sqrt{\lambda}) \frac{d(\varphi/\sqrt{\lambda})}{dz} \right)^{-\frac{1}{2}}$$

$$\times \exp\left\{ -i \left[\sum_{n=0}^{N} \left(\int_{(\vartheta_1)}^{\varphi/\sqrt{\lambda}} q_c^{(2n+1)}(\varphi/\sqrt{\lambda}) d(\varphi/\sqrt{\lambda}) + \delta_c^{(2n+1)} \right) \right. \right.$$

$$\left. \left. - \pi/4 - \frac{1}{4i} \ln \frac{\Gamma(\frac{1}{2} + i\overline{K}/\lambda)}{\Gamma(\frac{1}{2} - i\overline{K}/\lambda)} \right] \right\}$$

$$+ i\mathcal{A} \exp(\pi\overline{K}/\lambda)[1 + \exp(2\pi\overline{K}/\lambda)]^{\frac{1}{2}} \left(\sum_{n=0}^{N} q_c^{(2n+1)}(\varphi/\sqrt{\lambda}) \frac{d(\varphi/\sqrt{\lambda})}{dz} \right)^{-\frac{1}{2}}$$

$$\times \exp\left\{+i\left[\sum_{n=0}^{N}\left(\int_{(\vartheta_1)}^{\varphi/\sqrt{\lambda}} q_c^{(2n+1)}(\varphi/\sqrt{\lambda})d(\varphi/\sqrt{\lambda}) + \delta_c^{(2n+1)}\right)\right.\right.$$

$$\left.\left. - \pi/4 - \frac{1}{4i}\ln\frac{\Gamma(\frac{1}{2}+i\overline{K}/\lambda)}{\Gamma(\frac{1}{2}-i\overline{K}/\lambda)}\right]\right\}, \tag{5.5.11b}$$

valid in particular when z is the point z_1 in Figure 5.2. The quantities $\delta_c^{(2n+1)}$ are, according to (5.5.9) and (5.5.5), defined by

$$\delta_c^{(1)} = -\lim_{\varphi/\sqrt{\lambda}\to\infty}\left(\int_{\vartheta_1}^{\varphi/\sqrt{\lambda}} q_c^{(1)}(\varphi/\sqrt{\lambda})d(\varphi/\sqrt{\lambda})\right.$$

$$\left. + (\varphi/\sqrt{\lambda})^2/4 - (\overline{K}/\lambda)\ln(\varphi/\sqrt{\lambda})\right) \tag{5.5.12}$$

$$\delta_c^{(2n+1)} = -\int_{(\vartheta_1)}^{\infty} q_c^{(2n+1)}(\varphi/\sqrt{\lambda})d(\varphi/\sqrt{\lambda}); \quad n > 0. \tag{5.5.12'}$$

We emphasize that $\varphi/\sqrt{\lambda}$ in (5.5.11a,b) plays a different role than $\varphi/\sqrt{\lambda}$ does in (5.5.12) and (5.5.12'); viz. that in (5.5.11a,b) $\varphi/\sqrt{\lambda}$ is a function of z and λ according to (5.4.22) and (5.4.15a–c), whereas in (5.5.12) and (5.5.12') $\varphi/\sqrt{\lambda}$ is merely an integration variable.

Heretofore, the lower "limit of integration" ϑ_1 in the integrals in (5.5.4), (5.5.9), (5.5.10a,b), (5.5.11a,b), (5.5.12) and (5.5.12') is an unspecified point in the complex $(\varphi/\sqrt{\lambda})$-plane; henceforth, we will let ϑ_1 be the transition zero, associated with the square of the base function (5.5.2), which approaches $\tau_1/\sqrt{\lambda}$, with τ_1 defined by (5.4.13a), when $\lambda \to 0$. Letting the other transition zero associated with the square of the base function (5.5.2) be ϑ_2, we set

$$\vartheta_1 = 2(\overline{K}/\lambda)^{\frac{1}{2}}\exp(+i\pi), \tag{5.5.13a}$$

$$\vartheta_2 = 2(\overline{K}/\lambda)^{\frac{1}{2}}, \tag{5.5.13b}$$

where the phase of $(\overline{K}/\lambda)^{\frac{1}{2}}$ is chosen so as to approach the phase of $\overline{K}_0^{\frac{1}{2}}$ in (5.4.13a,b) when $\lambda \to 0$. Using this convention, we obtain from (5.5.13a,b), (5.4.13a,b), and (5.4.23)

$$\vartheta_\nu = \frac{\tau_\nu}{\sqrt{\lambda}}\left(\sum_{\beta=0}^{N}\frac{\overline{K}_{2\beta}}{\overline{K}_0}\lambda^{2\beta}\right)^{\frac{1}{2}}, \quad \nu = 1,2, \tag{5.5.14}$$

λ and $\sqrt{\lambda}$ being positive.

The transition zeros t_1 and t_2 in the complex z-plane are mapped by the function $\varphi_0(z)$ onto the points τ_1 and τ_2, respectively, in the complex φ_0-plane; i.e., $\varphi_0(t_\nu) = \tau_\nu$, $\nu = 1, 2$. However, in general, the transition zeros t_1 and t_2 in the z-plane are not mapped by the function $\varphi(z)/\sqrt{\lambda}$

onto the zeros ϑ_1 and ϑ_2 of $Q_c^2(\varphi/\sqrt{\lambda})$ except in the limit when $\lambda \to 0$. We see this by using (5.4.22) to obtain

$$\frac{1}{\sqrt{\lambda}}\varphi(t_\nu) = \frac{1}{\sqrt{\lambda}}\sum_{\beta=0}^{N}\varphi_{2\beta}(t_\nu)\lambda^{2\beta} = \frac{\tau_\nu}{\sqrt{\lambda}}\sum_{\beta=0}^{N}\frac{\varphi_{2\beta}(t_\nu)}{\varphi_0(t_\nu)}\lambda^{2\beta}, \quad \nu = 1, 2,$$

$$(5.5.15)$$

and by comparing this formula with (5.5.14).

The actual path of integration in the definition of the integral in (5.5.11a,b) is a nonclosed loop in the complex $\varphi/\sqrt{\lambda}$-plane enclosing the point ϑ_1, but not the point ϑ_2. This contour corresponds in the complex z-plane to a similar contour that can be deformed so that it encircles t_1 but not t_2, and so that, for sufficiently small values of λ, it lies so far away from t_1 and t_2 that the phase-integral approximation, which one obtains by expansion in powers of λ, is valid along the whole contour. The integrand of the integral in (5.5.11a) and (5.5.11b) has the branch points ϑ_1 and ϑ_2 in the $\varphi/\sqrt{\lambda}$-plane, but after z has been introduced as integration variable, and the expansion in powers of λ has been performed, the integrand has the branch points t_1 and t_2 in the z-plane, although the points t_1 and t_2 in the z-plane are not mapped by the function $\varphi(z)/\sqrt{\lambda}$ on the points ϑ_1 and ϑ_2 in the $\varphi/\sqrt{\lambda}$-plane.

To continue in a direct and straightforward fashion, as mentioned in the beginning of the present section, we should as the next step in our calculations use the contour just discussed and expand the z-dependent function $\sum_{n=0}^{N} q_c^{(2n+1)}(\varphi/\sqrt{\lambda})d(\varphi/\sqrt{\lambda})/dz$ as well as the z-independent phase $\sum_{n=0}^{N} \delta_c^{(2n+1)}$ in powers of λ in order to bring (5.5.11a) and (5.5.11b) into the form of linear combinations of the phase-integral functions (5.2.4a,b) associated with the auxiliary differential equation (5.2.3). These somewhat tedious calculations are presented in the Appendix. We shall present below an alternative, more elegant procedure, which corresponds to that originally given in [25] for another problem.

Recalling the truncated expansions (5.4.22) and (5.4.23) of φ and $\overline{K}$ in powers of λ, and the general structure of the expressions (5.5.6a–c) along with (5.5.2) and (5.5.7), we realize that

$$\sum_{n=0}^{N} q_c^{(2n+1)}(\varphi/\sqrt{\lambda})\frac{d(\varphi/\sqrt{\lambda})}{dz} = \sum_{n=0}^{N} q^{(2n+1)}(z)\lambda^{2n-1} + O(\lambda^{2N+1}) \quad (5.5.16)$$

and, also considering (5.5.12) and (5.5.12'),

$$\sum_{n=0}^{N} \delta_c^{(2n+1)} = \sum_{n=0}^{N} \delta^{(2n+1)}\lambda^{2n-1} + O(\lambda^{2N+1}) - \frac{\overline{K}/\lambda}{2}\ln\lambda, \quad (5.5.17)$$

where the λ-independent functions $q^{(2n+1)}(z)$ and phase constants $\delta^{(2n+1)}$ are as yet unknown. It will, however, soon be shown that the functions

$q^{(2n+1)}(z)$ introduced here are the same as the functions $q^{(2n+1)}(z)$ in (5.2.6). The term $-\frac{1}{2}(\overline{K}/\lambda)\ln\lambda$ in (5.5.17) derives from the last term in (5.5.12). Inserting (5.5.16) and (5.5.17), neglecting the terms $O(\lambda^{(2N+1)})$, into (5.5.11a), and choosing the normalization factor $\mathcal{A}$ to be

$$\mathcal{A} = \exp\left[\frac{1}{2}i(\pi/2 - \phi) - \pi\overline{K}/\lambda\right] \tag{5.5.18}$$

with ϕ defined by

$$\phi = \frac{1}{2i}\ln\frac{\Gamma(\frac{1}{2}+i\overline{K}/\lambda)}{\Gamma(\frac{1}{2}-i\overline{K}/\lambda)} + (\overline{K}/\lambda)(\ln\lambda + 2\pi i) - \sum_{n=0}^{N}2\delta^{(2n+1)}\lambda^{2n-1}, \tag{5.5.19}$$

we obtain the formula

$$\psi = \left(\sum_{n=0}^{N}q^{(2n+1)}(z)\lambda^{2n-1}\right)^{-\frac{1}{2}}\exp\left(-i\int_{(t_1)}^{z}\sum_{n=0}^{N}q^{(2n+1)}(z)\lambda^{2n-1}dz\right), \tag{5.5.20a}$$

which in particular is valid when z is the point z_{-1} in Figure 5.2. From (5.5.11b) we similarly obtain the formula

$$\psi = i[1 + \exp(-2\pi\overline{K}/\lambda]^{\frac{1}{2}}\exp(-i\phi)$$

$$\times\left(\sum_{n=0}^{N}q^{(2n+1)}(z)\lambda^{2n-1}\right)^{-\frac{1}{2}}\exp\left(+i\int_{(t_1)}^{z}\sum_{n=0}^{N}q^{(2n+1)}(z)\lambda^{2n-1}dz\right)$$

$$+\left(\sum_{n=0}^{N}q^{(2n+1)}(z)\lambda^{2n-1}\right)^{-\frac{1}{2}}\exp\left(-i\int_{(t_1)}^{z}\sum_{n=0}^{N}q^{(2n+1)}(z)\lambda^{2n-1}dz\right), \tag{5.5.20b}$$

which in particular is valid when z is the point z_1 in Figure 5.2.

So far the functions $q^{(2n+1)}(z)$ introduced in (5.5.16) are not known. However, since the parameter λ appears in the same way in (5.5.20a,b) as in the phase-integral solution of the auxiliary differential equation (5.2.3); i.e., in the phase-integral functions (5.2.4a,b) along with (5.2.5) and (5.2.6); the functions $q^{(2n+1)}(z)$ in the present section must be given by the phase-integral formula (5.2.7); that is,

$$q^{(2n+1)}(z) = Q(z)Y_{2n}. \tag{5.5.21}$$

In the above $Q(z)$ is the base function chosen for solving the original differential equation (5.2.1); it appears squared in the auxiliary differential equation (5.2.3); the first few functions Y_{2n} are given by (5.2.8a–c).

In the definition (5.5.19) of ϕ, the quantity $\overline{K}$ is a function of λ given originally by the infinite series (5.4.4); in order to obtain an expression for

$\overline{K}$ corresponding to the $(2N+1)$th-order phase-integral approximation, one must use instead the truncated series (5.4.23), in which the quantities $\overline{K}_{2\beta}$ are given by (5.4.21).

Using the reflection formula for the Γ-function, we can rewrite the expression (5.5.19) for ϕ into the form

$$\phi = \frac{1}{i} \left\{ \ln[1 + \exp(-2\pi \overline{K}/\lambda)]^{\frac{1}{2}} + \ln \Gamma \left(\tfrac{1}{2} + i\overline{K}/\lambda \right) - \ln(2\pi)^{\frac{1}{2}} + \frac{\pi \overline{K}/\lambda}{2} \right\}$$

$$+ (\overline{K}/\lambda)(\ln \lambda + 2\pi i) - \sum_{n=0}^{N} 2\delta^{(2n+1)} \lambda^{2n-1}. \qquad (5.5.22)$$

From (5.2.13) and (5.4.26) it follows for the usual case in Figure 5.2(a) that $\mathrm{Im}(i\overline{K}) > 0$; for the exceptional case in Figure 5.2(b) it follows that $\mathrm{Re}(i\overline{K}) > 0$. In both cases the absolute value of $\arg(i\overline{K}/\lambda)$ is, therefore, smaller than π (a small, positive, λ-independent quantity is deducted); hence we have, according to Eq. (5) on p. 32 in [30], for large values of $|z|$, which in our notation corresponds to $|\overline{K}/\lambda|$, the asymptotic formula

$$\ln \Gamma \left(\tfrac{1}{2} + i\overline{K}/\lambda \right) = \ln(2\pi)^{\frac{1}{2}} - \frac{\pi \overline{K}/\lambda}{2} + i\left((\overline{K}/\lambda) \ln(\overline{K}/\lambda) \right.$$

$$\left. - \overline{K}/\lambda + \frac{1}{24\,\overline{K}/\lambda} + \frac{7}{2880(\overline{K}/\lambda)^3} + \cdots \right), \qquad (5.5.23)$$

where, in agreement with the choice of branch for the logarithm function in (5.5.1a,b), the principal branch of $\ln(\overline{K}/\lambda)$ is to be used; i.e., $\ln(\overline{K}/\lambda)$ is chosen to be real when $\overline{K}/\lambda$ is real and positive. Inserting (5.5.23) into (5.5.22) and using for $\overline{K}$ the truncated series (5.4.23), we obtain

$$\phi = \frac{1}{i} \ln[1 + \exp(-2\pi \overline{K}/\lambda)]^{\frac{1}{2}} + \sum_{n=0}^{N} \left[\overline{K}_{2n}(\ln \overline{K}_0 + 2\pi i) \right.$$

$$\left. - \phi^{(2n+1)} - 2\delta^{(2n+1)} \right] \lambda^{2n-1} - \phi^{(2N+3)} \lambda^{2N+1} + O(\lambda^{2N+3}) \qquad (5.5.24)$$

where

$$\phi^{(1)} = \overline{K}_0 \qquad (5.5.25a)$$

$$\phi^{(3)} = -\frac{1}{24\,\overline{K}_0} \qquad (5.5.25b)$$

$$\phi^{(5)} = -\frac{\overline{K}_2^2}{2\overline{K}_0} + \frac{\overline{K}_2}{24\,\overline{K}_0^2} - \frac{7}{2880\,\overline{K}_0^3}. \qquad (5.5.25c)$$

The reason for the choice of the notations in the left-hand members of (5.5.25a–c) will be made clear below, *viz.* in connection with (5.5.30).

Considering the solution (5.5.20a,b), and recalling that, in particular, (5.5.20a) is valid when z is the point z_{-1} in Figure 5.2, and (5.5.20b) is valid when z is the point z_1 in Figure 5.2; we obtain with the use of (5.2.14)

$$a_1 = i[1 + \exp(-2\pi \overline{K}/\lambda)]^{\frac{1}{2}} \exp(-i\phi). \tag{5.5.26}$$

With the aid of (5.5.24) we can write (5.5.26) as

$$a_1 = i \exp\left\{ -i \sum_{n=0}^{N} \left[\overline{K}_{2n}(\ln \overline{K}_0 + 2\pi i) - \phi^{(2n+1)} \right. \right.$$

$$\left. \left. - 2\delta^{(2n+1)} \right] \lambda^{2n-1} + i\phi^{(2N+3)}\lambda^{2N+1} + O(\lambda^{2N+3}) \right\}. \tag{5.5.27}$$

If we let λ tend to zero, we realize (assuming that $t_1 \neq t_2$) that the absolute value of the integral $\int_{t_1}^{t_2}[Q(z)/\lambda]dz$, associated with the first-order phase-integral solution of the auxiliary differential equation (5.2.3), approaches infinity. The corresponding integral for the phase-integral approximation of an arbitrary order $2N+1$ must also approach infinity. Consequently, the phase-integral formulas for well separated transition zeros become applicable; in particular, for the matrix $\mathbf{F}(z_1, z_0)$ we obtain the formula that is valid when the transition zero t_1 is well isolated (see Figure 5.2). To derive this formula we proceed as in the derivation of Eq. (7.5a) in [13] and note that the μ-integral that appears is of the order of magnitude λ^{2N+1}. Thus, we find that

$$a_1 = i + O(\lambda^{2N+1}) \quad \text{as} \quad \lambda \to 0. \tag{5.5.28}$$

Comparing this formula with (5.5.27), we see that the λ-independent quantities $2\delta^{(2n+1)}$ in (5.5.19) are given by the formula

$$2\delta^{(2n+1)} = \overline{K}_{2n}(\ln \overline{K}_0 + 2\pi i) - \phi^{(2n+1)}. \tag{5.5.29}$$

With the aid of (5.5.29) and (5.4.23) we can write (5.5.19) in the form

$$\phi = \frac{1}{2i} \ln \frac{\Gamma(\frac{1}{2} + i\overline{K}/\lambda)}{\Gamma(\frac{1}{2} - i\overline{K}/\lambda)} - \frac{\overline{K}}{\lambda} \ln(\overline{K}_0/\lambda) + \sum_{n=0}^{N} \phi^{(2n+1)}\lambda^{2n-1}. \tag{5.5.30}$$

The formula (5.5.30) along with (5.5.25a–c) is valid for all values of the energy, even those close to the top of the barrier. Although (5.5.30) is valid also when the barrier is thick, it may be useful to have a simpler formula for this case. One can obtain such a formula by inserting (5.5.29) into (5.5.24). The result is

$$\phi = -\phi^{(2N+3)}\lambda^{2N+1} + O(\lambda^{2N+3}). \tag{5.5.31}$$

The quantities $\phi^{(2n+1)}$ are related to quantities $D^{(2n+1)}$, which were introduced in Regge pole theory in [20] and which are also considered in §5.8, in accordance with the formula

$$\phi^{(2n+1)} = iD^{(2n+1)}. \tag{5.5.32}$$

Expressions for $D^{(2n+1)}$ in terms of the quantities $\overline{\gamma}_0, \overline{\gamma}_2, \overline{\gamma}_4, \ldots$ related to $\overline{K}_0, \overline{K}_2, \overline{K}_4, \ldots$ according to (5.8.3) were obtained up to $2n+1 = 13$ in [2]; therefore, one can obtain expressions for $\phi^{(2n+1)}$ up to $2n+1 = 13$. Thus, one encounters (5.5.25a–c) again; in addition, one finds the formulas

$$\phi^{(7)} = -\frac{\overline{K}_2\overline{K}_4}{\overline{K}_0} + \frac{4\overline{K}_2^3 + \overline{K}_4}{24\,\overline{K}_0^2} - \frac{\overline{K}_2^2}{24\,\overline{K}_0^3} + \frac{7\overline{K}_2}{960\,\overline{K}_0^4} - \frac{31}{40320\,\overline{K}_0^5} \quad (5.5.25d)$$

$$\phi^{(9)} = -\frac{2\overline{K}_2\overline{K}_6 + \overline{K}_4^2}{2\overline{K}_0} + \frac{12\,\overline{K}_2^2\overline{K}_4 + \overline{K}_6}{24\,\overline{K}_0^2} - \frac{\overline{K}_2^4 + \overline{K}_2\overline{K}_4}{12\,\overline{K}_0^3}$$

$$+ \frac{40\,\overline{K}_2^3 + 7\overline{K}_4}{960\,\overline{K}_0^4} - \frac{7\overline{K}_2^2}{480\,\overline{K}_0^5} + \frac{31\,\overline{K}_2}{8064\,\overline{K}_0^6} - \frac{127}{215040\,\overline{K}_0^7} \quad (5.5.25e)$$

$$\phi^{(11)} = -\frac{\overline{K}_2\overline{K}_8 + \overline{K}_4\overline{K}_6}{\overline{K}_0} + \frac{12\,\overline{K}_2^2\overline{K}_6 + 12\,\overline{K}_2\overline{K}_4^2 + \overline{K}_8}{24\,\overline{K}_0^2}$$

$$- \frac{8\overline{K}_2^3\overline{K}_4 + 2\overline{K}_2\overline{K}_6 + \overline{K}_4^2}{24\,\overline{K}_0^3} + \frac{48\,\overline{K}_2^5 + 120\,\overline{K}_2^2\overline{K}_4 + 7\overline{K}_6}{960\,\overline{K}_0^4} - \frac{10\,\overline{K}_2^4 + 7\overline{K}_2\overline{K}_4}{240\,\overline{K}_0^5}$$

$$+ \frac{196\,\overline{K}_2^3 + 31\,\overline{K}_4}{8064\,\overline{K}_0^6} - \frac{31\,\overline{K}_2^2}{2688\,\overline{K}_0^7} + \frac{127\,\overline{K}_2}{30720\,\overline{K}_0^8} - \frac{511}{608256\,\overline{K}_0^9} \quad (5.5.25f)$$

$$\phi^{(13)} = -\frac{2\overline{K}_2\overline{K}_{10} + 2\overline{K}_4\overline{K}_8 + \overline{K}_6^2}{2\overline{K}_0} + \frac{12\,\overline{K}_2^2\overline{K}_8 + 24\,\overline{K}_2\overline{K}_4\overline{K}_6 + 4\overline{K}_4^3 + \overline{K}_{10}}{24\,\overline{K}_0^2}$$

$$- \frac{4\overline{K}_2^3\overline{K}_6 + 6\overline{K}_2^2\overline{K}_4^2 + \overline{K}_2\overline{K}_8 + \overline{K}_4\overline{K}_6}{12\,\overline{K}_0^3}$$

$$+ \frac{240\,\overline{K}_2^4\overline{K}_4 + 120\,\overline{K}_2^2\overline{K}_6 + 120\,\overline{K}_2\overline{K}_4^2 + 7\overline{K}_8}{960\,\overline{K}_0^4}$$

$$- \frac{16\,\overline{K}_2^6 + 80\,\overline{K}_2^3\overline{K}_4 + 14\,\overline{K}_2\overline{K}_6 + 7\overline{K}_4^2}{480\,\overline{K}_0^5} + \frac{336\,\overline{K}_2^5 + 588\,\overline{K}_2^2\overline{K}_4 + 31\,\overline{K}_6}{8064\,\overline{K}_0^6}$$

$$- \frac{49\,\overline{K}_2^4 + 31\,\overline{K}_2\overline{K}_4}{1344\,\overline{K}_0^7} + \frac{2480\,\overline{K}_2^3 + 381\,\overline{K}_4}{92160\,\overline{K}_0^8} - \frac{127\,\overline{K}_2^2}{7680\,\overline{K}_0^9}$$

$$+ \frac{511\,\overline{K}_2}{67584\,\overline{K}_0^{10}} - \frac{1414477}{738017280\,\overline{K}_0^{11}}. \quad (5.5.25g)$$

Comparing the formulas (5.5.30) and (5.5.25a–c) for ϕ and $\phi^{(2n+1)}$, $n = 0, 1, 2$, with Eqs. (10) and (10a–c) in [27], which were derived for a real barrier; we realize that ϕ in this chapter corresponds to -2σ in [27]. The formulas published in [27], which thus represent a specific case of the general results in this chapter, have been used during the past two decades in a number of applications, where barrier transmission through a real barrier is essential [3, 4, 5, 10, 12, 18, 31, 35, 36]. After generalization, the formulas in [27] have, in fact, also been applied in a situation corresponding to a complex barrier, *viz.* in the calculation of phase shifts for a heavy-ion optical potential by Linnaeus [29]. The results were found to be in good agreement with previously published numerical results.

5.6   Stokes Constants

The solution of the connection problem, associated with anti-Stokes lines emerging from the cluster in four different directions, is expressed in terms of four Stokes constants, which determine the connection matrices (5.2.11a–d). From (5.5.26), (5.2.12a–c), and (5.4.26) one obtains the following expressions for these Stokes constants:

$$a_{-1} = i[1 + \exp(-2\pi\overline{K}/\lambda)]^{\frac{1}{2}} \exp[-(2\pi\overline{K}/\lambda + i\phi)], \qquad (5.6.1a)$$

$$b_0 = i[1 + \exp(-2\pi\overline{K}/\lambda)]^{\frac{1}{2}} \exp[+(2\pi\overline{K}/\lambda + i\phi)], \qquad (5.6.1b)$$

$$a_1 = i[1 + \exp(-2\pi\overline{K}/\lambda)]^{\frac{1}{2}} \exp[-i\phi], \qquad (5.6.1c)$$

$$b_2 = i[1 + \exp(-2\pi\overline{K}/\lambda)]^{\frac{1}{2}} \exp[+i\phi]. \qquad (5.6.1d)$$

From (5.5.30) we obtain

$$\exp(-i\phi) = \left(\frac{\Gamma(\frac{1}{2} - i\overline{K}/\lambda)}{\Gamma(\frac{1}{2} + i\overline{K}/\lambda)}\right)^{\frac{1}{2}} \exp\left(i\frac{\overline{K}}{\lambda} \ln(\overline{K}_0/\lambda)\right.$$

$$\left. - i\sum_{n=0}^{N} \phi^{(2n+1)}\lambda^{2n-1}\right). \qquad (5.6.2)$$

The factor in (5.6.2) containing the gamma functions can, with the aid of the reflection formula for the gamma function, be written as

$$\left(\frac{\Gamma(\frac{1}{2} - i\overline{K}/\lambda)}{\Gamma(\frac{1}{2} + i\overline{K}/\lambda)}\right)^{\frac{1}{2}} = \exp\left(\frac{1}{2}\pi\overline{K}/\lambda\right)[1 + \exp(-2\pi\overline{K}/\lambda)]^{\frac{1}{2}}$$

$$\times \Gamma(\frac{1}{2} - i\overline{K}/\lambda)/(2\pi)^{\frac{1}{2}} \qquad (5.6.3)$$

$$= \frac{(2\pi)^{\frac{1}{2}}}{\exp(\frac{1}{2}\pi\overline{K}/\lambda)[1 + \exp(-2\pi\overline{K}/\lambda)]^{\frac{1}{2}}\Gamma(\frac{1}{2} + i\overline{K}/\lambda)}. \qquad (5.6.3')$$

Using (5.6.2) and (5.6.3′) in (5.6.1a,c), and using (5.6.2) and (5.6.3) in (5.6.1b,d); we obtain

$$a_{-1} = i(2\pi)^{\frac{1}{2}} \exp\left(+\frac{i\overline{K}}{\lambda} \ln(\overline{K}_0/\lambda) - \frac{5\pi}{2} \frac{\overline{K}}{\lambda}\right.$$

$$\left. - i \sum_{n=0}^{N} \phi^{(2n+1)} \lambda^{2n-1}\right) / \Gamma\left(\tfrac{1}{2} + i\overline{K}/\lambda\right), \tag{5.6.4a}$$

$$b_0 = i(2\pi)^{\frac{1}{2}} \exp\left(-\frac{i\overline{K}}{\lambda} \ln(\overline{K}_0/\lambda) + \frac{3\pi}{2} \frac{\overline{K}}{\lambda}\right.$$

$$\left. + i \sum_{n=0}^{N} \phi^{(2n+1)} \lambda^{2n-1}\right) / \Gamma\left(\tfrac{1}{2} - i\overline{K}/\lambda\right), \tag{5.6.4b}$$

$$a_1 = i(2\pi)^{\frac{1}{2}} \exp\left(+\frac{i\overline{K}}{\lambda} \ln(\overline{K}_0/\lambda) - \frac{\pi}{2} \frac{\overline{K}}{\lambda}\right.$$

$$\left. - i \sum_{n=0}^{N} \phi^{(2n+1)} \lambda^{2n-1}\right) / \Gamma\left(\tfrac{1}{2} + i\overline{K}/\lambda\right), \tag{5.6.4c}$$

$$b_2 = i(2\pi)^{\frac{1}{2}} \exp\left(-\frac{i\overline{K}}{\lambda} \ln(\overline{K}_0/\lambda) - \frac{\pi}{2} \frac{\overline{K}}{\lambda}\right.$$

$$\left. + i \sum_{n=0}^{N} \phi^{(2n+1)} \lambda^{2n-1}\right) / \Gamma\left(\tfrac{1}{2} - i\overline{K}/\lambda\right). \tag{5.6.4d}$$

Since the gamma function is an analytic function without zeros, it is immediately apparent from (5.6.4a–d) that the Stokes constants are always finite. Taking into account the expressions (5.6.4a–d) for the Stokes constants, the elementary matrices (5.2.11a–d) are completely determined; one is thus able to trace a given solution from one anti-Stokes line receding from the cluster in one direction to anti-Stokes lines receding from the cluster in other directions.

We remark that with the aid of (5.4.13a,b) we can express the quantity $\ln(\overline{K}_0/\lambda)$ in (5.6.4a–d) in terms of τ_1 and τ_2 as

$$\ln(\overline{K}_0/\lambda) = 2\ln\frac{\tau_1 \exp(-i\pi)}{2\sqrt{\lambda}} = 2\ln\frac{\tau_2}{2\sqrt{\lambda}}. \tag{5.6.5}$$

For values of $\overline{K}/\lambda$ such that $\frac{1}{2} \pm i\overline{K}/\lambda$ is not equal to zero or a negative integer, all four Stokes constants are different from zero, according to (5.6.4a–d). If, however, $\frac{1}{2} + i\overline{K}/\lambda$ is equal to zero or a negative integer, then the function $\Gamma(\frac{1}{2} + i\overline{K}/\lambda)$ is infinite; hence, a_{-1} and a_1 are equal to zero. By analogy, if $\frac{1}{2} - i\overline{K}/\lambda$ is equal to zero or a negative integer, the function $\Gamma(\frac{1}{2} - i\overline{K}/\lambda)$ is infinite; hence, b_0 and b_2 are equal to zero. For the situation in Figure 5.2(b) it was shown below (5.2.15) that Ω is positive, and from (5.4.26) it then follows that $\frac{1}{2} + i\overline{K}/\lambda$ is also positive; therefore, a_{-1} and a_1 cannot be equal to zero. Thus, when the phase of $Q(z)$ and the cuts are chosen according to Figure 5.2(b), the only possible "quantization condition" is $b_0 = b_2 = 0$.

5.7   Application to Complex Potential Barrier

For an arbitrary order of the phase-integral approximation we shall use the terminology that an anti-Stokes line "emerges" from a simple transition point t if $\int_{(t)}^{z} q(z)dz$ is real on the anti-Stokes line; similarly we shall say that two simple transition points t_1 and t_2 are "joined" by an anti-Stokes line if an anti-Stokes line emerging from t_1 coincides with an anti-Stokes line emerging from t_2. In the specific case when the integral of $q(z)$ along the closed loop Λ is real; which is illustrated in Figures 5.1(c), 5.1(g) and 5.2(b); the transition zeros t_1 and t_2 are thus joined by an anti-Stokes line. There are therefore in this case only four anti-Stokes lines (instead of six in the general case), which emerge from the two transition zeros and recede from the cluster. We shall characterize a cluster for which the particular situation described above prevails as a complex potential well, which may support a localized state. In all other cases all six anti-Stokes lines emerging from the two transition zeros recede from the cluster, as shown in Figure 5.1 and Figure 5.2(a). We shall then characterize the cluster as a complex potential barrier. Whether a cluster of two transition zeros is to be characterized as a potential barrier or not thus depends, not only on the parameters in the base function, but, in general, also on the order of the phase-integral approximation used. However, in the special case when t_1 and t_2 lie on the real axis, and the function $Q^2(z)$ is real there, the characterization of a cluster of two transition zeros as a potential well or a potential barrier is independent of the order of the approximation used.

We shall now consider wave transmission through a complex barrier from the point z_1 to the point z_{-1}; see Figure 5.2(a). In physical applications the anti-Stokes line on which z_1 is located is approximately parallel to the negative real axis when z_1 is large (negative); the anti-Stokes line on which z_{-1} lies is approximately parallel to the positive real axis when z_{-1} is large (positive). With the choice of phase for $Q(z)$ introduced in §5.2, and the complex z-plane cut as described there and illustrated in Figure 5.2(a); the phase-integral functions $f_1(z_1)$ and $f_1(z_{-1})$ represent waves propagating towards the barrier, whereas $f_2(z_1)$ and $f_2(z_{-1})$ represent waves propagating away from the barrier. Hence, the solution (5.5.20a,b) represents an incident wave plus a reflected wave at z_1 and a transmitted wave at z_{-1}. The connection matrix for transmission through the cluster is given by

(5.2.14) with (5.4.26), that is,

$$\mathbf{F}(z_1, z_{-1}) = \begin{pmatrix} -\exp(2\pi\overline{K}/\lambda) & a_1 \\ b_0 & 1 \end{pmatrix}, \tag{5.7.1}$$

where the Stokes constants b_0 and a_1 are given by (5.6.1b,c) or (5.6.4b,c). With the aid of (5.6.1b,c) we can write (5.7.1) as

$$\mathbf{F}(z_1, z_{-1}) =$$
$$\begin{pmatrix} -\exp(2K) & i[1 + \exp(-2K)]^{\frac{1}{2}}\exp(-i\phi) \\ i[1 + \exp(-2K)]^{\frac{1}{2}}\exp(2K + i\phi) & 1 \end{pmatrix}, \tag{5.7.2}$$

where K is given by (5.4.27), and ϕ is given by (5.5.30) in addition to (5.5.25a–g). For a thick barrier one can use the simple formula (5.5.31) instead of (5.5.30).

5.8   Application to Regge Pole Theory

An arbitrary-order phase-integral formula for the residue of the S-matrix at a Regge pole was derived by Fröman and Fröman [20] on the assumption that there are two relevant transition zeros that may lie at an arbitrary distance from each other—possibly close together—but far away from all other transition points, including the pole at the origin. It was also shown that the already known phase-integral formula for the positions of the Regge poles is valid under the same general conditions. Amaha and Thylwe [2] have applied these formulas with great success to particular potentials. The derivation of the residue formula mentioned above was at one point, *viz.* the justification of the expression for the Stokes constant b_2, given by Eqs. (2.15) and (2.14) in [20], done in very sketchy terms. Therefore, we shall briefly describe how this expression can be obtained from formulas in the present chapter.

With $w(r)$ defined according to Eq. (2.1) in [20], with the contour of integration $\Gamma_{t_2}(r)$ shown in Figure 1 of [20]; we have instead of (5.6.4d), which is based on the definition (5.2.5) of $w(z)$, involving a nonclosed contour of integration encircling the transition zero t_1, the formula

$$b_2 = \exp(+2\pi\overline{K}/\lambda)i(2\pi)^{\frac{1}{2}}\exp\left(-\frac{i\overline{K}}{\lambda}\ln(\overline{K}_0/\lambda)\right.$$
$$\left. -\frac{\pi}{2}\frac{\overline{K}}{\lambda} + i\sum_{n=0}^{N}\phi^{(2n+1)}\lambda^{2n-1}\right)/\Gamma(\tfrac{1}{2} - i\overline{K}/\lambda). \tag{5.8.1}$$

With the aid of the reflection formula for the gamma function this can be written as

$$b_2 = i(2\pi)^{-\frac{1}{2}}[1 + \exp(2\pi\overline{K}/\lambda)]\exp\left(-\frac{i\overline{K}}{\lambda}\ln(\overline{K}_0/\lambda)\right.$$

$$+ \frac{\pi}{2} \frac{\overline{K}}{\lambda} + i \sum_{n=0}^{N} \phi^{(2n+1)} \lambda^{2n-1} \bigg) \Gamma(\tfrac{1}{2} + i\overline{K}/\lambda). \qquad (5.8.2)$$

According to Eq. (2.9) in [20] and (5.4.21), the quantities $\overline{\gamma}_{2n}$ in [20] are the same as the quantities $i\overline{K}_{2n}$ in this chapter; that is,

$$\overline{K}_{2n} = \overline{\gamma}_{2n} \exp(-i\pi/2). \qquad (5.8.3)$$

Hence,

$$\overline{K}/\lambda = \overline{\gamma} \exp(-i\pi/2), \qquad (5.8.4)$$

where $\overline{K}$ is given by (5.4.23), and $\overline{\gamma}$ is given by Eq. (2.13a) in [20]; that is,

$$\overline{\gamma} = \sum_{n=0}^{N} \overline{\gamma}_{2n} \lambda^{2n-1}. \qquad (5.8.5)$$

We now define

$$D = \sum_{n=0}^{N} D^{(2n+1)} \lambda^{2n-1}, \qquad (5.8.6)$$

where according to (5.5.32)

$$D^{(2n+1)} = -i\phi^{(2n+1)}. \qquad (5.8.7)$$

From (5.8.7), (5.5.25a–c) and (5.8.3) it follows that

$$D^{(1)} = -\overline{\gamma}_0 \qquad (5.8.8a)$$

$$D^{(3)} = -\frac{1}{24\,\overline{\gamma}_0} \qquad (5.8.8b)$$

$$D^{(5)} = \frac{\overline{\gamma}_2^2}{2\overline{\gamma}_0} + \frac{\overline{\gamma}_2}{24\,\overline{\gamma}_0^2} + \frac{7}{2880\,\overline{\gamma}_0^3}. \qquad (5.8.8c)$$

Using (5.8.3), (5.8.4), (5.8.6), and (5.8.7), we can now write (5.8.2) as

$$b_2 = i(2\pi)^{-\frac{1}{2}} [1 + \exp[-2\pi i\overline{\gamma}]] \exp[-\overline{\gamma} \ln(\overline{\gamma}_0/\lambda) - D] \Gamma(\tfrac{1}{2} + \overline{\gamma}). \qquad (5.8.9)$$

Setting $\lambda = 1$, we obtain for b_2 the formula given by (2.15), (2.14), (2.13b), and (2.11a–c) in [20].

Appendix: Phase-Integral Solution Obtained from the Comparison Equation Solution by Straightforward Calculation

In this Appendix we describe a tedious, but straightforward, calculation of the functions $q^{(2n+1)}(z)$ in (5.5.16) and the phases $\delta^{(1)}$, $\delta^{(3)}$, and $\delta^{(5)}$ in (5.5.17), which in §5.5 were obtained in a simpler and more elegant way.

From (5.5.2), (5.5.7), and (5.5.6a–c) we obtain

$$Q_c(\varphi/\sqrt{\lambda}) = \lambda^{-1/2}(\varphi^2/4 - \overline{K})^{1/2} = -\frac{1}{2}\lambda^{-1/2}\varphi\,(1 - 4\overline{K}/\varphi^2)^{1/2}, \quad (5.A.1)$$

$$\varepsilon_{0c}(\varphi/\sqrt{\lambda}) = \frac{1}{16}\left(\frac{3}{(\varphi^2/4 - \overline{K})^2} + \frac{5\overline{K}}{(\varphi^2/4 - \overline{K})^3}\right)\lambda^2, \qquad (5.A.2)$$

$$q_c^{(1)}(\varphi/\sqrt{\lambda})\frac{d(\varphi/\sqrt{\lambda})}{d\varphi} = (\varphi^2/4 - \overline{K})^{1/2}\lambda^{-1} \qquad (5.A.3a)$$

$$= \frac{d}{d\varphi}\left[\frac{1}{2}\varphi(\varphi^2/4 - \overline{K})^{1/2} + \overline{K}\,\ln[\varphi/2 - (\varphi^2/4 - \overline{K})^{1/2}]\right]\lambda^{-1}, \quad (5.A.3a')$$

$$q_c^{(3)}(\varphi/\sqrt{\lambda})\frac{d(\varphi/\sqrt{\lambda})}{d\varphi} = \frac{1}{32}(\varphi^2/4 - \overline{K})^{1/2}\left(\frac{3}{(\varphi^2/4 - \overline{K})^2} + \frac{5\overline{K}}{(\varphi^2/4 - \overline{K})^3}\right)\lambda$$
$$\hspace{11cm}(5.A.3b)$$

$$= \frac{1}{96}\frac{d}{d\varphi}\left(\frac{\varphi}{\overline{K}(\varphi^2/4 - \overline{K})^{1/2}} - \frac{5\varphi}{(\varphi^2/4 - \overline{K})^{3/2}}\right)\lambda, \qquad (5.A.3b')$$

$$q_c^{(5)}(\varphi/\sqrt{\lambda})\frac{d(\varphi/\sqrt{\lambda})}{d\varphi} = -(\varphi^2/4 - \overline{K})^{1/2}\Bigg\{\frac{1}{2048}\left(\frac{3}{(\varphi^2/4 - \overline{K})^2} + \frac{5\overline{K}}{(\varphi^2/4 - \overline{K})^3}\right)^2$$

$$+ \frac{1}{128(\varphi^2/4 - \overline{K})^{1/2}}\frac{d}{d\varphi}\left[\frac{1}{(\varphi^2/4 - \overline{K})^{1/2}}\frac{d}{d\varphi}\left(\frac{3}{(\varphi^2/4 - \overline{K})^2}\right.\right.$$

$$\left.\left.+ \frac{5\overline{K}}{(\varphi^2/4 - \overline{K})^3}\right)\right]\Bigg\}\lambda^3, \qquad (5.A.3c)$$

$$= \frac{d}{d\varphi}\left[\frac{7}{3840}\left(\frac{\varphi}{3\overline{K}^3(\varphi^2/4 - \overline{K})^{1/2}} - \frac{\varphi}{6\overline{K}^2(\varphi^2/4 - \overline{K})^{3/2}}\right.\right.$$

$$\left.+ \frac{\varphi}{8\overline{K}(\varphi^2/4 - \overline{K})^{5/2}}\right) + \frac{221}{9216}\left(\frac{\varphi}{(\varphi^2/4 - \overline{K})^{7/2}}\right.$$

$$\left.\left.+ \frac{5\overline{K}\varphi}{2(\varphi^2/4 - \overline{K})^{9/2}}\right)\right]\lambda^3. \qquad (5.A.3c')$$

Here, and in the following, the precise meaning of $(\varphi^2/4 - \overline{K})^{1/2}$ is given by the second equality in (5.A.1), where the square root in the last member is to tend to unity as φ tends to infinity.

Derivation of (5.5.16) with Explicit Expressions for $q^{(2n+1)}$

With the aid of (5.4.14) and (5.4.11) we obtain

$$q_c^{(2n+1)}(\varphi/\sqrt{\lambda})\frac{d(\varphi/\sqrt{\lambda})}{dz}$$

$$= q_c^{(2n+1)}(\varphi/\sqrt{\lambda})\frac{d(\varphi/\sqrt{\lambda})}{d\varphi}\frac{d\varphi}{d\varphi_0}\frac{Q(z)}{(\varphi_0^2/4 - \overline{K}_0)^{1/2}}. \tag{5.A.4}$$

Using (5.A.4), (5.A.3a,b,c), (5.4.3), (5.4.4), and the definition (5.4.20), we obtain after straightforward calculations

$$q_c^{(1)}(\varphi/\sqrt{\lambda})\frac{d(\varphi/\sqrt{\lambda})}{dz} = Q(z)\lambda^{-1} + \left(\frac{1}{(\varphi_0^2/4 - \overline{K}_0)^{1/2}}\frac{d[(\varphi_0^2/4 - \overline{K}_0)^{1/2}\varphi_2]}{d\varphi_0}\right.$$

$$\left. - \frac{\overline{K}_2}{2(\varphi_0^2/4 - \overline{K}_0)}\right)Q(z)\lambda + \left\{\frac{1}{(\varphi_0^2/4 - \overline{K}_0)^{1/2}}\frac{d[(\varphi_0^2/4 - \overline{K}_0)^{1/2}\varphi_4]}{d\varphi_0}\right.$$

$$- \left[\frac{\overline{K}_4}{2(\varphi_0^2/4 - \overline{K}_0)} + \frac{1}{2}\left(\frac{\varphi_0\varphi_2 - 2\overline{K}_2}{4(\varphi_0^2/4 - \overline{K}_0)}\right)^2\right.$$

$$\left. - \frac{1}{8(\varphi_0^2/4 - \overline{K}_0)}\frac{d(\varphi_0\varphi_2^2)}{d\varphi_0} + \frac{\overline{K}_2}{2(\varphi_0^2/4 - \overline{K}_0)}\frac{d\varphi_2}{d\varphi_0}\right]\right\}Q(z)\lambda^3 + \cdots, \tag{5.A.5a}$$

$$q_c^{(3)}(\varphi/\sqrt{\lambda})\frac{d(\varphi/\sqrt{\lambda})}{dz} = \frac{1}{2}h_0(\varphi_0, \overline{K}_0)Q(z)\lambda + \left(\frac{1}{2}\frac{d[h_0(\varphi_0, \overline{K}_0)\varphi_2]}{d\varphi_0}\right.$$

$$+ \frac{\varphi_0 h_0(\varphi_0, \overline{K}_0)}{8(\varphi_0^2/4 - \overline{K}_0)}\varphi_2 + \frac{19\overline{K}_2}{64(\varphi_0^2/4 - \overline{K}_0)^3}$$

$$\left. + \frac{25\overline{K}_0\overline{K}_2}{64(\varphi_0^2/4 - \overline{K}_0)^4}\right)Q(z)\lambda^3 + \cdots, \tag{5.A.5b}$$

$$q_c^{(5)}(\varphi/\sqrt{\lambda})\frac{d(\varphi/\sqrt{\lambda})}{dz} = -\frac{1}{8}\left[[h_0(\varphi_0, \overline{K}_0)]^2\right.$$

$$+ \frac{1}{(\varphi_0^2/4 - \overline{K}_0)^{1/2}}\frac{d}{d\varphi_0}\left(\frac{1}{(\varphi_0^2/4 - \overline{K}_0)^{1/2}}\frac{dh_0(\varphi_0, \overline{K}_0)}{d\varphi_0}\right)\right]Q(z)\lambda^3 + \cdots. \tag{5.A.5c}$$

To obtain a more convenient expression for $q_c^{(1)}(\varphi/\sqrt{\lambda})d(\varphi/\sqrt{\lambda})/dz$ we shall now rewrite two of the terms in the right-hand member of (5.A.5a). From (5.4.15b,c), (5.4.14), and (5.4.11) we obtain

$$\frac{1}{(\varphi_0^2/4 - \overline{K}_0)^{1/2}}\frac{d[(\varphi_0^2/4 - \overline{K}_0)^{1/2}\varphi_2]}{d\varphi_0} = Y_2 - \frac{1}{2}h, \tag{5.A.6a}$$

$$\frac{1}{(\varphi_0^2/4 - \overline{K}_0)^{1/2}}\frac{d[(\varphi_0^2/4 - \overline{K}_0)^{1/2}\varphi_4]}{d\varphi_0} = Y_4 + \frac{1}{8}h^2$$

$$+ \frac{1}{8(\varphi_0^2/4 - \overline{K}_0)^{1/2}} \frac{d}{d\varphi_0} \left(\frac{1}{(\varphi_0^2/4 - \overline{K}_0)^{1/2}} \frac{dh}{d\varphi_0} \right)$$

$$+ \frac{\overline{K}_4}{2(\varphi_0^2/4 - \overline{K}_0)} - \frac{1}{2(\varphi_0^2/4 - \overline{K}_0)^{1/2}}$$

$$\times \frac{d}{d\varphi_0} \left(h(\varphi_0^2/4 - \overline{K}_0)^{1/2}\varphi_2 + \frac{\varphi_0\varphi_2^2}{4(\varphi_0^2/4 - \overline{K}_0)^{1/2}} \right). \tag{5.A.6b}$$

Using (5.4.19), we can rewrite (5.A.6a,b) as

$$\frac{1}{(\varphi_0^2/4 - \overline{K}_0)^{1/2}} \frac{d[(\varphi_0^2/4 - \overline{K}_0)^{1/2}\varphi_2]}{d\varphi_0} = Y_2 - \frac{1}{2}h_0(\varphi_0,\overline{K}_0) + \frac{\overline{K}_2}{2(\varphi_0^2/4 - \overline{K}_0)}, \tag{5.A.7a}$$

$$\frac{1}{(\varphi_0^2/4 - \overline{K}_0)^{1/2}} \frac{d[(\varphi_0^2/4 - \overline{K}_0)^{1/2}\varphi_4]}{d\varphi_0} = Y_4 + \left[\frac{\overline{K}_4}{2(\varphi_0^2/4 - \overline{K}_0)} \right.$$

$$+ \frac{1}{2} \left(\frac{\varphi_0\varphi_2 - 2\overline{K}_2}{4(\varphi_0^2/4 - \overline{K}_0)} \right)^2 - \frac{1}{8(\varphi_0^2/4 - \overline{K}_0)} \frac{d(\varphi_0\varphi_2^2)}{d\varphi_0}$$

$$+ \frac{\overline{K}_2}{2(\varphi_0^2/4 - \overline{K}_0)} \frac{d\varphi_2}{d\varphi_0} - \left[\frac{1}{2} \frac{d[h_0(\varphi_0,\overline{K}_0)\varphi_2]}{d\varphi_0} + \frac{\varphi_0 h_0(\varphi_0,\overline{K}_0)}{8(\varphi_0^2/4 - \overline{K}_0)}\varphi_2 \right.$$

$$\left. + \frac{19\overline{K}_2}{64(\varphi_0^2/4 - \overline{K}_0)^3} + \frac{25\overline{K}_0\overline{K}_2}{64(\varphi_0^2/4 - \overline{K}_0)^4} \right]$$

$$+ \frac{1}{8} \left[[h_0(\varphi_0,\overline{K}_0)]^2 + \frac{1}{(\varphi_0^2/4 - \overline{K}_0)^{1/2}} \frac{d}{d\varphi_0} \left(\frac{dh_0(\varphi_0,\overline{K}_0)/d\varphi_0}{(\varphi_0^2/4 - \overline{K}_0)^{1/2}} \right) \right]. \tag{5.A.7b}$$

Inserting (5.A.7a,b) into (5.A.5a) and recalling that $Y_0 = 1$, we obtain

$$q_c^{(1)}(\varphi/\sqrt{\lambda}) \frac{d(\varphi/\sqrt{\lambda})}{dz} = Y_0 Q(z)\lambda^{-1} + \left[Y_2 - \frac{1}{2}h_0(\varphi_0,\overline{K}_0) \right] Q(z)\lambda$$

$$+ \left\{ Y_4 - \left[\frac{1}{2} \frac{d[h_0(\varphi_0,\overline{K}_0)\varphi_2]}{d\varphi_0} + \frac{\varphi_0 h_0(\varphi_0,\overline{K}_0)}{8(\varphi_0^2/4 - \overline{K}_0)}\varphi_2 \right. \right.$$

$$+ \frac{19\overline{K}_2}{64(\varphi_0^2/4 - \overline{K}_0)^3} + \frac{25\overline{K}_0\overline{K}_2}{64(\varphi_0^2/4 - \overline{K}_0)^4} \left] + \frac{1}{8} \left[[h_0(\varphi_0,\overline{K}_0)]^2 \right. \right.$$

$$\left. \left. + \frac{1}{(\varphi_0^2/4 - \overline{K}_0)^{1/2}} \frac{d}{d\varphi_0} \left(\frac{dh_0(\varphi_0,\overline{K}_0)/d\varphi_0}{(\varphi_0^2/4 - \overline{K}_0)^{1/2}} \right) \right] \right\} Q(z)\lambda^3 + \cdots. \tag{5.A.5a'}$$

For $0 \leq N \leq 2$ we obtain from (5.A.5a',b,c) the formula (5.5.16) with $q^{(2n+1)}(z)$ given by (5.2.7). This formula is valid also for $N > 2$.

Derivation of (5.5.17) with Explicit Expressions for $\delta^{(2n+1)}$

Using (5.A.3a′–c′), and recalling that according to (5.5.13a) $\varphi^2/4 - \overline{K} = 0$ when $\varphi/\sqrt{\lambda} = \vartheta_1$, and that according to (5.A.1) $(\varphi^2/4 - \overline{K})^{1/2} = -\frac{1}{2}\varphi(1 - 4\overline{K}/\varphi^2)^{1/2}$, where the last square root tends to unity as φ tends to infinity, we obtain from (5.5.12) and (5.5.12′), where $\varphi/\sqrt{\lambda}$ is merely an integration variable (not to be expanded in powers of λ),

$$\delta_c^{(1)} = \left(\overline{K}\ln\frac{\vartheta_1}{2} - \frac{1}{2}\overline{K}\right)\lambda^{-1} \tag{5.A.8a}$$

$$\delta_c^{(3)} = \frac{1}{48\overline{K}}\lambda \tag{5.A.8b}$$

$$\delta_c^{(5)} = \frac{7}{5760\overline{K}^3}\lambda^3. \tag{5.A.8c}$$

Using the expansion (5.4.4) of $\overline{K}$ in powers of λ and (5.5.13a), we obtain from (5.A.8a–c)

$$\delta_c^{(1)} = \frac{\overline{K}}{2\lambda}\left(\ln\frac{\overline{K}_0}{\lambda} + 2\pi i\right) - \frac{1}{2}\overline{K}_0\lambda^{-1} + \frac{\overline{K}_2}{4\overline{K}_0}\lambda^3 + \cdots \tag{5.A.9a}$$

$$\delta_c^{(3)} = \frac{1}{48\overline{K}_0}\lambda - \frac{\overline{K}_2}{48\overline{K}_0^2}\lambda^3 + \cdots \tag{5.A.9b}$$

$$\delta_c^{(5)} = \frac{7}{5760\overline{K}_0^3}\lambda^3 + \cdots . \tag{5.A.9c}$$

For $0 \leq N \leq 2$ we obtain from (5.A.9a–c) and (5.4.4) the formula (5.5.17) with $\delta^{(2n+1)}$ given by

$$\delta^{(1)} = \frac{\overline{K}_0}{2}(\ln\overline{K}_0 + 2\pi i) - \frac{\overline{K}_0}{2} \tag{5.A.10a}$$

$$\delta^{(3)} = \frac{\overline{K}_2}{2}(\ln\overline{K}_0 + 2\pi i) + \frac{1}{48\overline{K}_0} \tag{5.A.10b}$$

$$\delta^{(5)} = \frac{\overline{K}_4}{2}(\ln\overline{K}_0 + 2\pi i) + \frac{\overline{K}_2^2}{4\overline{K}_0} - \frac{\overline{K}_2}{48\overline{K}_0^2} + \frac{7}{5760\overline{K}_0^3}. \tag{5.A.10c}$$

These formulas can also be obtained from (5.5.29) with the aid of (5.5.25a–c).

References

[1] M. Abramowitz and I.A. Stegun (Editors), *Handbook of Mathematical Functions*, Dover, New York, 1972.

[2] A. Amaha and K.-E. Thylwe, Regge-pole positions and residues calculated from phase-integral formulas, *Phys. Rev.* **A44** (1991), 4203–4209.

[3] Ö. Dammert, Transmission through a system of potential barries. I. Transmission coefficient, *J. Math. Phys.* **24** (1983), 2163–2175.

[4] Ö. Dammert, Transmission through a system of potential barriers. II. Necessary condition for complete transparency. A maximum transmission problem, *J. Math. Phys.* **27** (1986), 461–470.

[5] G. Drukarev, N. Fröman, and P.O. Fröman, The Jost function treated by the F-matrix phase integral method, *J. Phys. A: Math. Gen.* **12** (1979), 171–186.

[6] K.W. Ford, D.L. Hill, M. Wakano, and J.A. Wheeler, Quantum effects near a barrier maximum, *Ann. Phys. (N.Y.)* **7** (1959), 239–258.

[7] N. Fröman, Connection formulas for certain higher-order phase-integral approximations, *Ann. Phys. (N.Y.)* **61** (1970), 451–464.

[8] N. Fröman, A simple formula for calculating quantal expectation values without the use of wave functions, *Phys. Lett.* **48A** (1974), 137–139.

[9] N. Fröman, Phase-integral formulas for level densities, normalization factors, and quantal expectation values, not involving wave functions, *Phys. Rev.* **A17** (1978), 493–504.

[10] N. Fröman, Dispersion relation for energy bands and energy gaps derived by the use of a phase-integral method, with an application to the Mathieu equation, *J. Phys. A: Math. Gen.* **12** (1979), 2355–2371.

[11] N. Fröman, Semiclassical and higher-order approximations: Properties. Solution of connection problems. In: *Semiclassical Methods in Molecular Scattering and Spectroscopy* (Reidel, 1980), pp. 1–44.

[12] N. Fröman, Relation, expressed in terms of elliptic integrals, for determining characteristic values and characteristic exponents within stable and unstable regions (bands and gaps) associated with the Mathieu potential, *Phys. Rev.* **D23** (1981), 1756–1759.

[13] N. Fröman and P.O. Fröman, *JWKB Approximation. Contributions to the theory*, (North-Holland Publishing Company, Amsterdam, 1965). Russian translation: MIR, Moscow, 1967.

[14] N. Fröman and P.O. Fröman, Transmission through a real potential barrier treated by means of certain phase-integral approximations, *Nucl. Phys.* **A147** (1970), 606–626.

[15] N. Fröman and P.O. Fröman, On Wentzel's proof of the quantization condition for a single-well potential, *J. Math. Phys.* **18** (1977), 96–99.

[16] N. Fröman and P.O. Fröman, Phase-integral calculation of quantal matrix elements without the use of wavefunctions, *J. Math. Phys.* **18** (1977), 903–906.

[17] N. Fröman and P.O. Fröman, On phase-integral quantization conditions for bound states in one-dimensional smooth single-well potentials, *J. Math. Phys.* **19** (1978), 1830–1837.

[18] N. Fröman and P.O. Fröman, Phase-integral calculation with very high accuracy of the Stark effect in a hydrogen atom. *Méthodes Semi-Classiques en Mécanique Quantique, Colloque du 10 au 15 septembre 1984, CIRM (Luminy), p. 45, Publications de l'Université de Nantes, Institut de Mathématiques et d'Informatiques, 2, rue de la Houssinière, 44072 Nantes CEDEX, France.*

[19] N. Fröman and P.O. Fröman, Exact formulas and phase-integral formulas, not involving wavefunctions, for expectation values pertaining to general potentials, *Ann. Phys. (N.Y.)* **163** (1985), 215–226.

[20] N. Fröman and P.O. Fröman, New two-turning-point phase-integral formula for the residue of the S-matrix at a Regge pole, *Phys. Rev.* **A43** (1991), 3563–3566.

[21] N. Fröman and P.O. Fröman, Phase-integral approximation of arbitrary order generated from an unspecified base function. Review article in: *Forty More Years of Ramifications: Spectral Asymptotics and its Applications,* edited by S.A. Fulling and F.J. Narcowich (Discourses in Mathematics and Its Applications, No. 1, Department of Mathematics, Texas A & M University, College Station, Texas, 1991). This paper is reprinted, with minor changes, as Chapter 1 of the present monograph.

[22] N. Fröman and P.O. Fröman, Technique of the comparison equation adapted to the phase-integral method. This is Chapter 2 of the present monograph.

[23] N. Fröman and P.O. Fröman, Problem involving one transition zero. This is Chapter 3 of the present monograph.

[24] N. Fröman, P.O. Fröman, and F. Karlsson, Phase-integral calculation of quantal matrix elements between unbound states, without the use of wavefunctions, *Molec. Phys.* **38** (1979), 749–767.

[25] N. Fröman, P.O. Fröman, and S. Linnaeus, New phase-integral formulas for the regular wave function when there are turning points close to a pole of the potential. This is Chapter 6 in this monograph.

[26] N. Fröman, P.O. Fröman, and B. Lundborg, The Stokes constants for a cluster of transition points, *Math. Proc. Camb. Phil. Soc.* **104** (1988), 153–179.

[27] N. Fröman, P.O. Fröman, U. Myhrman, and R. Paulsson, On the quantal treatment of the double-well potential problem by means of certain phase-integral approximations, *Ann. Phys. (N.Y.)* **74** (1972), 314–323.

[28] J. Heading, The approximate use of the complex gamma function in some wave propagation problems, *J. Phys. A: Math. Nucl. Gen.* **6** (1973), 958–973.

[29] S. Linnaeus, Phase-integral calculation of phase shifts for a heavy-ion optical potential, *Phys. Rev.* **C34** (1986), 1274–1277.

[30] Y.L. Luke, *The Special Functions and Their Approximations*, Vol. I. (Academic Press, New York and London, 1969).

[31] B. Lundborg, Phase-integral treatment of transmission through an inverted Morse potential, *Math. Proc. Camb. Phil. Soc.* **81** (1977), 463–483.

[32] B. Lundborg and B. Thidé, Standing wave pattern of HF radio waves in the ionospheric reflection region, 1. General formulas, *Radio Sci.* **20** (1985), 947–958.

[33] B. Lundborg and B. Thidé, Standing wave pattern of HF radio waves in the ionospheric reflection region, 2. Applications, *Radio Sci.* **21** (1986), 486–500.

[34] H. Moriguchi, An improvement of the WKB method in the presence of turning points and the asymptotic solutions of a class of Hill equations, *J. Phys. Soc. Japan* **14** (1959), 1771–1796.

[35] R. Paulsson, Calculation of expectation values and matrix elements for symmetric double-well potentials. II. Investigation of the accuracy of phase-integral formulas, *Ann. Phys. (N.Y.)* **163** (1985), 245–251.

[36] R. Paulsson and N. Fröman, Calculation of expectation values and matrix elements for symmetric double-well potentials. I. General theory according to phase-integral method, *Ann. Phys. (N.Y.)* **163** (1985), 227–244.

6

Phase-Integral Formulas for the Regular Wave Function When There Are Turning Points Close to a Pole of the Potential

Nanny Fröman, Per Olof Fröman, and Staffan Linnaeus

Abstract. Comparison equation technique, with the Coulomb potential plus the centrifugal barrier as comparison potential, is applied to the Schrödinger equation when there are turning points close to a pole of the potential. By asymptotic expansion, with respect to a "small" parameter, of the resulting formal solution one obtains the arbitrary-order phase-integral approximation generated from an unspecified base function. The phase and the amplitude of the phase-integral solution thus obtained remain accurate also when the turning points lie close to the pole. For a solution that is regular at the pole of the potential the results yield new phase-integral formulas in the classically allowed region. These formulas contain as special limiting cases previously known formulas, valid when the turning points recede from the pole. The normalization of the wave function is also considered.

6.1 Introduction

When, for a differential equation of the Schrödinger type, there are turning points close to a pole of the potential, a convenient kind of comparison equation technique is most appropriate for obtaining the correct phase and amplitude of the phase-integral wave function at some distance from the pole and the turning points. The radial Schrödinger equation was considered by Good [15], who used the pure centrifugal barrier as comparison potential; this yields a comparison equation with two turning points situated symmetrically on opposite sides of the pole at the origin. However, due to certain series expansions performed in [15], the treatment does not yield an optimal result. For instance, the author mentions in connection with the nonbound case that a difficulty arises in the neighborhood of the positive turning point, so that no acceptable solution is obtained. A fully adequate treatment would remove this deficiency, but in the asymptotic region (i.e., sufficiently far from the origin) one would still not obtain any improvement of the phase yielded by the usual connection formula. Erdélyi [4] later treated a more general class of problems by means of a comparison equation with only one turning point and a second-order pole. The solution that he obtained, valid uniformly through the transition points, yields asymptotically in the classically allowed region the same phase as the usual one-turning point connection formula. Thus, in the asymptotic region, neither of the two comparison equations mentioned yields an improvement of the WKB expression obtained from the usual connection formula for tracing the wave function from the origin to the classically allowed region.

In the present chapter we use the Coulomb potential plus the centrifugal barrier as comparison potential. We use comparison equation technique for solving the Schrödinger equation with the aim of extending the applicability of the phase-integral method to problems where transition points may approach each other. We shall treat the cases in which the effective potential has a pole of the first or second order and in which there are thus one or two real turning points, which may lie arbitrarily close to the pole. All other turning points or singularities of the potential are assumed to be situated well outside the region of the complex z-plane where the comparison equation technique is to be used. We push the calculations far enough to obtain final analytical results, simplified as far as possible. These results yield significant corrections to the usual phase-integral formulas [5, 6, 8, 9, 10, 12] for the phase and amplitude of the wave function. The latter formulas are obtained as special limiting cases for the situation in which the turning points and the pole in question do not lie close to each other.

The comparison equation technique that we shall use is described by Fröman and Fröman [13]. It is adapted to yield automatically the arbitrary-order phase-integral approximation generated from an unspecified base function [9, 10, 12] with correct values of phase and amplitude, when appropriate expansions in terms of a "small" bookkeeping parameter are made in the comparison equation solution.

In §6.2 we present general formulas concerning the comparison equation treatment.

In §6.3 we consider the case in which the comparison equation corresponds to a *scattering problem*, and derive the corresponding phase-integral wave function in the classically allowed region at some distance from the closely lying transition points. Although the comparison equation in §6.3 corresponds to unbound states, primarily yielding phase shifts in scattering problems; one can also apply the results obtained there in the treatment of bound-state problems. The regular solution traced from the pole into the neighboring classically allowed region must then be joined to a phase-integral wave function regular at the other end (which possibly may also be a pole) of a physically relevant interval, which may contain several wells separated by barriers. In the simplest case of a single-well bound-state problem the solution that is regular at the pole can be joined to the solution that one obtains by using the usual connection formula at the right-hand turning point for tracing the bound-state solution from the classically forbidden region to the classically allowed region. In this way one obtains new quantization conditions in which transition points, lying close to poles occurring in the potential, are taken into account, in addition to the transition points delimiting the classically allowed regions.

Using a comparison equation corresponding to a *single-well potential*, we derive in §6.4 already familiar phase-integral quantization conditions. This is an example of the fact that the use of comparison equation technique sometimes provides only an alternative derivation of already available formulas that can be obtained in a much simpler way without the use of comparison equation technique. There are also situations in which the comparison equation tool may turn out to be superfluous unless it is fully exploited [3]; see the discussion at the end of §7.4. In §6.4 we also consider the normalized wave function of a bound state.

For the radial Schrödinger equation the results obtained in the present chapter are very useful when the angular momentum quantum number ℓ is small. When ℓ is sufficiently large, the already available phase-integral connection formulas [5, 6, 8, 9, 10, 12], derived directly without the use of comparison equation technique; yield the wave function with the appropriate phase and amplitude.

In Chapter 7 the results obtained in this chapter are used for investigating in detail the behavior of the conveniently normalized wave function in the neighborhood of the pole.

6.2 Definitions and Preparatory Calculations

For the sake of simplicity we shall place the pole at the origin. The differential equation then takes the form of the radial Schrödinger equation; it is natural to use the terminology associated with that equation. This is, of course, not an essential restriction.

The radial Schrödinger equation can be written in the form of Eq. (2.2.1); that is,

$$\frac{d^2\psi}{dz^2} + R(z)\psi = 0, \tag{6.2.1}$$

where, using obvious notation,

$$R(z) = \frac{2m}{\hbar^2}[E - V(z)] - \frac{\ell(\ell+1)}{z^2}. \tag{6.2.2}$$

$V(z)$ is an analytic function of z that is regular except for a possible Coulomb singularity at $z = 0$. We emphasize that in this chapter it is not necessary for ℓ to be an integer. From (6.2.2) it follows that

$$\lim_{z \to 0} z^2 R(z) = -\ell(\ell + 1). \tag{6.2.3}$$

We now introduce a formal, "small" expansion parameter λ by considering, instead of the original Schrödinger equation (6.2.1), the auxiliary differential equation (2.2.2); that is,

$$\frac{d^2\psi}{dz^2} + \left(\frac{Q^2(z)}{\lambda^2} + [R(z) - Q^2(z)] \right) \psi = 0, \tag{6.2.4}$$

where the *base function* $Q(z)$ is an analytic function that, to some extent, can be chosen arbitrarily. At the end λ is set equal to unity, so that the original differential equation (6.2.1) is restored. The differential equation (6.2.4) will be solved by the comparison equation technique described in Chapter 2. The solution of (6.2.1) thus obtained, when λ is set equal to unity, is constructed so that, sufficiently far away from the origin and the transition points, it can be transformed into the phase-integral approximation of arbitrary order generated from the base function $Q(z)$ (see Chapter 1). We shall assume that $Q^2(z)$ has a first- or second-order pole at the origin. Therefore,

$$\lim_{z \to 0} z^2 Q^2(z) = -\xi_0^2 \tag{6.2.5}$$

where ξ_0 is an unspecified constant that, for the sake of simplicity, we assume to be real and nonnegative.

To solve (6.2.4) by the comparison equation technique described in Chapter 2, we let, according to Eq. (2.2.3),

$$\psi = \left(\frac{d\varphi}{dz} \right)^{-\frac{1}{2}} \Phi(\varphi), \tag{6.2.6}$$

where φ is a function of z and λ that is to be determined from the differential equation (6.2.4). The function $\Phi(\varphi)$ is a known solution of the differential equation (2.2.5); that is,

$$\frac{d^2\Phi}{d\varphi^2} + \frac{\Pi(\varphi)}{\lambda^2} \Phi = 0, \tag{6.2.7}$$

where $\Pi(\varphi)$ is a function of φ and λ that, in the region of the complex φ-plane corresponding to the region of the complex z-plane under consideration, has similar zeros and poles as does the function $Q^2(z) + [R(z) - Q^2(z)]\lambda^2$ in (6.2.4). Inserting (6.2.6) into (6.2.4), using (6.2.7), and assuming that $\Phi(\varphi)$ is not identically equal to zero, we obtain the differential equation (2.2.8).

For the purpose of the present chapter the function $\Pi(\varphi)$ is chosen to be

$$\Pi(\varphi) = \frac{A_0 + A_1\varphi + A_2\varphi^2}{\varphi^2}, \tag{6.2.8}$$

where the coefficients A_0, A_1, and A_2 are independent of φ, but may depend on λ. In (6.2.8) the coefficient A_0 may be different from zero [second-order pole of $\Pi(\varphi)$] or equal to zero [first-order pole of $\Pi(\varphi)$ if $A_1 \neq 0$]. Our formula (6.2.8) is the particular case of Eqs. (2.2.9a–c) that one obtains by setting

$$A_n = 0 \quad \text{for} \quad n > 2 \tag{6.2.9}$$

and

$$B_q = \delta_{q,2}. \tag{6.2.10}$$

In order for the comparison equation technique to be valid in the neighborhood of the origin, the condition (2.2.6), particularized to the present case of the radial Schrödinger equation, must be fulfilled. Using (6.2.3), (6.2.5), and (6.2.8) in that condition, one obtains

$$A_0 = -\left(\xi_0^2 + [\ell(\ell+1) - \xi_0^2]\lambda^2\right). \tag{6.2.11}$$

When z is positive and sufficiently large (but still lies within the region of the complex z-plane considered in the comparison equation treatment), the functions $Q^2(z)$ and $\Pi(\varphi)$ must have the same sign in order that $\psi(z)$ and $\Phi(\varphi)$ have similar behavior. The sign of $\Pi(\varphi)$ for large values of φ is determined by A_2. Therefore, A_2 must be positive in the cases shown in Figures 6.1–6.3, but negative in the cases shown in Figures 6.4–6.5. By introducing in (6.2.7) along with (6.2.8) a new independent variable, $(\pm A_2)^{\frac{1}{2}}\varphi$, instead of the original independent variable φ, one realizes that it is no restriction to set

$$A_2 = \pm 1, \tag{6.2.12}$$

where the sign depends on whether, for the largest positive values of φ that correspond to points in the region of the complex z-plane under consideration in our comparison equation treatment, the function $\Pi(\varphi)$, and hence $Q^2(z)$, is positive or negative.

From the differential equation (2.2.8) together with (6.2.8) we shall determine φ as a function of z and λ. For this purpose we expand φ formally according to Eq. (2.2.11); that is,

$$\varphi = \sum_{\beta=0}^{\infty} \varphi_{2\beta}\lambda^{2\beta}, \tag{6.2.13}$$

where the quantities $\varphi_{2\beta}$ depend on z but are independent of λ; we assume that the quantities A_n depend on λ according to the formal series (2.2.12); that is,

$$A_n = \sum_{\beta=0}^{\infty} A_{n,2\beta}\lambda^{2\beta}, \tag{6.2.14}$$

the coefficients $A_{n,2\beta}$ being independent of λ. Using (6.2.11), (6.2.12), and (6.2.9), we obtain

$$A_{0,0} = -\xi_0^2 \tag{6.2.15a}$$

$$A_{0,2} = \xi_0^2 - \ell(\ell+1) \tag{6.2.15b}$$

$$A_{0,2\beta} = 0 \quad \text{for} \quad 2\beta > 2 \tag{6.2.15c}$$

$$A_{2,2\beta} = \pm\delta_{0,\beta} \tag{6.2.16}$$

$$A_{n,2\beta} = 0 \quad \text{for} \quad n > 2. \tag{6.2.17}$$

From the definitions (2.2.16a), (2.2.31), and (2.2.16c); that is,

$$P^2(\varphi_0) = \frac{S_0(\varphi_0)}{T(\varphi_0)} \tag{6.2.18}$$

$$S_{2\beta}(\varphi_0) = \sum_{n=0}^{\infty} A_{n,2\beta}\varphi_0^n \tag{6.2.19}$$

$$T(\varphi_0) = \sum_{q=0}^{\infty} B_q\varphi_0^q, \tag{6.2.20}$$

and (6.2.10), (6.2.15a–c), (6.2.16), and (6.2.17); we obtain

$$P(\varphi_0) = \left(\pm 1 + \frac{A_{1,0}}{\varphi_0} - \frac{\xi_0^2}{\varphi_0^2}\right)^{\frac{1}{2}} \tag{6.2.21}$$

and

$$\frac{S_{2\beta}}{T} = \frac{A_{1,2\beta}}{\varphi_0} - \frac{\ell(\ell+1) - \xi_0^2}{\varphi_0^2}\delta_{\beta,1}, \quad \beta > 0. \tag{6.2.22}$$

In (6.2.21) the double sign arises from the double sign in (6.2.12); in addition, the sign of the square root must be chosen so as to correspond to that of $Q(z)$ in relevant corresponding regions of the complex φ_0- and z-planes.

6.2.1 Determination of φ_0 and $A_{1,0}$

In order for the comparison equation technique for solving the differential equation (6.2.4) to work satisfactorily, the mapping of the z-plane and the φ_0-plane onto each other must, according to Chapter 2, be conformal in the region under consideration. This means that φ_0 must be a regular analytic function of z, the derivative of which has no zeros in the region of the complex z-plane under consideration. As a consequence of the conformal mapping it follows that, at corresponding points in the complex z- and φ_0-planes, the functions $Q^2(z)$ and $P^2(\varphi_0)$ must have the same kinds of zeros and poles (i.e., with the same multiplicities).

The function φ_0 must satisfy the differential equation (2.2.17); that is,

$$\frac{d\varphi_0}{dz} = \frac{Q(z)}{P(\varphi_0)}. \tag{6.2.23}$$

When $\xi_0 > 0$ (Figures 6.1 and 6.4) the square of the base function; i.e., $Q^2(z)$, as well as $P^2(\varphi_0)$; has a second-order pole at the origin, and in the neighborhood of the origin the mapping of the z-plane and the φ_0-plane on each other is, because of (6.2.15a), conformal for any choice of the integration constant appearing by the solution of (6.2.23). Because $\xi_0 \neq 0$ there are two zeros (τ_1 and τ_2) of $P^2(\varphi_0)$. The zeros of $Q^2(z)$ corresponding to τ_1 and τ_2 are called t_1 and t_2. It follows from (6.2.21) that when $A_2 = +1$ one of the zeros of $P^2(\varphi_0)$, which we denote by τ_1, is positive; the other, τ_2, is negative (see Figure 6.1). When $A_2 = -1$ and $A_{1,0} > 2\xi_0$, it follows that both zeros τ_1 and τ_2 ($> \tau_1$) are positive (see Figure 6.4). We disregard the less common cases in which both τ_1 and τ_2 are negative ($A_2 = -1$, $A_{1,0} < -2\xi_0$) and in which τ_1 and τ_2 are complex conjugate ($A_2 = -1$, $|A_{1,0}| < 2\xi_0$).

When $\xi_0 = 0$ (Figures 6.2, 6.3, and 6.5) there is a first-order pole at the origin and one zero, τ_1, of $P^2(\varphi_0)$, the position of which (according to [6.2.21]) is given by $\tau_1 = \mp A_{1,0}$ when $A_2 = \pm 1$. The corresponding zero of $Q^2(z)$ is called t_1 (> 0 or < 0).

From (6.2.23) and the requirement that the turning point t_1 be mapped into the zero τ_1 of $P^2(\varphi_0)$ we get both for $\xi_0 > 0$ and for $\xi_0 = 0$ the relation

$$\int_{\tau_1}^{\varphi_0} P(\varphi_0)d\varphi_0 = \int_{t_1}^{z} Q(z)\, dz, \qquad (6.2.24)$$

which determines φ_0 as a function of z.

The function $\varphi_0(z)$ must be regular analytic and hence single-valued in the whole region containing the pole and the zero(s) of $Q^2(z)$. Therefore, to determine the parameter $A_{1,0}$ we impose the condition that $\varphi_0(z)$ be single-valued on a curve Λ enclosing the origin and the turning point(s); see Figures 6.1(b)–6.5(b). Recalling that $Y_0 = 1$ according to Eq. (2.1.5a), we obtain from (6.2.24)

$$\int_{\Lambda} P(\varphi_0)\, d\varphi_0 = \int_{\Lambda} Y_0 Q(z)\, dz, \qquad (6.2.25)$$

where Λ is the closed contour of integration in Figures 6.1(b)–6.5(b).

In Figures 6.1, 6.2, and 6.3 we choose $Q(z)$ to be positive when z is positive and lies to the right of t_1; thus $P(\varphi_0)$ approaches $+1$ as $\varphi_0 \to +\infty$. In Figures 6.4 and 6.5 we choose $Q(z)$, and hence $P(\varphi_0)$, to be positive in the classically allowed region of the positive z-axis and φ_0-axis, respectively; this means that $P(\varphi_0) \to -i$ as $\varphi_0 \to +\infty$. For large values of $|\varphi_0|$ it then follows from (6.2.21) that

$$P(\varphi_0) = \begin{cases} 1 + \frac{1}{2}A_{1,0}/\varphi_0 + \cdots & \text{when } A_2 = +1 \text{ (Figs. 6.1–6.3)} \\ -i(1 - \frac{1}{2}A_{1,0}/\varphi_0 + \cdots) & \text{when } A_2 = -1 \text{ (Figs. 6.4, 6.5).} \end{cases}$$
$$(6.2.26)$$

Therefore, one can evaluate the integral in the left-hand member of (6.2.25) by deforming the contour Λ into a very large circle. This yields

$$A_{1,0} = \begin{cases} -\frac{i}{\pi} \int_\Lambda Y_0 Q(z)dz & \text{when } A_2 = +1 \text{ (Figs. 6.1–6.3)} \\ -\frac{1}{\pi} \int_\Lambda Y_0 Q(z)dz & \text{when } A_2 = -1 \text{ (Figs. 6.4, 6.5).} \end{cases} \qquad (6.2.27)$$

6.2.2 Determination of $\varphi_{2\beta}$ and $A_{1,2\beta}$ for $\beta > 0$

As described in Chapter 2, one obtains for $\varphi_{2\beta}$ differential equations that after integration yield (see Eqs. (2.2.25) and (2.2.25′))

$$\varphi_{2\beta} = \frac{1}{P(\varphi_0)} \int_{\tau_1}^{\varphi_0} I_{2\beta} P(\varphi_0)\, d\varphi_0, \quad \beta > 0, \qquad (6.2.28)$$

where the choice of τ_1 as the constant lower limit of integration ensures that the functions $\varphi_{2\beta}$ will be regular analytic at τ_1; and where according to Eqs. (2.2.33), (2.2.46), (2.1.5b,c), (2.2.41), (2.1.6), and (2.2.34)

$$I_2 = Y_2 - \frac{1}{2}h \qquad (6.2.29a)$$

$$I_4 = Y_4 + \frac{1}{8}h^2 - \frac{S_4/T}{2P^2} + \frac{d}{d\zeta}\left(\frac{1}{8}\frac{dh}{d\zeta} - \frac{1}{2}hP\varphi_2 - \frac{1}{2}\frac{dP}{d\varphi_0}\varphi_2^2\right) \qquad (6.2.29b)$$

with

$$Y_2 = \frac{1}{2}\varepsilon_0 \qquad (6.2.30a)$$

$$Y_4 = -\frac{1}{8}\varepsilon_0^2 - \frac{1}{8}\frac{d^2\varepsilon_0}{d\zeta^2} \qquad (6.2.30b)$$

$$\frac{d}{d\zeta} = \frac{1}{Q}\frac{d}{dz} = \frac{1}{P}\frac{d}{d\varphi_0} \qquad (6.2.31)$$

$$\varepsilon_0 = Q^{-\frac{3}{2}}\frac{d^2}{dz^2}Q^{-\frac{1}{2}} + \frac{R - Q^2}{Q^2} \qquad (6.2.32)$$

$$h = P^{-\frac{3}{2}}\frac{d^2 P^{-\frac{1}{2}}}{d\varphi_0^2} + \frac{S_2/T}{P^2}, \qquad (6.2.33)$$

S_2/T and S_4/T being given by (6.2.22).

The quantities $A_{1,2\beta}$ are determined from the condition that $\varphi_{2\beta}$ must be regular analytic in the region under consideration and hence single-valued on the previously mentioned curve Λ, which by means of (6.2.28) yields

$$\int_\Lambda I_{2\beta} P(\varphi_0)\, d\varphi_0 = 0, \quad \beta > 0, \qquad (6.2.34)$$

where Λ is the contour of integration shown in Figures 6.1(b)–6.5(b). Inserting the expressions (6.2.29a,b) for I_2 and I_4 into (6.2.34) with $\beta = 1$

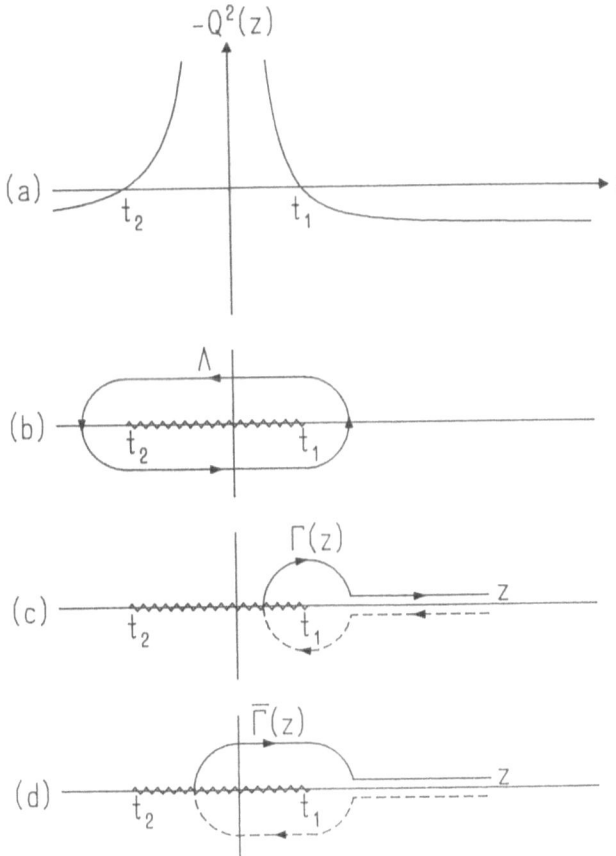

FIGURE 6.1. (a) Qualitative behavior of $-Q^2(z)$ in the case when $A_2 = +1$, $\xi_0 > 0$, and $\eta_0 > 0$ or $\eta_0 < 0$. The zeros of $Q^2(z)$ are denoted by t_1 and t_2. (b) Closed contour of integration Λ in the complex z-plane. The wavy line indicates a branch cut. The base function $Q(z)$, as well as $P(\varphi_0)$, is chosen to be positive on the real axis to the right of t_1 [τ_1]. (c) Open contour of integration $\Gamma(z)$ in the complex z-plane [φ_0-plane]. The dashed part of the contour $\Gamma(z)$ lies on a Riemann sheet adjacent to the complex z-plane under consideration. (d) Alternative open contour of integration $\overline{\Gamma}(z)$.

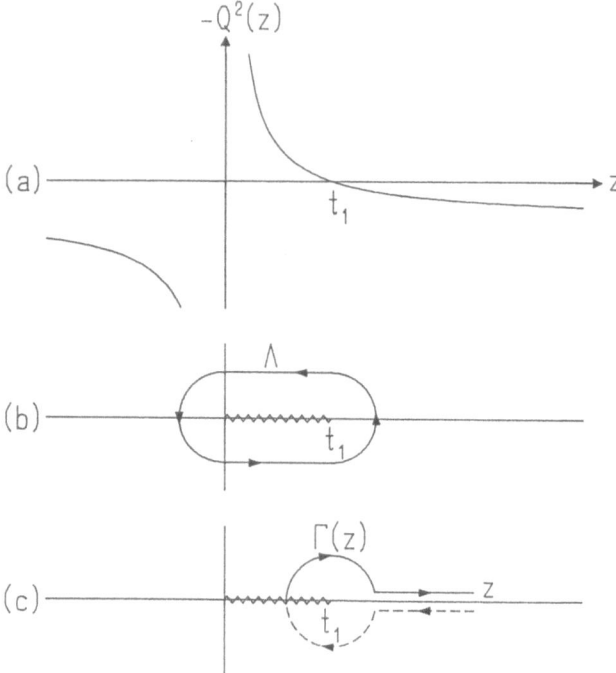

FIGURE 6.2. (a) Qualitative behavior of $-Q^2(z)$ in the case when $A_2 = +1$, $\xi_0 = 0$, and $\eta_0 > 0$. The zero of $Q^2(z)$ is denoted by t_1. (b) Closed contour of integration Λ in the complex z-plane. The wavy line indicates a branch cut. The base function $Q(z)$, as well as $P(\varphi_0)$, is chosen to be positive on the real axis to the right of t_1 [τ_1]. (c) Open contour of integration $\Gamma(z)$ in the complex z-plane [φ_0-plane]. The dashed part of the contour $\Gamma(z)$ lies on a Riemann sheet adjacent to the complex z-plane under consideration.

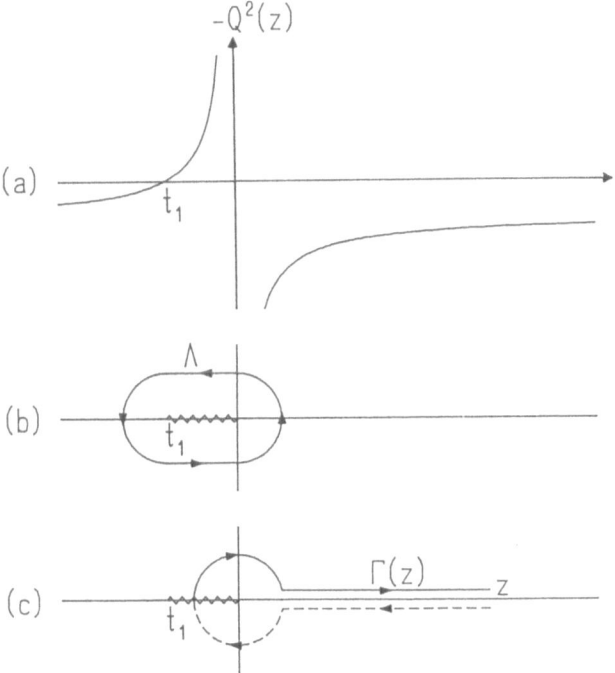

FIGURE 6.3. (a) Qualitative behavior of $-Q^2(z)$ for the case when $A_2 = +1$, $\xi_0 = 0$ and $\eta_0 < 0$. The zero of $Q^2(z)$ is denoted by t_1. (b) Closed contour of integration in the complex z-plane. The wavy line indicates a branch cut. The base function $Q(z)$, as well as $P(\varphi_0)$, is chosen to be positive on the positive real axis. (c) Open contour of integration $\Gamma(z)$ in the complex z-plane [φ_0-plane]. The dashed part of the contour $\Gamma(z)$ lies on a Riemann sheet adjacent to the z-plane under consideration.

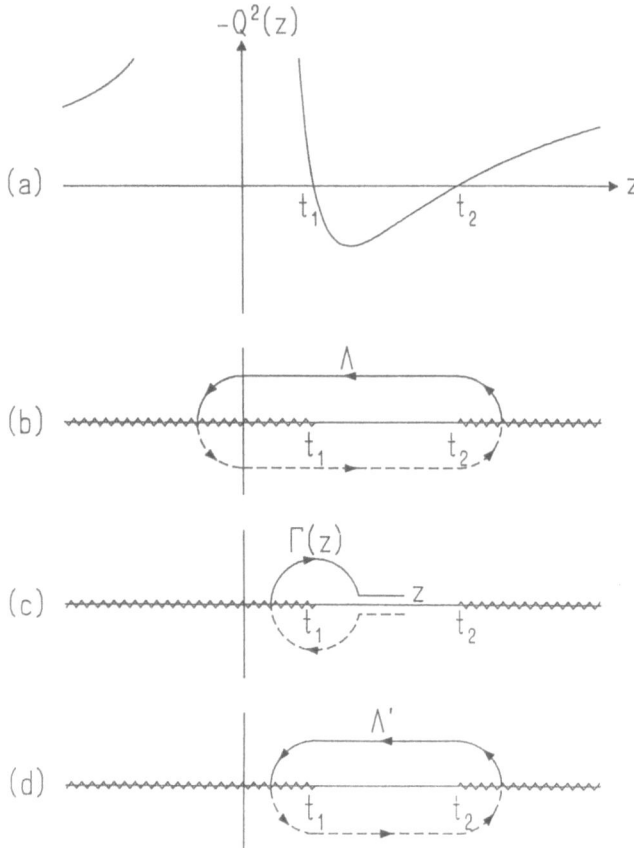

FIGURE 6.4. (a) Qualitative behavior of $-Q^2(z)$ for the case when $A_2 = -1$ and $\kappa_0 > \xi_0 > 0$. The zeros of $Q^2(z)$ are denoted by t_1 and t_2. (b) Closed contour of integration Λ in the complex z-plane. The wavy lines indicate branch cuts. The base function $Q(z)$, as well as $P(\varphi_0)$, is chosen to be positive on the real axis between $t_1[\tau_1]$ and $t_2[\tau_2]$. (c) Open contour of integration $\Gamma(z)$ in the complex z-plane [φ_0-plane]. The dashed part of the contour $\Gamma(z)$ lies on a Riemann sheet adjacent to the complex z-plane under consideration. (d) Closed contour of integration Λ' in the complex z-plane. The wavy lines indicate branch cuts.

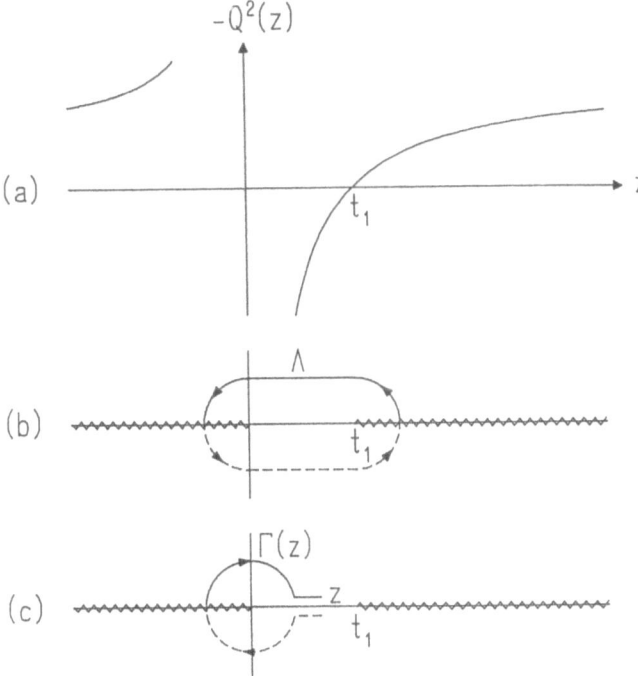

FIGURE 6.5. (a) Qualitative behavior of $-Q^2(z)$ for the case when $A_2 = -1$, $\xi_0 = 0$, and $\kappa_0 > 0$. The zero of $Q^2(z)$ is denoted by t_1. (b) Closed contour in the complex z-plane. The branch cuts are indicated by wavy lines. The base function $Q(z)$, as well as $P(\varphi_0)$, is chosen to be positive on the real axis between the origin and t_1 [τ_1]. (c) Open contour of integration $\Gamma(z)$ in the complex z-plane [φ_0-plane]. The dashed part of the contour $\Gamma(z)$ lies on a Riemann sheet adjacent to the complex z-plane under consideration.

and $\beta = 2$, respectively, and using (6.2.23), we obtain, in accordance with Eqs. (2.2.59a,b),

$$\int_\Lambda \frac{1}{2} h P(\varphi_0) \, d\varphi_0 = \int_\Lambda Y_2 Q(z) \, dz \qquad (6.2.35a)$$

and

$$\int_\Lambda \left(-\frac{1}{8} h^2 + \frac{S_4/T}{2P^2} \right) P(\varphi_0) \, d\varphi_0 = \int_\Lambda Y_4 Q(z) \, dz. \qquad (6.2.35b)$$

The integrals in the left-hand members of (6.2.35a,b) are easily evaluated when the contour Λ is deformed into a very large circle and when use is made of the formulas

$$\frac{1}{2} h P = \begin{cases} \frac{1}{2} A_{1,2}/\varphi_0 + \dots, & \varphi_0 \to \infty, \quad \text{when } A_2 = +1 \text{ (Figs. 6.1--6.3)} \\ \frac{1}{2} i A_{1,2}/\varphi_0 + \dots, & \varphi_0 \to \infty, \quad \text{when } A_2 = -1 \text{ (Figs. 6.4, 6.5),} \end{cases}$$
$$(6.2.36a)$$

and

$$\left(-\frac{1}{8} h^2 + \frac{S_4/T}{2P^2} \right) P =$$

$$\begin{cases} \frac{1}{2} A_{1,4}/\varphi_0 + \dots, & \varphi_0 \to \infty, \quad \text{when } A_2 = +1 \text{ (Figs. 6.1--6.3)} \\ \frac{1}{2} i A_{1,4}/\varphi_0 + \dots, & \varphi_0 \to \infty, \quad \text{when } A_2 = -1 \text{ (Figs. 6.4, 6.5).} \end{cases}$$
$$(6.2.36b)$$

Thus, one obtains

$$A_{1,2\beta} = \begin{cases} -\frac{i}{\pi} \int_\Lambda Y_{2\beta} Q(z) dz & \text{when } A_2 = +1 \text{ (Figs. 6.1--6.3)} \\ -\frac{1}{\pi} \int_\Lambda Y_{2\beta} Q(z) dz & \text{when } A_2 = -1 \text{ (Figs. 6.4, 6.5)} \end{cases} \qquad (6.2.37)$$

for $\beta = 1$ and $\beta = 2$. This formula is also valid for $\beta = 0$, according to (6.2.27); it should be valid for any nonnegative integer β. The direction of integration on the contour Λ in (6.2.37) is indicated in Figs. 6.1(b)--6.5(b).

6.3   Comparison Equation Corresponding to Scattering States

In this section we shall treat the situations corresponding to scattering states illustrated in Figures 6.1--6.3. Some of the formulas that we shall give here are, however, also relevant for the situations corresponding to bound states (Figures 6.4--6.5), and will therefore be referred to in §6.4.

6.3.1   Comparison Equation Solution

It is obviously the plus sign in (6.2.12) that is appropriate to use now. With the aid of (6.2.11) the comparison equation (6.2.7) along with (6.2.8) can then be written

$$\frac{d^2\Phi}{d\rho^2} + \left(1 - \frac{2\eta}{\rho} - \frac{L(L+1)}{\rho^2}\right)\Phi = 0 \tag{6.3.1}$$

with

$$\rho = \varphi/\lambda \tag{6.3.2}$$

$$\eta = -\frac{1}{2}A_1/\lambda \tag{6.3.3}$$

$$L = \left(\frac{\xi_0^2}{\lambda^2} + \left[\left(\ell + \frac{1}{2}\right)^2 - \xi_0^2\right]\right)^{\frac{1}{2}} - 1/2, \tag{6.3.4}$$

where the sign of the square root can be chosen such that $L + \frac{1}{2} \geq 0$; since we assume that $\xi_0 \geq 0$ and $\ell + \frac{1}{2} \geq 0$, and that λ is a "small" parameter ($\lambda \leq 1$). We note that the λ-dependence of L is quite different in the two cases $\xi_0 \neq 0$ and $\xi_0 = 0$. We also note that for $\lambda = 1$ we have, in both cases, $L = \ell$.

Since $L + \frac{1}{2}$ is nonnegative, the appropriate regular solution of the comparison equation (6.3.1) that for small values of ρ behaves as ρ^{L+1} times a power series in ρ (the constant term of which is different from zero); is proportional to the regular Coulomb wave function $F_L(\eta, \rho)$. From (6.2.6) and (6.3.2) we therefore obtain

$$\psi = A\left(\frac{d\rho}{dz}\right)^{-\frac{1}{2}} F_L(\eta, \rho), \tag{6.3.5}$$

where A is a normalization factor which is independent of ρ.

We define the $(2N+1)$th-order comparison equation solution (valid uniformly through the transition points) as the solution obtained by truncating the expansions of φ and A_1 [i.e., (6.2.13) and (6.2.14) with $n = 1$] at $\beta = N$, but retaining the expression (6.3.4) for L. By means of (6.3.2) and (6.3.3) we then obtain

$$\rho = \frac{1}{\lambda}\sum_{\beta=0}^{N}\varphi_{2\beta}\lambda^{2\beta} \tag{6.3.6}$$

and

$$\eta = \frac{1}{\lambda}\sum_{\beta=0}^{N}\eta_{2\beta}\lambda^{2\beta}, \tag{6.3.7}$$

where

$$\eta_{2\beta} = -\frac{1}{2}A_{1,2\beta}. \tag{6.3.8}$$

According to (6.2.37) along with $A_2 = +1$ and (6.3.8) we have

$$\eta_{2\beta} = \frac{i}{2\pi}\int_\Lambda Y_{2\beta}Q(z)\,dz, \quad \beta = 0, 1, 2, \ldots. \tag{6.3.9}$$

We recall that the functions $\varphi_{2\beta}$ are obtained from (6.2.24) and (6.2.28)–(6.2.33), where, according to (6.2.21), (6.2.22), and (6.3.8), we have to introduce

$$P(\varphi_0) = \left(1 - \frac{2\eta_0}{\varphi_0} - \frac{\xi_0^2}{\varphi_0^2}\right)^{\frac{1}{2}} \tag{6.3.10}$$

and

$$\frac{S_{2\beta}}{T} = -\frac{2\eta_{2\beta}}{\varphi_0} - \frac{\ell(\ell+1) - \xi_0^2}{\varphi_0^2}\delta_{\beta,1}, \quad \beta > 0. \tag{6.3.11}$$

6.3.2   Phase-Integral Approximation Obtained from the Comparison Equation Solution

When $\rho \to +\infty$, while L and η remain fixed, we have the well known asymptotic formula

$$F_L(\eta, \rho) \sim \sin[\rho - \eta \ln(2\rho) - L\pi/2 + \sigma_L], \tag{6.3.12}$$

where

$$\sigma_L = \frac{1}{2i} \ln \frac{\Gamma(L+1+i\eta)}{\Gamma(L+1-i\eta)} = \arg \Gamma(L+1+i\eta). \tag{6.3.13}$$

The phase-integral approximation representing $F_L(\eta, \rho)$, which agrees with (6.3.12) when $\rho = +\infty$ in any order of approximation, is given by

$$F_L(\eta, \rho) = q_c^{-\frac{1}{2}}(\rho) \sin\left(\frac{1}{2}\int_{\Gamma(\rho)} q_c(\rho)\, d\rho - \delta_c - L\pi/2 + \sigma_L\right), \tag{6.3.14}$$

where $\Gamma(\rho)$ is an open contour enclosing the positive zero of $P^2(\varphi_0)$ or the origin, corresponding to the contour $\Gamma(z)$ in the z-plane shown in Figures 6.1(c), 6.2(c), and 6.3(c); the constant phase δ_c is defined by

$$\delta_c = \lim_{\rho \to +\infty}\left(\frac{1}{2}\int_{\Gamma(\rho)} q_c(\rho)\, d\rho - \rho + \eta \ln(2\rho)\right). \tag{6.3.15}$$

The function $q_c(\rho)$, which appears in (6.3.14) as well as in (6.3.15), is given in the $(2N+1)$th-order approximation by

$$q_c(\rho) = \sum_{n=0}^{N} q_c^{(2n+1)}(\rho), \tag{6.3.16}$$

where the expressions for the first few functions in the right-hand member of (6.3.16), according to Chapter 1, are

$$q_c^{(1)}(\rho) = Q_c(\rho) \tag{6.3.17a}$$

$$q_c^{(3)}(\rho) = Q_c(\rho)\frac{1}{2}\varepsilon_{0c}(\rho) \tag{6.3.17b}$$

$$q_c^{(5)}(\rho) = Q_c(\rho)\left[-\frac{1}{8}\varepsilon_{0c}^2(\rho) - \frac{1}{8}\frac{1}{Q_c(\rho)}\frac{d}{d\rho}\left(\frac{1}{Q_c(\rho)}\frac{d}{d\rho}\varepsilon_{0c}(\rho)\right)\right] \tag{6.3.17c}$$

with

$$\varepsilon_{0c}(\rho) = Q_c^{-\frac{3}{2}}(\rho)\frac{d^2}{d\rho^2}Q_c^{-\frac{1}{2}}(\rho) + \frac{[1 - 2\eta/\rho - L(L+1)/\rho^2] - Q_c^2(\rho)}{Q_c^2(\rho)}. \tag{6.3.18}$$

For our present purposes the base function $Q_c(\rho)$ can be chosen to be

$$Q_c(\rho) = (1 - 2\eta/\rho - \Lambda^2/\rho^2)^{\frac{1}{2}}, \tag{6.3.19}$$

with the parameter Λ specified only with regard to its dependence on λ, which is assumed to be such that $\Lambda^2 = \xi_0^2/\lambda^2 +$ terms containing nonnegative powers of λ^2. The use of the notation Λ with two quite different meanings, $viz.$ for a contour of integration in Figures 6.1–6.5 and for a parameter in (6.3.19), should not cause confusion. The subscript c in (6.3.14)–(6.3.19) indicates that the quantities in question pertain to the comparison equation (6.3.1). Unlike the asymptotic formula (6.3.12), the phase-integral formula (6.3.14) with (6.3.15)–(6.3.19) is a good approximation for all values of ρ well to the right of the positive turning point (Figures 6.1 and 6.2) or the origin (Figure 6.3).

Inserting the phase-integral approximation (6.3.14) with (6.3.15) and (6.3.16) for the Coulomb wave function $F_L(\eta, \rho)$ into the comparison equation solution (6.3.5), we obtain

$$\psi = A\left(\sum_{n=0}^{N} q_c^{(2n+1)}(\rho)\frac{d\rho}{dz}\right)^{-\frac{1}{2}} \sin\left[\sum_{n=0}^{N}\left(\frac{1}{2}\int_{\Gamma(z)} q_c^{(2n+1)}(\rho)\frac{d\rho}{dz}dz - \delta_c^{(2n+1)}\right)\right.$$

$$\left. - L\pi/2 + \sigma_L\right], \quad \rho = \rho(z), \tag{6.3.20}$$

where

$$\delta_c^{(1)} = \lim_{\rho \to +\infty}\left(\frac{1}{2}\int_{\Gamma(\rho)} q_c^{(1)}(\rho)\,d\rho - \rho + \eta\ln(2\rho)\right) \tag{6.3.21a}$$

$$= \int_b^\infty \left(Q_c(\rho) - 1 + \frac{\eta}{\rho}\right)d\rho - b + \eta\ln(2b). \tag{6.3.21a'}$$

Here $b = \eta + (\eta^2 + \Lambda^2)^{\frac{1}{2}}$, which according to (6.3.19) is the positive zero of $Q_c^2(\rho)$, and

$$\delta_c^{(2n+1)} = \frac{1}{2}\int_{\Gamma(+\infty)} q_c^{(2n+1)}(\rho)\,d\rho, \quad n > 0, \tag{6.3.21b}$$

and $\Gamma(z)$ is an open contour encircling the positive turning point or the origin as shown in Figures 6.1(c), 6.2(c), and 6.3(c); the corresponding contour in the complex ρ-plane is denoted by $\Gamma(\rho)$. The formula (6.3.20) with (6.3.21a,b) is valid when z does not lie too close to the positive zero of $Q^2(z)$ or the origin.

The next step in our calculation is to expand the z-dependent function $\sum_{n=0}^{N} q_c^{(2n+1)}(\rho)d\rho/dz$ as well as the z-independent phase $\sum_{n=0}^{N} \delta_c^{(2n+1)}$, defined with the aid of (6.3.21a,b), in powers of λ in order to bring the solution (6.3.20) into the form of the phase-integral solution to the differential equation (6.2.4). This can be done in a concrete and straightforward way; however, the calculations are very tedious. Instead, we shall present an alternative, more elegant procedure for obtaining the same result. This procedure is simple in the sense that it requires a moderate amount of calculations, but it is based on much more delicate arguments; the calculations mentioned above are sketched in the Appendix.

By recalling the expansions (6.3.6) and (6.3.7) of $\rho(z)$ and η in powers of λ, the λ-dependence assumed for Λ^2 in (6.3.19), and the general structure of the expressions (6.3.17a,b,c), (6.3.18), and (6.3.19) for $q_c^{(2n+1)}(\rho)$; we realize that

$$\sum_{n=0}^{N} q_c^{(2n+1)}(\rho)\frac{d\rho}{dz} = \frac{1}{\lambda}\sum_{n=0}^{N} q^{(2n+1)}(z)\lambda^{2n} + O(\lambda^{2N+1}), \quad \rho = \rho(z), \quad (6.3.22)$$

where $q^{(2n+1)}(z)$ are certain λ-independent functions of z that, for the moment, need not be further specified. Similarly, by recalling once more the expansion of η in powers of λ, the λ-dependence of Λ^2, and the general structure of the functions $q_c^{(2n+1)}(\rho)$; we realize that the definitions (6.3.21a,b) give

$$\sum_{n=0}^{N} \delta_c^{(2n+1)} = \frac{1}{\lambda}\sum_{n=0}^{N} \delta^{(2n+1)}\lambda^{2n} + O(\lambda^{2N+1}) - \eta \ln \lambda, \quad (6.3.23)$$

where $\delta^{(2n+1)}$ are certain λ-independent phase constants that, for the moment, need not be further specified. The term $\eta \ln \lambda$ in (6.3.23) derives from the last term in (6.3.21a′), since $b = \eta + (\eta^2 + \Lambda^2)^{\frac{1}{2}}$ depends on λ in accordance with the expansion (6.3.7) of η and the expansion of Λ^2 mentioned in the text below Eq. (6.3.19). Inserting (6.3.22) and (6.3.23) into (6.3.20), and neglecting terms of the order $O(\lambda^{2N+1})$; we obtain

$$\psi = Aq^{-\frac{1}{2}}(z)\sin\left(\frac{1}{2}\int_{\Gamma(z)} q(z)\,dz + \Delta\right) \quad (6.3.24)$$

where

$$q(z) = \frac{1}{\lambda}\sum_{n=0}^{N} q^{(2n+1)}(z)\lambda^{2n} \quad (6.3.25)$$

and

$$\Delta = \eta \ln \lambda - L\pi/2 + \sigma_L - \frac{1}{\lambda} \sum_{n=0}^{N} \delta^{(2n+1)} \lambda^{2n}. \qquad (6.3.26)$$

We emphasize that in the definition (6.3.26) of Δ the quantities L and σ_L are functions of ℓ, ξ_0, η, and λ, given by the original definitions (6.3.4) and (6.3.13), respectively. In order to obtain an expression for η one uses the truncated series (6.3.7) with (6.3.9) which is thus to be inserted into (6.3.13).

Since the parameter λ appears in the same way in (6.3.24) with (6.3.25) as in the phase-integral solution of the differential equation (6.2.4); the functions $q^{(2n+1)}(z)$ must, according to Chapter 1, be given by the phase-integral formula

$$q^{(2n+1)}(z) = Y_{2n}Q(z), \qquad (6.3.27)$$

where $Q(z)$ is the base function which is chosen for solving the original differential equation (6.2.1) and which appears squared in the auxiliary differential equation (6.2.4), and $Y_0 = 1$, whereas Y_2 and Y_4 are given by (6.2.30a,b).

The calculations required for determining the phases $\delta^{(2n+1)}$ can be greatly simplified if one uses the following knowledge, which is available in the phase-integral method, concerning the λ-dependence of the phase Δ in (6.3.24) when λ approaches zero. When we permit $\lambda \to 0$, we realize that the absolute value of the integral $\int_{\frac{1}{2}t_1}^{t_1} [Q(z)/\lambda]dz$, associated with the first-order phase-integral solution of the auxiliary differential equation (6.2.4) that we are now considering, approaches infinity and that, consequently, the previously discovered phase-integral connection formulas in [6, 8] become applicable. Thus, for the examples in Figures 6.1 and 6.2 we have

$$\Delta = \pi/4 + O(\lambda^{2N+1}) \quad \text{as} \quad \lambda \to 0, \quad \text{Figures 6.1 and 6.2}, \qquad (6.3.28a)$$

according to one of the usual phase-integral connection formulas for a single isolated turning point [6, 8]. Similarly, for the case in Figure 6.3, when $\eta_0 \neq 0$ and, hence, $|\eta| \to \infty$ as $\lambda \to 0$; we have according to formula (7.22) in [8], generalized to the phase-integral approximation of arbitrary order,

$$\Delta = -(L + 1/4)\pi + O(\lambda^{2N+1}) \quad \text{as} \quad \lambda \to 0, \quad \text{Figure 6.3}. \qquad (6.3.28b)$$

Although we are concerned with situations, where it is *not* justified to simply neglect the O-symbols in (6.3.28a,b), when λ is made equal to unity, these formulas, combined with the fact that the quantities $\delta^{(2n+1)}$ are independent of λ, can be used in the derivation of expressions for $\delta^{(2n+1)}$.

To calculate the quantities $\delta^{(2n+1)}$, we let $\lambda \to 0$ so that (6.3.28a,b) can be used. Inserting (6.3.28a,b) into (6.3.26), we thus obtain

$$\mp \left(L + \frac{1}{2} \right) \pi/2 + \sigma_L + \eta \ln \lambda = \frac{1}{\lambda} \sum_{n=0}^{N} \delta^{(2n+1)} \lambda^{2n} + O(\lambda^{2N+1}), \qquad (6.3.29)$$

where the upper sign pertains to Figures 6.1 and 6.2, whereas the lower sign pertains to Figure 6.3. Hence, because of the fact that the phases $\delta^{(2n+1)}$ are independent of λ, we obtain the explicit expressions for $\delta^{(2n+1)}$ simply by expanding the left-hand member of (6.3.29) in powers of λ. To make this expansion we note that for $L + \frac{1}{2} \geq 0$, η real, and large values of $|L + \frac{1}{2} + i\eta|$ we have, according to Eq. (5) on p. 32 and Eq. (4) on p. 20 in [16], the asymptotic expansion

$$\ln \Gamma(L+1\pm i\eta) = \left(L + \frac{1}{2} \pm i\eta\right) \ln \left(L + \frac{1}{2} \pm i\eta\right) - \left(L + \frac{1}{2} \pm i\eta\right) + \ln(2\pi)^{\frac{1}{2}}$$

$$- \frac{1}{24\left(L + \frac{1}{2} \pm i\eta\right)} + \frac{7}{2880\left(L + \frac{1}{2} \pm i\eta\right)^3} + \cdots . \tag{6.3.30}$$

Inserting (6.3.30) into (6.3.13), when L and η are real and at least one of the quantities $L + \frac{1}{2}$ and $|\eta|$ is large, we obtain

$$\sigma_L = \left(L + \frac{1}{2}\right) \arctan \left[\eta / \left(L + \frac{1}{2}\right)\right] + \frac{1}{2}\eta \ln \left[\left(L + \frac{1}{2}\right)^2 + \eta^2\right] - \eta$$

$$+ \frac{\eta}{24\left[\left(L + \frac{1}{2}\right)^2 + \eta^2\right]} - \frac{7\eta}{960\left[\left(L + \frac{1}{2}\right)^2 + \eta^2\right]^2} + \frac{7\eta^3}{720\left[\left(L + \frac{1}{2}\right)^2 + \eta^2\right]^3} + \cdots . \tag{6.3.31}$$

We shall now treat the cases $\xi_0 > 0$ (Figure 6.1) and $\xi_0 = 0$, but $\eta_0 \neq 0$, (Figures 6.2 and 6.3) individually.

The Case When $\xi_0 > 0$ (Figure 6.1). For any sufficiently small value of λ we obtain from (6.3.4)

$$L + \frac{1}{2} = \frac{1}{\lambda} \sum_{\beta=0}^{\infty} \xi_{2\beta} \lambda^{2\beta}, \quad \text{when } \xi_0 \neq 0, \tag{6.3.32}$$

where

$$\xi_{2\beta} = \xi_0 \binom{\frac{1}{2}}{\beta} \left(\frac{(\ell + \frac{1}{2})^2 - \xi_0^2}{\xi_0^2}\right)^{\beta}, \quad \beta \geq 0; \tag{6.3.33}$$

hence,

$$\xi_2 = \frac{(\ell + \frac{1}{2})^2 - \xi_0^2}{2\xi_0} \tag{6.3.34a}$$

$$\xi_4 = -\frac{\left[(\ell + \frac{1}{2})^2 - \xi_0^2\right]^2}{8\xi_0^3}. \tag{6.3.34b}$$

Inserting the expansion (6.3.31) of σ_L into (6.3.29) with the upper sign, and using the expansions (6.3.7) and (6.3.32) of η and $L + \frac{1}{2}$, respectively, we obtain

$$\delta^{(1)} = \xi_0 \left(\arctan\frac{\eta_0}{\xi_0} - \frac{\pi}{2} \right) + \eta_0 \ln(\xi_0^2 + \eta_0^2)^{\frac{1}{2}} - \eta_0, \qquad (6.3.35a)$$

$$\delta^{(3)} = \xi_2 \left(\arctan\frac{\eta_0}{\xi_0} - \frac{\pi}{2} \right) + \eta_2 \ln(\xi_0^2 + \eta_0^2)^{\frac{1}{2}} + \frac{\eta_0}{24(\xi_0^2 + \eta_0^2)}, \qquad (6.3.35b)$$

$$\delta^{(5)} = \xi_4 \left(\arctan\frac{\eta_0}{\xi_0} - \frac{\pi}{2} \right) + \eta_4 \ln(\xi_0^2 + \eta_0^2)^{\frac{1}{2}}$$

$$+ \frac{\eta_2 + 12(\xi_0\eta_2 - \eta_0\xi_2)\xi_2 + 12(\xi_0\xi_2 + \eta_0\eta_2)\eta_2}{24(\xi_0^2 + \eta_0^2)}$$

$$- \frac{7\eta_0 + 80\eta_0(\xi_0\xi_2 + \eta_0\eta_2)}{960(\xi_0^2 + \eta_0^2)^2} + \frac{7\eta_0^3}{720(\xi_0^2 + \eta_0^2)^3} \qquad (6.3.35c)$$

Now that we have determined the λ-independent phases $\delta^{(2n+1)}$ we will suspend our assumption that $\lambda \to 0$. Let us recall that σ_L is given by (6.3.13) and that L is given by (6.3.4) and is therefore equal to ℓ when $\lambda = 1$. Inserting (6.3.35a–c) into (6.3.26) and letting $\lambda = 1$; we obtain the following final formula for Δ, pertaining to the solution of the original differential equation (6.2.1):

$$\Delta = -\ell\pi/2 + \arg \Gamma(\ell + 1 + i\eta) - \xi \left(\arctan\frac{\eta_0}{\xi_0} - \frac{\pi}{2} \right)$$

$$- \eta \ln(\xi_0^2 + \eta_0^2)^{\frac{1}{2}} + \sum_{n=0}^{N} \Delta^{(2n+1)} \qquad (6.3.36)$$

where

$$\Delta^{(1)} = \eta_0 \qquad (6.3.37a)$$

$$\Delta^{(3)} = -\frac{\eta_0}{24(\xi_0^2 + \eta_0^2)} \qquad (6.3.37b)$$

$$\Delta^{(5)} = -\frac{\eta_2 + 12(\xi_0\eta_2 - \eta_0\xi_2)\xi_2 + 12(\xi_0\xi_2 + \eta_0\eta_2)\eta_2}{24(\xi_0^2 + \eta_0^2)}$$

$$+ \frac{7\eta_0 + 80\eta_0(\xi_0\xi_2 + \eta_0\eta_2)}{960(\xi_0^2 + \eta_0^2)^2} - \frac{7\eta_0^3}{720(\xi_0^2 + \eta_0^2)^3} \qquad (6.3.37c)$$

and

$$\xi = \sum_{\beta=0}^{N} \xi_{2\beta} \qquad (6.3.38)$$

$$\eta = \sum_{\beta=0}^{N} \eta_{2\beta}. \qquad (6.3.39)$$

For the sake of completeness we mention that, in the case when $\xi_0 > 0$, the contour $\Gamma(z)$ in (6.3.24) [see Figure 6.1(c)] can be replaced by the contour $\overline{\Gamma}(z)$ also enclosing the origin as shown in Figure 6.1(d); provided that $\Delta^{(2n+1)}$ in (6.3.36) is simultaneously replaced by $\overline{\Delta}^{(2n+1)}$, which is defined as follows

$$\overline{\Delta}^{(2n+1)} = \Delta^{(2n+1)} + i\pi \mathrm{Res}_{z=0}[Y_{2n}Q(z)]. \tag{6.3.40}$$

Using some general properties of the functions Y_{2n} given in [2], one can prove the formula

$$\mathrm{Res}_{z=0}[Y_{2n}Q(z)] = i\xi_{2n}, \tag{6.3.41}$$

which for $n = 0, 1$, and 2 may be also obtained from Eqs. (32a–c) in [11], where α is the same as our ξ_0. Inserting (6.3.41) into (6.3.40), we obtain

$$\overline{\Delta}^{(2n+1)} = \Delta^{(2n+1)} - \pi\xi_{2n}. \tag{6.3.42}$$

The Case When $\xi_0 = 0$ But $\eta_0 \neq 0$ (Figures 6.2 and 6.3). According to (6.3.4) we have

$$L = \ell. \tag{6.3.43}$$

Inserting the expansion (6.3.31) of σ_L into (6.3.29), using (6.3.43) and the expansion (6.3.7) for η, and noting that the upper and lower signs in (6.3.29) correspond to Figure 6.2 ($\eta_0 > 0$) and Figure 6.3 ($\eta_0 < 0$), respectively; we realize that the double sign in (6.3.29) disappears, and we obtain

$$\delta^{(1)} = \eta_0 \ln|\eta_0| - \eta_0 \tag{6.3.44a}$$

$$\delta^{(3)} = \eta_2 \ln|\eta_0| - \frac{3(2\ell+1)^2 - 1}{24\eta_0} \tag{6.3.44b}$$

$$\delta^{(5)} = \eta_4 \ln|\eta_0| + \frac{\eta_2^2}{2\eta_0} + \frac{[3(2\ell+1)^2 - 1]\eta_2}{24\eta_0^2}$$
$$+ \frac{15(2\ell+1)^4 - 30(2\ell+1)^2 + 7}{2880\eta_0^3}. \tag{6.3.44c}$$

Inserting (6.3.13), (6.3.43), and (6.3.44a–c) into (6.3.26) and letting $\lambda = 1$, we obtain the final expression for Δ, which pertains to the solution of the original differential equation (6.2.1):

$$\Delta = -\ell\pi/2 + \arg\Gamma(\ell + 1 + i\eta) - \eta \ln|\eta_0| + \sum_{n=0}^{N} \Delta^{(2n+1)}, \tag{6.3.45}$$

where

$$\Delta^{(1)} = \eta_0 \tag{6.3.46a}$$

$$\Delta^{(3)} = \frac{3(2\ell+1)^2 - 1}{24\eta_0} \tag{6.3.46b}$$

$$\Delta^{(5)} = -\frac{\eta_2^2}{2\eta_0} - \frac{[3(2\ell+1)^2 - 1]\eta_2}{24\eta_0^2} - \frac{15(2\ell+1)^4 - 30(2\ell+1)^2 + 7}{2880\eta_0^3}$$

$$(6.3.46c)$$

and η is given by (6.3.39).

6.3.3 Behavior of the Wave Function Close to the Origin

From the formulas 14.1.3–14.1.7 in [1] it follows that

$$\lim_{\rho \to 0} F_L(\eta, \rho)/\rho^{L+1} = C_L(\eta) \qquad (6.3.47)$$

where

$$C_L(\eta) = \frac{2^L \exp(-\eta\pi/2)|\Gamma(L+1+i\eta)|}{\Gamma(2L+2)}. \qquad (6.3.48)$$

From (6.3.5) and (6.3.47) and the fact that $d\rho/dz \to \rho/z$ as $z \to 0$, we obtain

$$\lim_{z \to 0} \frac{\psi}{z^{L+1}} = A C_L(\eta) \left(\lim_{z \to 0} \frac{\rho}{z}\right)^{L+\frac{1}{2}}. \qquad (6.3.49)$$

We obtain the $(2N+1)$th-order approximation of (6.3.49) with (6.3.48) by introducing there the truncated expansions (6.3.6) and (6.3.7) for ρ and η, respectively, and letting $\lambda = 1$.

6.3.4 Summary of Formulas in §6.3

As a result of the calculations in this section we found that the regular solution of the original differential equation (6.2.1) with (6.2.2) well to the right of the positive turning point (Figures 6.1 and 6.2) or the origin (Figure 6.3) can be represented by the phase-integral formula (6.3.24) with $\lambda = 1$; that is,

$$\psi = Aq^{-\frac{1}{2}}(z) \sin\left(\frac{1}{2}\int_{\Gamma(z)} q(z)\,dz + \Delta\right), \qquad (6.3.50)$$

where A is an arbitrary constant factor, $q(z)$ is given by (6.3.25) with $\lambda = 1$ and (6.3.27), and Δ is given by (6.3.36) with (6.3.37a–c), (6.3.38), (6.3.39), (6.2.5), and (6.3.34a,b), and (6.3.9) when $\xi_0 > 0$ (Figure 6.1); and by (6.3.45) with (6.3.46a–c), (6.3.39) and (6.3.9) when $\xi_0 = 0$ but $\eta_0 \neq 0$ (Figures 6.2 and 6.3) (see also [6.2.5]).

One can easily check that the formulas for Δ that result from our present comparison equation treatment are consistent with previously obtained formulas that are valid when the relevant turning point recedes from the pole. In fact, from the expressions for Δ referred to above, and the asymptotic formula for the Γ-function, we find that

$$\Delta \to \pi/4 \quad \text{when} \quad \xi_0 \to +\infty \text{ (Figure 6.1)} \qquad (6.3.51)$$

while η_0 is kept fixed, and that

$$\Delta \to \begin{cases} \pi/4 & \text{when } \eta_0 \to +\infty \text{ (Figs. 6.1 and 6.2)} \\ [1/4 + \xi_0 - (\ell + 1/2)]\pi & \text{when } \eta_0 \to -\infty \text{ (Figs. 6.1 and 6.3)} \end{cases}$$

(6.3.52)

while ξ_0 is fixed and $\eta - \eta_0$ remains finite. When $\xi_0 \to +\infty$ (Figure 6.1) the barrier immediately to the right of the origin becomes very thick, and the usual connection formula corresponding to $\Delta = \pi/4$ becomes applicable, in agreement with (6.3.51). The same situation occurs when $\eta_0 \to +\infty$ both when $\xi_0 > 0$ (Figure 6.1) and $\xi_0 = 0$ (Figure 6.2), which is in agreement with (6.3.52). When $\xi_0 > 0$ and $\eta_0 \to -\infty$, the zero τ_1 approaches the origin; according to (6.3.52), only the choice $\xi_0 = \ell + \frac{1}{2}$, which makes the wave function good at the origin, ensures that the usual connection formula corresponding to $\Delta = \pi/4$ is valid. When $\xi_0 = 0$ and $\eta_0 \to -\infty$ (Figure 6.3) we have $\Delta = -(\ell + 1/4)$ according to (6.3.52), which, as expected, is in agreement with formulas (7.22) and (7.28) in [8], since the conditions for the validity of those formulas are fulfilled.

6.4   Comparison Equation Corresponding to Bound States

In this section we shall treat the situation shown in Figures 6.4–6.5. Thus, we choose the minus sign in (6.2.12). With the aid of (6.2.11) the comparison equation (6.2.7) with (6.2.8) can then be written

$$\frac{d^2\Phi}{d\rho^2} + \left(-1 + \frac{2\kappa}{\rho} - \frac{L(L+1)}{\rho^2}\right)\Phi = 0 \qquad (6.4.1)$$

with ρ defined by (6.3.2) and L defined by (6.3.4), whereas κ is defined by

$$\kappa = \frac{1}{2}A_1/\lambda. \qquad (6.4.2)$$

We have already mentioned that some of the formulas in §6.3 can also be used in the present section. In regard to the sign and λ-dependence of $L + \frac{1}{2}$ we refer to the text below Eq. (6.3.4).

6.4.1   Quantization Condition

The differential equation (6.4.1) can have solutions corresponding to bound states only if κ is positive. According to 13.1.31, 13.1.32, and 13.1.33 in [1] two solutions which vanish at $\rho = 0$ and $\rho = +\infty$, respectively, for every positive value of κ are given by

$$\Phi_1(\rho) = A\lambda^{\frac{1}{2}}M_{\kappa,L+\frac{1}{2}}(2\rho), \qquad (6.4.3a)$$

$$\Phi_2(\rho) = A\lambda^{\frac{1}{2}} \exp[(\kappa - L - 1)\pi i] \frac{\Gamma(2L+2)}{\Gamma(\kappa+L+1)} W_{\kappa, L+\frac{1}{2}}(2\rho), \qquad (6.4.3b)$$

respectively; where A is an unspecified constant normalization factor. The other ρ-independent factor in front of $W_{\kappa, L+\frac{1}{2}}(2\rho)$ in (6.4.3b) has been introduced in order to fulfill (6.4.5) when κ is given by (6.4.4). This factor is finite and different from zero, since κ and $L + \frac{1}{2}$ are assumed to be nonnegative. The functions $\Phi_1(\rho)$ and $\Phi_2(\rho)$ are linearly independent, unless κ satisfies the quantization condition

$$\kappa = L + 1 + k, \quad k = 0, 1, 2, \ldots, \qquad (6.4.4)$$

in which case formula 13.1.34 in [1] yields

$$\Phi_1(\rho) = \Phi_2(\rho). \qquad (6.4.5)$$

In the $(2N+1)$th-order approximation we use for ρ the truncated series (6.3.6), where the quantities $\varphi_{2\beta}$ are obtained from (6.2.24) and (6.2.28)–(6.2.33), and for κ, according to (6.2.14) with $n = 1$ and (6.4.2), we use the truncated series

$$\kappa = \frac{1}{\lambda} \sum_{\beta=0}^{N} \kappa_{2\beta} \lambda^{2\beta}, \qquad (6.4.6)$$

where

$$\kappa_{2\beta} = \frac{1}{2} A_{1,2\beta}. \qquad (6.4.7)$$

The quantities $\kappa_{2\beta}$ are obtained from (6.4.7) and (6.2.37) with $A_2 = -1$ as

$$\kappa_{2\beta} = -\frac{1}{2\pi} \int_\Lambda Y_{2\beta} Q(z)\, dz, \qquad (6.4.8)$$

where Λ is the contour of integration shown in Figure 6.4(b) or Figure 6.5(b). Inserting (6.4.8) into (6.4.6) and using (6.3.25) with (6.3.27), we obtain

$$\kappa = \frac{1}{2\pi} \int_{-\Lambda} q(z)\, dz, \qquad (6.4.9)$$

where $-\Lambda$ is the same contour as Λ but with the opposite direction of integration. With the aid of (6.4.9) we can rewrite the quantization condition (6.4.4) as

$$\frac{1}{2} \int_{-\Lambda} q(z)\, dz = (L + 1 + k)\pi, \quad k = 0, 1, 2, \ldots. \qquad (6.4.10)$$

Using (6.4.7), we obtain from (6.2.21)

$$P(\varphi_0) = \left(-1 + \frac{2\kappa_0}{\varphi_0} - \frac{\xi_0^2}{\varphi_0^2}\right)^{\frac{1}{2}} \qquad (6.4.11)$$

and from (6.2.22)

$$\frac{S_{2\beta}}{T} = \frac{2\kappa_{2\beta}}{\varphi_0} - \frac{\ell(\ell+1) - \xi_0^2}{\varphi_0^2}\delta_{\beta,1}, \quad \beta > 0. \qquad (6.4.12)$$

The Case When $\xi_0 > 0$ (Figure 6.4). With the aid of residue calculus we obtain

$$\frac{1}{2\pi}\int_{-\Lambda} q(z)\,dz = \frac{1}{2\pi}\int_{-\Lambda'} q(z)\,dz - i\,\mathrm{Res}_{z=0}q(z), \qquad (6.4.13)$$

where $-\Lambda'$ is a contour which encloses in the negative sense only the two turning points t_1 and t_2, but not the origin. The same contour with the opposite direction of integration (which we denote by Λ') is shown in Figure 6.4(d). From (6.3.25), (6.3.27), and (6.3.41) it follows that, in the $(2N + 1)$th-order approximation,

$$\mathrm{Res}_{z=0}q(z) = i\sum_{n=0}^{N}\xi_{2n}\lambda^{2n-1}. \qquad (6.4.14)$$

Inserting (6.4.13) with (6.4.14) into (6.4.10) and letting $\lambda = 1$ (which yields $L = \ell$ according to [6.3.4]), we obtain the quantization condition

$$\frac{1}{2}\int_{-\Lambda'} q(z)\,dz = \left[\left(\ell + \frac{1}{2} - \xi\right) + \left(k + \frac{1}{2}\right)\right]\pi, \quad k = 0, 1, 2, \ldots, \qquad (6.4.15)$$

where ξ is given by the truncated series (6.3.38) with (6.3.33).

One can also derive the quantization condition (6.4.15) without comparison equation technique by taking the origin into account, as is done in the derivation of Eqs. (33) and (34a,b,c) in [11], where α is the same as our ξ_0. If we can let the order $2N+1$ of approximation approach infinity, which is possible in the special case of the hydrogen-like ion treated in [11]; then it becomes obvious from (6.3.4) with $\lambda = 1$, (6.3.32) with $\lambda = 1$, and (6.3.38) with $N = \infty$, that $\ell + \frac{1}{2} - \xi = 0$ independently of the value of ξ_0 ($\neq 0$). Thus, for this special case the quantization condition (6.4.15) takes the form of an exact half-integer quantization condition with $q(z)$ pertaining to the infinite-order phase-integral approximation. In the general case, when the order of approximation used cannot be increased beyond a certain optimal order, ξ_0^2 must fulfill the requirement of not differing too much from $\ell(\ell+1)$. Let us therefore assume that ξ_0 and ℓ approach $+\infty$, while $\left(\ell + \frac{1}{2}\right)^2 - \xi_0^2$ is a finite constant. This assumption implies that $\ell + \frac{1}{2} - \xi_0 \to 0$ and, according to (6.3.33), that $\xi_{2\beta} \to 0$ for $\beta > 0$. Thus, recalling the definition (6.3.38), we conclude that $\ell + \frac{1}{2} - \xi \to 0$ as $\ell \to +\infty$. In the limit of very large values of ℓ the quantization condition (6.4.15) therefore goes over into the quantization condition given by Eqs. (19) and (20a) in [7], which is the arbitrary-order half-integer condition, which pertains to the situation when no zeros or singularities of the square of the base function (apart from the two zeros delimiting the classically allowed region) are taken into account. This is to be expected since, when ℓ becomes very large, the barrier becomes very thick, so that the presence of the pole at the origin can be disregarded. Let us next consider the important special case when $\xi_0 = \ell + \frac{1}{2}$. According to (6.3.33) we then have $\xi_{2\beta} = 0$ for $\beta > 0$ and hence $\xi = \xi_0 = \ell + \frac{1}{2}$, which means that (6.4.15) becomes the same as the quantization condition (19) with (20a) in [7] for all orders of approximation and for all values of ℓ. With regard to the special case of a hydrogen-like ion mentioned above, the quantization condition already becomes exact in the first-order approximation, since (when $\xi_0 = \ell + \frac{1}{2}$) the higher-order terms in $q(z)$ yield no contribution to the integral in (6.4.15).

The Case When $\xi_0 = 0$ But $\eta_0 \neq 0$ (Figure 6.5). We have $L = \ell$ regardless of the value of λ; see (6.3.4). Formula (6.4.10) then gives

$$\frac{1}{2}\int_\Lambda q(z)\,dz = (\ell + 1 + k)\pi, \quad k = 0, 1, 2, \dots . \tag{6.4.16}$$

This is in agreement with the quantization condition given by Eqs. (19) and (20b) in [7], although the derivation of that condition is unnecessarily specialized, in that the coefficients in the right-hand member of Eq. (18) in [7] are the same as those occurring in Eq. (17) in [7].

We have thus demonstrated that the application in the present section of the comparison equation technique only reproduces already determined quantization conditions that can be derived without the use of that technique [7, 11].

6.4.2 Normalized Wave Function

We shall now show how the $(2N+1)$th-order bound-state comparison equation wave function can be normalized by means of a formula devised by Yngve [17]. Replacing $\Phi(\varphi)$ in (6.2.6) by $\Phi_1(\rho)$ and $\Phi_2(\rho)$, respectively; we obtain, with the aid of (6.4.3a,b) and (6.3.2), the comparison equation solutions

$$\psi_1(\kappa, z) = \left(\frac{d\varphi}{dz}\right)^{-\frac{1}{2}} \Phi_1(\rho) = A \left(\frac{d\rho}{dz}\right)^{-\frac{1}{2}} M_{\kappa, L+\frac{1}{2}}(2\rho) \tag{6.4.17a}$$

and

$$\psi_2(\kappa, z) = \left(\frac{d\varphi}{dz}\right)^{-\frac{1}{2}} \Phi_2(\rho) = A \exp[(\kappa - L - 1)\pi i]\frac{\Gamma(2L+2)}{\Gamma(\kappa + L + 1)}$$

$$\times \left(\frac{d\rho}{dz}\right)^{-\frac{1}{2}} W_{\kappa, L+\frac{1}{2}}(2\rho), \tag{6.4.17b}$$

which approach zero as $z \to 0$ and $z \to +\infty$, respectively. The functions $\psi_1(\kappa, z)$ and $\psi_2(\kappa, z)$ are linearly independent unless the quantization condition (6.4.4) is fulfilled, in which case (6.4.5) yields

$$\psi_1(\kappa, z) = \psi_2(\kappa, z), \quad \kappa = L + 1 + k. \tag{6.4.18}$$

Writing the function in (6.4.18) as $\psi(z)$, we have according to Eq. (6) in [17] and (6.2.2)

$$\int_0^\infty [\psi(z)]^2 dz = \frac{\hbar^2}{2m} \left[\frac{\partial}{\partial E} \left(\psi_1 \frac{\partial \psi_2}{\partial z} - \psi_2 \frac{\partial \psi_1}{\partial z} \right) \right]_{\kappa = L+1+k} \tag{6.4.19}$$

From (6.4.17a,b) we obtain

$$\psi_1 \frac{\partial \psi_2}{\partial z} - \psi_2 \frac{\partial \psi_1}{\partial z} = 2A^2 \exp[(\kappa - L - 1)\pi i] \frac{\Gamma(2L+2)}{\Gamma(\kappa + L + 1)}$$

$$\times \left(M_{\kappa, L+\frac{1}{2}}(2\rho) W'_{\kappa, L+\frac{1}{2}}(2\rho) - W_{\kappa, L+\frac{1}{2}}(2\rho) M'_{\kappa, L+\frac{1}{2}}(2\rho) \right), \tag{6.4.20}$$

where the prime mark denotes differentiation with respect to 2ρ. Using the formulas 13.1.32, 13.1.33, 13.1.12, 13.1.16, and 13.1.22 in [1], we obtain from (6.4.20)

$$\psi_1 \frac{\partial \psi_2}{\partial z} - \psi_2 \frac{\partial \psi_1}{\partial z} = 2A^2 \exp[(\kappa - L)\pi i] \frac{[\Gamma(2L+2)]^2}{\Gamma(L+1+\kappa)\Gamma(L+1-\kappa)} \tag{6.4.21}$$

and hence, by means of Eq. (13) on p. 13 in [16],

$$\left[\frac{\partial}{\partial \kappa} \left(\psi_1 \frac{\partial \psi_2}{\partial z} - \psi_2 \frac{\partial \psi_1}{\partial z} \right) \right]_{\kappa = L+1+k} = \frac{2A^2 k! [\Gamma(2L+2)]^2}{\Gamma(2L+2+k)}. \tag{6.4.22}$$

Assuming the function $R(z) - Q^2(z)$ in (6.2.4) to be independent of the energy E, and letting $\lambda = 1$ in the expression (6.3.25) for $q(z)$, we obtain from (6.4.9), (6.2.2), and Eqs. (20) and (22a–c) in [14] the formula

$$\frac{\partial \kappa}{\partial E} = \frac{m}{2\pi \hbar^2} \int_{-\Lambda} \sum_{\beta=0}^N C_{2\beta} \frac{dz}{Q(z)} \tag{6.4.23}$$

where

$$C_0 = 1, \tag{6.4.24a}$$
$$C_2 = -Y_2 \tag{6.4.24b}$$
$$C_4 = -(Y_4 - Y_2^2). \tag{6.4.24c}$$

Calculating the energy derivative in (6.4.19) with the aid of (6.4.22) and (6.4.23), and requiring that the normalization integral be equal to unity, we obtain

$$A = \left(\frac{k! [\Gamma(2L+2)]^2}{2\pi \Gamma(2L+2+k)} \int_{-\Lambda} \sum_{\beta=0}^N C_{2\beta} \frac{dz}{Q(z)} \right)^{-\frac{1}{2}}. \tag{6.4.25}$$

It is sometimes convenient to rewrite (6.4.25) by means of the approximate relation (25) in [14]; that is,

$$\sum_{\beta=0}^{N} C_{2\beta} \frac{1}{Q(z)} = \frac{1}{q(z)},$$

(6.4.26)

where $q(z)$ is given by (6.3.25), with $\lambda = 1$, and (6.3.27).

6.4.3   The Behavior of the Wave Function Close to the Origin

According to formulas 13.1.32 and 13.1.2 in [1] the function $M_{\kappa, L+\frac{1}{2}}(2\rho)/\rho^{L+1}$ tends to 2^{L+1} as $\rho \to 0$, and therefore we obtain from (6.4.17a)

$$\lim_{z \to 0} \frac{\psi_1(\kappa, z)}{z^{L+1}} = A 2^{L+1} \left(\lim_{z \to 0} \frac{\rho}{z} \right)^{L+\frac{1}{2}}$$

(6.4.27)

We obtain the $(2N+1)$th-order approximation of (6.4.27) by introducing there the truncated expansion (6.3.6) for ρ and putting $\lambda = 1$.

Appendix: Calculation of $q(z)$ and $\delta^{(2n+1)}$

The procedure used in §6.3.2 for obtaining $q(z)$ and the phases $\delta^{(2n+1)}$ involves a minimum of calculations, but it is also instructive to present the following alternative, concrete, and quite straightforward procedure that yields $q(z)$ and $\delta^{(2n+1)}$ from (6.3.17a–c), (6.3.18), (6.3.20), and (6.3.21a,b) by direct, but tedious, calculations. For this purpose we choose the base function $Q_c(\rho)$ to be

$$Q_c(\rho) = \left(1 - \frac{2\eta}{\rho} - \frac{(L+\frac{1}{2})^2}{\rho^2} \right)^{\frac{1}{2}},$$

(6.A.1)

which implies that the functions $q_c^{(3)}(\rho)/Q_c(\rho)$ and $q_c^{(5)}(\rho)/Q_c(\rho)$ approach zero as $\rho \to 0$. The functions $q_c^{(2n+1)}(\rho)$, given by (6.3.17a–c) with (6.3.18) and (6.A.1), depend on λ, since ρ, η and L depend on λ according to (6.3.6), (6.3.7), and (6.3.4); and the quantities $\delta_c^{(2n+1)}$, given by (6.3.21a,b), depend on λ, since η and L depend on λ according to (6.3.7) and (6.3.4). Using the expression (6.A.1) for the base function $Q_c(\rho)$ we shall now show that by straightforward but tedious expansion in powers of λ it is possible to rewrite (6.3.20) into the form (6.3.24), with $q(z)$ given by (6.3.25) and (6.3.27) and Δ given by (6.3.26); where the λ-independent quantities $\delta^{(2n+1)}$ are given by (6.3.35a–c) when $\xi_0 > 0$, and by (6.3.44a–c) when $\xi_0 = 0$, but $\eta_0 \neq 0$.

Recalling that $\rho = \varphi/\lambda$ according to (6.3.2) and that $d\varphi_0/dz = Q(z)/P(\varphi_0)$ according to (6.2.23), we can rewrite the decisive function appearing in

(6.3.20) as follows

$$\sum_{n=0}^{N} q_c^{(2n+1)}(\rho)\frac{d\rho}{dz} = \frac{Q(z)}{\lambda}\sum_{n=0}^{N}\frac{1}{P(\varphi_0)}\frac{d\varphi}{d\varphi_0}q_c^{(2n+1)}(\rho). \tag{6.A.2}$$

Using (6.3.17a,b,c), (6.3.18), (6.A.1), (6.3.2), (6.3.6), (6.3.7), (6.3.4), (6.3.10), and (6.3.11), and recalling that $Pd\varphi_0 = d\zeta$ according to (6.2.31); we obtain after straightforward calculations

$$\frac{1}{P(\varphi_0)}\frac{d\varphi}{d\varphi_0}q_c^{(1)}(\rho) = 1 + \left[\frac{d(P\varphi_2)}{d\zeta} + \frac{1}{2}\alpha\right]\lambda^2$$

$$+ \left[\frac{d(P\varphi_4)}{d\zeta} - \frac{1}{8}\alpha^2 + \frac{S_4/T}{2P^2} + \frac{d}{d\zeta}\left(\frac{1}{2}\alpha P\varphi_2 + \frac{1}{2}\frac{dP}{d\varphi_0}\varphi_2^2\right)\right]\lambda^4 + \cdots,$$

$$\tag{6.A.3a}$$

$$\frac{1}{P(\varphi_0)}\frac{d\varphi}{d\varphi_0}q_c^{(3)}(\rho) = \frac{1}{2}\beta\lambda^2 + \left[-\frac{1}{4}\alpha\beta - \frac{1}{8}\frac{d^2\alpha}{d\zeta^2} + \frac{d}{d\zeta}\left(\frac{1}{2}\beta P\varphi_2\right)\right]\lambda^4 + \cdots,$$

$$\tag{6.A.3b}$$

$$\frac{1}{P(\varphi_0)}\frac{d\varphi}{d\varphi_0}q_c^{(5)}(\rho) = \left[-\frac{1}{8}\beta^2 - \frac{1}{8}\frac{d^2\beta}{d\zeta^2}\right]\lambda^4 + \cdots, \tag{6.A.3c}$$

where

$$\alpha = \frac{S_2/T}{P^2} - \frac{1}{4P^2\varphi_0^2} \tag{6.A.4a}$$

and

$$\beta = P^{-\frac{3}{2}}\frac{d^2}{d\varphi_0^2}P^{-\frac{1}{2}} + \frac{1}{4P^2\varphi_0^2}. \tag{6.A.4b}$$

Noting that, according to (6.A.4a,b) and (6.2.33),

$$\alpha + \beta = h, \tag{6.A.5}$$

we obtain from (6.A.3a,b,c)

$$\sum_{n=0}^{N}\frac{1}{P(\varphi_0)}\frac{d\varphi}{d\varphi_0}q_c^{(2n+1)}(\rho) = 1 + \left[\frac{d(P\varphi_2)}{d\zeta} + \frac{1}{2}h\right]\lambda^2$$

$$+ \left[\frac{d(P\varphi_4)}{d\zeta} - \frac{1}{8}h^2 + \frac{S_4/T}{2P^2} + \frac{d'}{d\zeta}\left(-\frac{1}{8}\frac{dh}{d\zeta} + \frac{1}{2}hP\varphi_2 + \frac{1}{2}\frac{dP}{d\varphi_0}\varphi_2^2\right)\right]\lambda^4 + \cdots,$$

$$\tag{6.A.6}$$

where terms of the order $O(\lambda^{2N+1})$ in the right-hand member are to be neglected. Recalling that $d\zeta = Pd\varphi_0$ according to (6.2.31) and that $Y_0 = 1$; we can, with the aid of (6.2.28) and (6.2.29a,b), write (6.A.6) as

$$\sum_{n=0}^{N}\frac{1}{P(\varphi_0)}\frac{d\varphi}{d\varphi_0}q_c^{(2n+1)}(\rho) = \sum_{n=0}^{N}Y_{2n}\lambda^{2n}, \tag{6.A.7}$$

where terms of the order $O(\lambda^{2N+1})$ in the right-hand member have been neglected. Inserting (6.A.7) into (6.A.2), we obtain

$$\sum_{n=0}^{N} q_c^{(2n+1)}(\rho)\frac{d\rho}{dz} = q(z), \tag{6.A.8}$$

where $q(z)$ is given by (6.3.25) and (6.3.27). When (6.A.8) is inserted into (6.3.20), one obtains ψ in the form of (6.3.24), but we have not yet determined the appropriate constant phase Δ in (6.3.24). To do this we proceed as follows.

According to an unpublished investigation by Yngve concerning phase-integral formulas for Coulomb wave functions, the evaluation of the integrals in (6.3.21a,b), when the functions $q_c^{(2n+1)}(\rho)$ are given by (6.3.17a,b,c), (6.3.18), and (6.A.1) yields

$$\delta_c^{(1)} = \left(L+\frac{1}{2}\right)\left(\arctan\frac{\eta}{L+\frac{1}{2}} - \frac{\pi}{2}\right) + \eta\ln\left[\left(L+\frac{1}{2}\right)^2 + \eta^2\right]^{\frac{1}{2}} - \eta \tag{6.A.9a}$$

$$\delta_c^{(3)} = \frac{\eta}{24[(L+\frac{1}{2})^2 + \eta^2]} \tag{6.A.9b}$$

$$\delta_c^{(5)} = -\frac{7\eta}{960[(L+\frac{1}{2})^2 + \eta^2]^2} + \frac{7\eta^3}{720[(L+\frac{1}{2})^2 + \eta^2]^3}. \tag{6.A.9c}$$

We remark that with the expressions (6.3.16), (6.3.17a,b,c), (6.3.18), and (6.A.1) for $q_c(\rho)$ and (6.A.9a–c) for $\delta_c = \sum_{n=0}^{N}\delta_c^{(2n+1)}$ we obtain (6.3.12) from (6.3.14) in any order of approximation by letting $\rho \to +\infty$ while L and η are kept constant. We also remark that by using the expression (6.3.13) for σ_L we can confirm that, when η is fixed and $L \to \infty$, as well as when L is fixed and $|\eta| \to \infty$; we obtain from (6.A.9a–c) the expected limiting values

$$-\delta_c^{(1)} \to L\pi/2 + \sigma_L \to \pi/4, \tag{6.A.10a}$$

$$\delta_c^{(3)} \to 0, \quad \delta_c^{(5)} \to 0. \tag{6.A.10b}$$

Using (6.A.9a–c), we shall now separately consider the cases when $\xi_0 > 0$ and $\xi_0 = 0$ but $\eta_0 \neq 0$.

The Case When $\xi_0 > 0$ (Figure 6.1). Inserting the expansions (6.3.7) and (6.3.32) for η and $L + \frac{1}{2}$, respectively, in powers of λ into (6.A.9a–c), we obtain

$$\delta_c^{(1)} = -(\eta_0\lambda^{-1} + \eta_2\lambda + \eta_4\lambda^3 + \cdots)\ln\lambda$$

$$+ \left[\xi_0\left(\arctan\frac{\eta_0}{\xi_0} - \frac{\pi}{2}\right) + \eta_0\ln(\xi_0^2 + \eta_0^2)^{\frac{1}{2}} - \eta_0\right]\lambda^{-1}$$

$$+ \left[\xi_2\left(\arctan\frac{\eta_0}{\xi_0} - \frac{\pi}{2}\right) + \eta_2\ln(\xi_0^2 + \eta_0^2)^{\frac{1}{2}}\right]\lambda$$

$$+ \left[\xi_4 \left(\arctan \frac{\eta_0}{\xi_0} - \frac{\pi}{2} \right) + \eta_4 \ln(\xi_0^2 + \eta_0^2)^{\frac{1}{2}} \right.$$

$$\left. + \frac{(\xi_0 \eta_2 - \eta_0 \xi_2)\xi_2 + (\xi_0 \xi_2 + \eta_0 \eta_2)\eta_2}{2(\xi_0^2 + \eta_0^2)} \right] \lambda^3 + \cdots, \qquad (6.A.11a)$$

$$\delta_c^{(3)} = \frac{\eta_0}{24(\xi_0^2 + \eta_0^2)} \lambda + \left[\frac{\eta_2}{24(\xi_0^2 + \eta_0^2)} - \frac{\eta_0(\xi_0 \xi_2 + \eta_0 \eta_2)}{12(\xi_0^2 - \eta_0^2)^2} \right] \lambda^3 + \cdots,$$
$$\qquad (6.A.11b)$$

$$\delta_c^{(5)} = \left[-\frac{7\eta_0}{960(\xi_0^2 + \eta_0^2)^2} + \frac{7\eta_0^3}{720(\xi_0^2 + \eta_0^2)^3} \right] \lambda^3 + \cdots. \qquad (6.A.11c)$$

Summing the expressions for $\delta_c^{(2n+1)}$ from $n = 0$ to $n = N$, truncating the series in front of $\ln \lambda$ according to (6.3.7), and neglecting terms of the order $O(\lambda^{2N+1})$; we obtain

$$\sum_{n=0}^{N} \delta_c^{(2n+1)} = -\eta \ln \lambda + \frac{1}{\lambda} \sum_{n=0}^{N} \delta^{(2n+1)} \lambda^{2n}, \qquad (6.A.12)$$

where $\delta^{(1)}$, $\delta^{(3)}$, and $\delta^{(5)}$ are given by (6.3.35a,b,c), and η is the truncated series (6.3.7).

Inserting (6.A.8) and (6.A.12) into (6.3.20), we obtain (6.3.24) along with (6.3.25), (6.3.27), and (6.3.26).

The Case When $\xi_0 = 0$ But $\eta_0 \neq 0$ (Figures 6.2 and 6.3). Inserting $L = \ell$, according to (6.3.4) with $\xi_0 = 0$, and the expansion (6.3.7) of η in powers of λ into (6.A.9a–c), we obtain

$$\delta_c^{(1)} = -(\eta_0 \lambda^{-1} + \eta_2 \lambda + \eta_4 \lambda^3 + \cdots) \ln \lambda$$

$$+ (\eta_0 \ln |\eta_0| - \eta_0)\lambda^{-1} + \left(\eta_2 \ln |\eta_0| - \frac{(2\ell + 1)^2}{8\eta_0} \right) \lambda$$

$$+ \left(\eta_4 \ln |\eta_0| + \frac{\eta_2^2}{2\eta_0} + \frac{(2\ell + 1)^2 \eta_2}{8\eta_0^2} + \frac{(2\ell + 1)^4}{192\eta_0^3} \right) \lambda^3 + \cdots, \qquad (6.A.13a)$$

$$\delta_c^{(3)} = \frac{1}{24\eta_0} \lambda - \left(\frac{\eta_2}{24\eta_0^2} + \frac{(2\ell + 1)^2}{96\eta_0^3} \right) \lambda^3 + \cdots, \qquad (6.A.13b)$$

$$\delta_c^{(5)} = \frac{7}{2880\eta_0^3} \lambda^3 + \cdots. \qquad (6.A.13c)$$

Summing the expression for $\delta_c^{(2n+1)}$ from $n = 0$ to $n = N$, truncating the series in front of $\ln \lambda$ according to (6.3.7), and neglecting terms of the order $O(\lambda^{2N+1})$, we obtain

$$\sum_{n=0}^{N} \delta_c^{(2n+1)} = -\eta \ln \lambda + \frac{1}{\lambda} \sum_{n=0}^{N} \delta^{(2n+1)} \lambda^{2n}, \qquad (6.A.14)$$

where $\delta^{(1)}$, $\delta^{(3)}$, and $\delta^{(5)}$ are now given by (6.3.44a–c), and η is the truncated series (6.3.7).

Inserting (6.A.8) and (6.A.14) into (6.3.20), we obtain (6.3.24) with (6.3.25), (6.3.27), and (6.3.26).

References

[1] Abramowitz, M. and Stegun, I.A., (Editors), *Handbook of Mathematical Functions*, (National Bureau of Standards Applied Mathematics Series, 55, Seventh printing, May 1968, with corrections.)

[2] Campbell, J.A., Computation of a class of functions useful in the phase-integral approximation. I. Results, *J. Comp. Phys.* **10** (1972), 308–315.

[3] Durand, B. and Durand, L., Improved Fermi–Segrè formula for $|(R_{n,\ell}/r^{\ell})(0)|^2$ for singular and nonsingular potentials, *Phys. Rev. A* **33** (1986), 2887–2898.

[4] Erdélyi, A., Asymptotic solutions of differential equations with transition points or singularities, *J. Math. Phys.* **1** (1960), 16–26.

[5] Fröman, N., Outline of a general theory for higher order approximations of the JWKB-type, *Ark. Fys.* **32** (1966), 541–548.

[6] Fröman, N., Connction formulas for certain higher order phase-integral approximations, *Ann. Phys. (N.Y.)* **61** (1970), 451–464.

[7] Fröman, N., Phase-integral formulas for level densities, normalization factors, and quantal expectation values, not involving wave functions, *Phys. Rev. A* **17** (1978), 493–504.

[8] Fröman, N. and Fröman, P.O., *JWKB Approximation, Contributions to the Theory*, (North-Holland Publishing Co., Amsterdam, 1965.) Russian translation: MIR, Moscow, 1967.

[9] Fröman, N. and Fröman, P.O., A direct method for modifying certain phase-integral approximations of arbitrary order, *Ann. Phys. (N.Y.)* **83** (1974), 103–107.

[10] Fröman, N. and Fröman, P.O., On modifications of phase integral approximations of arbitrary order, *Nuovo Cimento* **20B** (1974), 121–132.

[11] Fröman, N. and Fröman, P.O., On phase-integral quantization conditions for bound states in one-dimensional smooth single-well potentials, *J. Math. Phys.* **19** (1978), 1830–1837.

[12] Fröman, N. and Fröman, P.O., Phase-integral approximation of arbitrary order generated from an unspecified base function. Review article in: *Forty More Years of Ramifications: Spectral Asymptotics and Its Applications*, edited by S.A. Fulling and F.J. Narcowich, Discourses in Mathematics and Its Applications, No. 1, Department of Mathematics, Texas A & M University, College Station, Texas, 1991, pp. 121–159. After minor changes reprinted as Chapter 1 in present monograph.

[13] Fröman, N. and Fröman, P.O., Technique of the comparison equation adapted to the phase-integral method. This is Chapter 2 in the present monograph.

[14] Fröman, P.O., On the normalization of wave functions for bound states in a single-well potential when certain phase-integral approximations of arbitrary order are used, *Ann. Phys. (N.Y.)* **88** (1974), 621–630.

[15] Good, Jr., R.H., The generalizations of the WKB method to radial wave equations, *Phys. Rev.* **90** (1953), 131–137.

[16] Luke, Y.L., *The Special Functions and Their Approximations*, Vol. I, (Academic Press, New York and London, 1969.)

[17] Yngve, S., Normalization of certain higher-order phase-integral approximations for wave functions of bound states in a potential well, II, *J. Math. Phys.* **13** (1972), 324–331.

7

Normalized Wave Function of the Radial Schrödinger Equation Close to the Origin

Nanny Fröman, Per Olof Fröman, Erik Walles, and Staffan Linnaeus

Abstract. This chapter concerns the derivation of general phase-integral formulas, up to the fifth-order approximation, for the behavior of the normalized wave function of the radial Schrödinger equation close to the origin.

7.1 Introduction

N. Fröman and P.O. Fröman [1] have adapted comparison equation technique for the purpose of calculating supplementary quantities in the phase-integral method. The results have been applied by N. Fröman et al. [2] to questions concerning the radial Schrödinger equation and also to more general related questions. On the basis of these results we shall consider the behavior of the normalized wave function of the radial Schrödinger equation close to the origin. For the background and the notations we refer to Chapters 2 and 6.

We shall start by recalling the expressions for some quantities appearing in the comparison equation treatment. Consider Eq. (6.2.1); that is,

$$\frac{d^2\psi}{dz^2} + R(z)\psi = 0, \tag{7.1.1}$$

where $R(z)$ close to the origin is assumed to behave according to Eq. (6.2.3); that is,

$$\lim_{z \to 0} z^2 R(z) = -\ell(\ell + 1), \tag{7.1.2}$$

with $\mathrm{Re}\, \ell \geq -1/2$, which is no restriction, since $\ell(\ell + 1)$ is equal to $(\ell + \frac{1}{2})^2 - 1/4$ and thus remains unchanged when $\ell + \frac{1}{2}$ is replaced by $-(\ell + \frac{1}{2})$.

The arbitrary-order phase-integral approximation used for representing the wave function is generated from an unspecified base function $Q(z)$, which is assumed to fulfill the condition (6.2.5); that is,

$$\lim_{z \to 0} z^2 Q^2(z) = -\xi_0^2, \tag{7.1.3}$$

where ξ_0 is an unspecified constant which we, as in Chapter 6, assume to be real and nonnegative. The phase-integral approximation (up to the fifth order) can be expressed in terms of the quantities

$$Y_0 = 1 \tag{7.1.4a}$$

$$Y_2 = \frac{1}{2}\varepsilon_0 \tag{7.1.4b}$$

$$Y_4 = -\frac{1}{8}\varepsilon_0^2 - \frac{1}{8Q}\frac{d}{dz}\left(\frac{1}{Q}\frac{d\varepsilon_0}{dz}\right), \tag{7.1.4c}$$

where

$$\varepsilon_0 = Q^{-\frac{3}{2}}(z)\frac{d^2Q^{-\frac{1}{2}}(z)}{dz^2} + \frac{R(z) - Q^2(z)}{Q^2(z)}. \tag{7.1.5}$$

In the comparison equation technique developed in Chapter 2 and used in Chapter 6 one introduces a new independent variable φ_0, which is related to z according to Eq. (6.2.24); that is,

$$\int_{\tau_1}^{\varphi_0} P(\varphi_0)\, d\varphi_0 = \int_{t_1}^{z} Q(z)\, dz, \tag{7.1.6}$$

where t_1 is a zero of $Q^2(z)$, and τ_1 is the corresponding zero of the function $P^2(\varphi_0)$. The square root of this function is given by the following expression, which covers both the case of a scattering state and the case of a bound state for the *comparison* equation,

$$P(\varphi_0) = \left(p + 2\mu_0/\varphi_0 - \xi_0^2/\varphi_0^2\right)^{\frac{1}{2}}, \tag{7.1.7}$$

where, in accordance with Eqs. (6.3.10) and (6.4.11),

$$p = +1, \quad \mu_0 = -\eta_0, \quad \text{scattering state}, \tag{7.1.8a}$$

$$p = -1, \quad \mu_0 = \kappa_0, \quad \text{bound state}. \tag{7.1.8b}$$

Here the concepts "scattering state" and "bound state" refer to the comparison equation (6.2.7) but not necessarily to the actual radial Schrödinger equation (7.1.1) under consideration. The case $\xi_0 > 0$ is illustrated in Figures 6.1 and 6.4, while the case $\xi_0 = 0$ is illustrated in Figures 6.2, 6.3, and 6.5. The function $P(\varphi_0)$ is chosen to be positive in the regions corresponding to those where $Q(z)$ is positive; i.e., for $\xi_0 > 0$ to the right of t_1 in Figure 6.1 and between t_1 and t_2 in Figure 6.4; and for $\xi_0 = 0$ to the right of t_1 in Figure 6.2, to the right of the origin in Figure 6.3, and between the origin and t_1 in Figure 6.5. In particular, this implies that for $\xi_0 > 0$ the function $P(\varphi_0)$ behaves as $+i\xi_0/\varphi_0$ as $\varphi_0 \to +i0$; i.e., as φ_0 approaches zero on the upper lip of the cuts in Figures 6.1 and 6.4. Generalizing the definition of μ_0 in (7.1.8a,b), we introduce the quantities $\mu_{2\beta}$, which are related to the quantities $\eta_{2\beta}$ and $\kappa_{2\beta}$ used in Chapter 6 according to the general definition

$$\mu_{2\beta} = -\eta_{2\beta}, \quad \text{scattering state,} \qquad (7.1.9a)$$

$$\mu_{2\beta} = \kappa_{2\beta}, \quad \text{bound state.} \qquad (7.1.9b)$$

Explicit expressions for these quantities are obtained with the aid of Eqs. (6.3.9) and (6.4.8); that is,

$$\mu_{2\beta} = -\eta_{2\beta} = \frac{1}{2\pi i} \int_\Lambda Y_{2\beta} Q(z)\, dz, \quad \text{scattering state,} \qquad (7.1.10a)$$

and

$$\mu_{2\beta} = \kappa_{2\beta} = -\frac{1}{2\pi} \int_\Lambda Y_{2\beta} Q(z)\, dz, \quad \text{bound state,} \qquad (7.1.10b)$$

where β is any nonnegative integer, and Λ is the contour of integration in Figures 6.1(b)–6.5(b). In the present chapter we need also the function $S_{2\beta}/T$ for $\beta > 0$, which is given by Eqs. (6.3.11) and (6.4.12); that is, taking (7.1.9a,b) into account,

$$\frac{S_{2\beta}}{T} = \frac{2\mu_{2\beta}}{\varphi_0} - \frac{\ell(\ell+1) - \xi_0^2}{\varphi_0^2}\delta_{\beta,1}, \quad \beta > 0, \qquad (7.1.11)$$

and the function h, which is given by Eq. (6.2.33); that is,

$$h = P^{-\frac{3}{2}} \frac{d^2 P^{-\frac{1}{2}}}{d\varphi_0^2} + \frac{S_2/T}{P^2}. \qquad (7.1.12)$$

The behavior of the comparison equation solution (6.3.5) close to the origin is given by Eqs. (6.3.49) with (6.3.48) and (6.4.27); that is,

$$\lim_{z\to0} \frac{\psi}{z^{L+1}} = A\frac{2^L \exp(-\eta\pi/2)|\Gamma(L+1+i\eta)|}{\Gamma(2L+2)} \left(\lim_{z\to0} \frac{\rho}{z}\right)^{L+\frac{1}{2}}, \text{ scattering state,} \qquad (7.1.13a)$$

$$\lim_{z\to0} \frac{\psi}{z^{L+1}} = A2^{L+1} \left(\lim_{z\to0} \frac{\rho}{z}\right)^{L+\frac{1}{2}}, \quad \text{bound state,} \qquad (7.1.13b)$$

where A is a normalization factor which is independent of z. We obtain the $(2N+1)$th-order approximation of (7.1.13a,b) by introducing the truncated expansion (6.3.6) for ρ in (7.1.13a,b); thereby obtaining

$$\lim_{z\to0} \frac{\rho}{z} = \frac{1}{\lambda}\left(\lim_{z\to0} \frac{\varphi_0}{z}\right) \sum_{\beta=0}^{N} \left(\lim_{\varphi_0\to0} \frac{\varphi_{2\beta}}{\varphi_0}\right) \lambda^{2\beta}. \qquad (7.1.14)$$

For η in (7.1.13a) we introduce the truncated expansion (6.3.7); that is

$$\eta = \frac{1}{\lambda} \sum_{\beta=0}^{N} \eta_{2\beta}\lambda^{2\beta}, \quad \text{scattering state.} \qquad (7.1.15)$$

For a scattering state of the comparison equation, the phase-integral expression for the wave function in the classically allowed region, either at some distance from the origin or from the classical turning point, is given by Eq. (6.3.24); that is,

$$\psi = Aq^{-\frac{1}{2}}(z) \sin \left(\frac{1}{2} \int_{\Gamma(z)} q(z)\, dz + \Delta \right), \qquad (7.1.16)$$

where $\Gamma(z)$ is an open contour encircling the positive turning point or the origin as shown in Figures 6.1(c), 6.2(c), and 6.3(c), and $q(z)$ and Δ are given by Eqs. (6.3.25) with (6.3.27) and (6.3.26) with (6.3.13); that is,

$$q(z) = \frac{1}{\lambda} \sum_{n=0}^{N} Y_{2n} Q(z) \lambda^{2n} \qquad (7.1.17)$$

and

$$\Delta = \eta \ln \lambda + \arg \Gamma(L + 1 + i\eta) - L\pi/2 - \frac{1}{\lambda} \sum_{n=0}^{N} \delta^{(2n+1)} \lambda^{2n}; \qquad (7.1.18)$$

L and η are assumed to be real. The expressions for $\delta^{(1)}$, $\delta^{(3)}$, and $\delta^{(5)}$ are given in Chapter 6; see Eqs. (6.3.35a,b,c), which are valid when $\xi_0 > 0$, and Eqs. (6.3.44a,b,c), which are valid when $\xi_0 = 0$.

For the occurence of a bound state of the radial Schrödinger equation one has to satisfy the quantization condition (6.4.10); that is,

$$\frac{1}{2} \int_{-\Lambda} q(z)\, dz = (L + 1 + k)\pi, \quad k = 0, 1, 2, \dots, \qquad (7.1.19)$$

where $-\Lambda$ denotes the closed contour of integration of Figures 6.4(b) and 6.5(b), but encircled in the opposite sense. As we require the wave function to be normalized so that

$$\int_0^\infty [\psi(z)]^2 dz = 1, \qquad (7.1.20)$$

we have to determine the normalization factor A in (7.1.13b) according to Eq. (6.4.25), where λ has already been set equal to unity. This formula reads

$$A = \left(\frac{k![\Gamma(2L + 2)]^2}{2\pi\Gamma(2L + 2 + k)} \int_{-\Lambda} \sum_{\beta=0}^{N} C_{2\beta} \frac{dz}{Q(z)} \right)^{-\frac{1}{2}}, \qquad (7.1.21)$$

where the first three of the quantities $C_{2\beta}$ are given by Eqs. (6.4.24a–c); that is,

$$C_0 = 1 \qquad (7.1.22a)$$

$$C_2 = -Y_2 \tag{7.1.22b}$$

$$C_4 = -(Y_4 - Y_2^2). \tag{7.1.22c}$$

In connection with (7.1.21) we also note the approximate formula (6.4.26); that is,

$$\sum_{\beta=0}^{N} C_{2\beta} \frac{1}{Q(z)} = \frac{1}{q(z)}, \quad \lambda = 1. \tag{7.1.23}$$

We will now consider separately the two cases $\xi_0 > 0$ and $\xi_0 = 0$.

7.2 $\xi_0 > 0$

In this case there appear, beside ξ_0, in the third- and fifth-order approximations, the quantities ξ_2 and ξ_4 defined by Eqs. (6.3.34a,b); that is,

$$\xi_2 = \frac{(\ell + \frac{1}{2})^2 - \xi_0^2}{2\xi_0} \tag{7.2.1a}$$

and

$$\xi_4 = -\frac{[(\ell + \frac{1}{2})^2 - \xi_0^2]^2}{8\xi_0^3}. \tag{7.2.1b}$$

For a scattering state of the comparison equation we have already stated the formulas (7.1.15) and (7.1.10a) for the expansion of η; for a bound state we correspondingly have Eq. (6.4.6); that is,

$$\kappa = \frac{1}{\lambda} \sum_{\beta=0}^{N} \kappa_{2\beta} \lambda^{2\beta}, \quad \text{bound state,} \tag{7.2.2}$$

and (7.1.10b).

Because we assume here that $\xi_0 > 0$, the functions $Q^2(z)$ and $P^2(\varphi_0)$ have second-order poles at $z = 0$ and $\varphi_0 = 0$, respectively. Considering for definiteness the upper lip of the cut through the origin in Figures 6.1 or 6.4, we let $zQ(z) = i|zQ(z)|$ and $\varphi_0 P(\varphi_0) = i|\varphi_0 P(\varphi_0)|$ and obtain from the condition (2.2.19) and (7.1.3)

$$\lim_{\varphi_0 \to +i0} \varphi_0 P(\varphi_0) = \lim_{z \to +i0} zQ(z) = i\xi_0, \quad \xi_0 > 0. \tag{7.2.3}$$

Introducing this limiting value into Eq. (2.2.62), with $t_j' = \tau_j' = 0$, we obtain

$$\lim_{z \to 0} \frac{\varphi_0}{z} = \exp\left[\lim_{z \to +i0} \left(\frac{1}{i\xi_0} \int_{t_1}^{z} Q(z)\, dz - \ln z \right) \right.$$

$$\left. - \lim_{\varphi_0 \to +i0} \left(\frac{1}{i\xi_0} \int_{\tau_1}^{\varphi_0} P(\varphi_0)\, d\varphi_0 - \ln \varphi_0 \right) \right], \quad \xi_0 > 0, \tag{7.2.4}$$

where each integral is to be performed on the upper lip of the cut through the origin. Recalling (7.1.7) and (7.1.8a,b), and evaluating the integral with respect to φ_0 in (7.2.4), we obtain

$$\lim_{z \to 0} \frac{\varphi_0}{z} = \frac{2\xi_0^2}{(\xi_0^2 + \eta_0^2)^{\frac{1}{2}}} \exp\left[\lim_{z \to +i0}\left(\frac{1}{i\xi_0}\int_{t_1}^z Q(z)\,dz - \ln z\right) - 1\right.$$

$$\left. + \frac{\eta_0}{\xi_0}\left(\arctan\frac{\eta_0}{\xi_0} + \frac{\pi}{2}\right)\right], \quad \text{scattering state, } \xi_0 > 0, \tag{7.2.5a}$$

and

$$\lim_{z \to 0} \frac{\varphi_0}{z} = \frac{2\xi_0^2}{(\kappa_0^2 - \xi_0^2)^{\frac{1}{2}}}\left(\frac{\kappa_0 - \xi_0}{\kappa_0 + \xi_0}\right)^{\kappa_0/(2\xi_0)}$$

$$\times \exp\left[\lim_{z \to +i0}\left(\frac{1}{i\xi_0}\int_{t_1}^z Q(z)\,dz - \ln z\right) - 1\right], \quad \text{bound state, } \xi_0 > 0, \tag{7.2.5b}$$

where the integral is still to be performed on the upper lip of the cut through the origin.

From the definitions (2.2.68a,b) of c_2 and c_4; that is,

$$c_2 = \lim_{z \to 0} Y_2 = \frac{1}{2}\lim_{z \to 0}\varepsilon_0 \tag{7.2.6a}$$

and

$$c_4 = \lim_{z \to 0} Y_4 = -\frac{1}{8}\lim_{z \to 0}\varepsilon_0^2, \tag{7.2.6b}$$

we obtain with the aid of (7.1.5), (7.1.2), (7.1.3), and (7.2.1a,b)

$$c_2 = \xi_2/\xi_0 \tag{7.2.7a}$$

and

$$c_4 = \xi_4/\xi_0. \tag{7.2.7b}$$

Inserting the last one of the equalities (7.2.3), together with (7.2.7a,b), into Eqs. (2.2.70a,b), with $t_j' = +i0$ and $\tau_j' = \varphi_0(t_j') = +i0$, and recalling the meaning of the short-hand notations for the integrals in Eqs. (2.2.70a,b), as explained in connection with Eq. (4.3.3), and using the same notation for the contours of integration as in Chapter 6; we obtain

$$\lim_{\varphi_0 \to 0} \frac{\varphi_2}{\varphi_0} = \frac{1}{2i\xi_0}\left[\int_{\Gamma(+i0)}\left(\frac{1}{2}\varepsilon_0 - \xi_2/\xi_0\right)Q(z)\,dz\right.$$

$$\left. - \int_{\Gamma(+i0)}\left(\frac{1}{2}h - \xi_2/\xi_0\right)P(\varphi_0)\,d\varphi_0\right] \tag{7.2.8a}$$

and

$$\lim_{\varphi_0 \to 0} \frac{\varphi_4}{\varphi_0} = \frac{1}{2i\xi_0}\left[\int_{\Gamma(+i0)}\left(-\frac{1}{8}\varepsilon_0^2 - \xi_4/\xi_0\right)Q(z)\,dz\right.$$

$$- \int_{\Gamma(+i0)} \left(-\frac{1}{8}h^2 + \frac{S_4/T}{2P^2} - \xi_4/\xi_0 \right) P(\varphi_0)\,d\varphi_0 \Bigg]$$

$$- \frac{\xi_2}{\xi_0} \lim_{\varphi_0 \to 0} \frac{\varphi_2}{\varphi_0} + \frac{1}{2} \left(\lim_{\varphi_0 \to 0} \frac{\varphi_2}{\varphi_0} \right)^2. \tag{7.2.8b}$$

Here $P(\varphi_0)$, h, and S_4/T are given by (7.1.7), (7.1.12), and (7.1.11), respectively, with (7.1.8a,b), (7.1.9a,b) and (7.1.10a,b); the contour $\Gamma(+i0)$ in the complex z-plane is obtained by turning the contour $\Gamma(z)$ in Figures 6.1(c) or 6.4(c) by the angle π in the positive sense and letting $z = +i0$. The corresponding contour in the complex φ_0-plane may also be denoted by $\Gamma(+i0)$.

We shall now evaluate the integrals with respect to φ_0 in (7.2.8a,b). To this purpose we will write (7.1.7) in the form

$$P(\varphi_0) = \frac{r^{\frac{1}{2}}(\varphi_0)}{\varphi_0} \tag{7.2.9}$$

where

$$r(\varphi_0) = p\varphi_0^2 + 2\mu_0\varphi_0 - \xi_0^2, \quad r^{\frac{1}{2}}(+i0) = +i\xi_0. \tag{7.2.10a, b}$$

Using (7.1.11), (7.1.12), (7.2.1a,b), (7.2.9), and (7.2.10a), we can rewrite (7.2.8a,b). We obtain the following expression for the integrand of the second integral in (7.2.8a):

$$\left(\frac{1}{2}h - \frac{\xi_2}{\xi_0} \right) P(\varphi_0) = -\xi_0\xi_2\varphi_0^{-1}r^{-\frac{1}{2}} - \frac{1}{8}r^{-\frac{3}{2}}r'$$

$$-\frac{1}{8}\varphi_0 r^{-\frac{3}{2}}r'' + \frac{5}{32}\varphi_0 r^{-5/2}(r')^2 - \frac{\xi_2}{\xi_0}\varphi_0^{-1}r^{+\frac{1}{2}} + \mu_2 r^{-\frac{1}{2}}. \tag{7.2.11a}$$

For the integrand of the second integral in (7.2.8b) we obtain

$$\left(-\frac{1}{8}h^2 + \frac{S_4/T}{2P^2} - \frac{\xi_4}{\xi_0} \right) P(\varphi_0) = \xi_0^3\xi_4\varphi_0^{-1}r^{-\frac{3}{2}} - \frac{\xi_0\xi_2}{8}r^{-\frac{3}{2}}r'$$

$$- \frac{\xi_0\xi_2}{8}\varphi_0 r^{-\frac{5}{2}}r'' + \frac{20\xi_0\xi_2 - 1}{128}\varphi_0 r^{-\frac{7}{2}}(r')^2$$

$$- \frac{1}{64}\varphi_0^2 r^{-\frac{7}{2}}r'r'' - \frac{1}{128}\varphi_0^3 r^{-\frac{7}{2}}(r'')^2 + \frac{5}{256}\varphi_0^2 r^{-\frac{9}{2}}(r')^3$$

$$+ \frac{5}{256}\varphi_0^3 r^{-\frac{9}{2}}(r')^2 r'' - \frac{25}{2048}\varphi_0^3 r^{-\frac{11}{2}}(r')^4$$

$$+ \mu_2 \left(\xi_0\xi_2 r^{-\frac{3}{2}} + \frac{1}{8}\varphi_0 r^{-\frac{5}{2}}r' + \frac{1}{8}\varphi_0^2 r^{-\frac{5}{2}}r'' - \frac{5}{32}\varphi_0^2 r^{-\frac{7}{2}}(r')^2 \right)$$

$$- \frac{\mu_2^2}{2}\varphi_0 r^{-\frac{3}{2}} + \mu_4 r^{-\frac{1}{2}} - \frac{\xi_4}{\xi_0}\varphi_0^{-1}r^{+\frac{1}{2}}, \tag{7.2.11b}$$

where the prime mark denotes differentiation with respect to φ_0. In these expressions all terms containing the first derivative of r can be "eliminated" by means of "integration by parts," which means that the first derivative of r ultimately appears only in expressions that can be written as derivatives with respect to φ_0. Thus, after rewriting the expressions in a convenient form and noting that $r'' = 2p$, $r''' = r'''' = 0$, and $p^2 = 1$; we obtain

$$\left(\frac{1}{2}h - \frac{\xi_2}{\xi_0}\right)P(\varphi_0) = -\frac{\xi_2}{\xi_0\varphi_0 r^{\frac{1}{2}}}(r + \xi_0^2) - \frac{p}{24}\varphi_0 r^{-\frac{3}{2}}$$

$$+ \mu_2 r^{-\frac{1}{2}} + \frac{d}{d\varphi_0}\left(\frac{1}{24}r^{-\frac{1}{2}} - \frac{5}{48}\varphi_0 r^{-\frac{3}{2}}r'\right) \qquad (7.2.12a)$$

and

$$\left(-\frac{1}{8}h^2 + \frac{S_4/T}{2P^2} - \frac{\xi_4}{\xi_0}\right)P(\varphi_0) = -\frac{\xi_4}{\xi_0\varphi_0 r^{\frac{3}{2}}}(r^2 - \xi_0^4) - \frac{p\xi_0\xi_2}{8}\varphi_0 r^{-\frac{5}{2}} - \frac{7}{384}\varphi_0^3 r^{-\frac{7}{2}}$$

$$+ \mu_2\left(\xi_0\xi_2 r^{-\frac{3}{2}} + \frac{p}{8}\varphi_0^2 r^{-\frac{5}{2}}\right) - \frac{\mu_2^2}{2}\varphi_0 r^{-\frac{3}{2}} + \mu_4 r^{-\frac{1}{2}}$$

$$+ \frac{d}{d\varphi_0}\left[\frac{25}{9216}\varphi_0^3 r^{-\frac{9}{2}}(r')^3 - \frac{5p}{768}\varphi_0^3 r^{-\frac{7}{2}}r' - \frac{5}{1536}\varphi_0^2 r^{-\frac{7}{2}}(r')^2\right.$$

$$- \frac{p}{1920}\varphi_0^2 r^{-\frac{5}{2}} + \frac{1}{8}\left(\frac{1}{240} - \frac{\xi_0\xi_2}{2}\right)\varphi_0 r^{-\frac{5}{2}}r'$$

$$\left.+ \left(\frac{1}{2880} + \frac{\xi_0\xi_2}{24}\right)r^{-\frac{3}{2}} + \frac{\mu_2}{16}\varphi_0^2 r^{-\frac{5}{2}}r'\right]. \qquad (7.2.12b)$$

With the aid of (7.2.10a) we can write the first term in the right-hand member of (7.2.12a) as

$$-\frac{\xi_2}{\xi_0\varphi_0 r^{\frac{1}{2}}}(r + \xi_0^2) = -\frac{\xi_2}{\xi_0}\left(p\varphi_0 r^{-\frac{1}{2}} + 2\mu_0 r^{-\frac{1}{2}}\right), \qquad (7.2.13a)$$

and we can write the first term in the right-hand member of (7.2.12b) as

$$-\frac{\xi_4}{\xi_0\varphi_0 r^{\frac{3}{2}}}(r^2 - \xi_0^4) = -\frac{\xi_4}{\xi_0}\left(p^2\varphi_0^3 r^{-\frac{3}{2}} + 4p\mu_0\varphi_0^2 r^{-\frac{3}{2}}\right.$$

$$\left.+ (4\mu_0^2 - 2p\xi_0^2)\varphi_0 r^{-\frac{3}{2}} - 4\mu_0\xi_0^2 r^{-\frac{3}{2}}\right). \qquad (7.2.13b)$$

Using (7.2.12a) with (7.2.13a), and (7.2.12b) with (7.2.13b), recalling (7.2.10b) and the fact that $p^2 = 1$; we obtain

$$-\frac{1}{2}\int_{\Gamma(+i0)}\left(\frac{1}{2}h - \frac{\xi_2}{\xi_0}\right)P(\varphi_0)d\varphi_0$$

$$= \frac{i}{24\xi_0} + \frac{p}{24}J_{1,3} + \frac{p\xi_2}{\xi_0}J_{1,1} + (2\mu_0\xi_2/\xi_0 - \mu_2)J_{0,1} \qquad (7.2.14a)$$

and

$$-\frac{1}{2}\int_{\Gamma(+i0)}\left(-\frac{1}{8}h^2+\frac{S_4/T}{2P^2}-\frac{\xi_4}{\xi_0}\right)P(\varphi_0)d\varphi_0$$

$$= -i\left(\frac{1}{2880\xi_0^3}+\frac{\xi_2}{24\xi_0^2}\right)+\frac{7}{384}J_{3,7}+\frac{\xi_4}{\xi_0}J_{3,3}-\frac{p\mu_2}{8}J_{2,5}+\frac{4p\mu_0\xi_4}{\xi_0}J_{2,3}$$

$$+\frac{p\xi_0\xi_2}{8}J_{1,5}+(4\mu_0^2\xi_4/\xi_0-2p\xi_0\xi_4+\mu_2^2/2)J_{1,3}$$

$$-\xi_0(4\mu_0\xi_4+\xi_2\mu_2)J_{0,3}-\mu_4J_{0,1}, \qquad (7.2.14b)$$

where, by definition,

$$J_{\mu,\nu}=\frac{1}{2}\int_{\Gamma(+i0)}\varphi_0^\mu r^{-\nu/2}d\varphi_0. \qquad (7.2.15)$$

Here μ is an arbitrary nonnegative integer, and ν is an arbitrary odd integer; there should be no risk of confusing the index μ appearing in (7.2.15) with the quantity $\mu_{2\beta}$ given by (7.1.10a,b). With the aid of (7.2.10a) we obtain the identities

$$(\nu-2)(\mu_0^2+p\xi_0^2)r^{-\nu/2}=-\frac{d}{d\varphi_0}\left[(p\varphi_0+\mu_0)r^{-(\nu-2)/2}\right]-p(\nu-3)r^{-(\nu-2)/2}$$

$$(7.2.16a)$$

$$p\varphi_0 r^{-\nu/2}=-\frac{1}{\nu-2}\frac{d}{d\varphi_0}\left(r^{-(\nu-2)/2}\right)-\mu_0 r^{-\nu/2} \qquad (7.2.16b)$$

$$p\varphi_0^\mu r^{-\nu/2}=-2\mu_0\varphi_0^{\mu-1}r^{-\nu/2}+\xi_0^2\varphi_0^{\mu-2}r^{-\nu/2}+\varphi_0^{\mu-2}r^{-(\nu-2)/2}. \qquad (7.2.16c)$$

Using the identities (7.2.16a–c) and recalling (7.2.10b) and the fact that ν is an odd integer, we find that the integrals $J_{\mu,\nu}$ defined by (7.2.15) satisfy the relations

$$(\nu-2)(\mu_0^2+p\xi_0^2)J_{0,\nu}=\frac{i(-1)^{(\nu+1)/2}\mu_0}{\xi_0^{\nu-2}}-p(\nu-3)J_{0,\nu-2} \qquad (7.2.17a)$$

$$pJ_{1,\nu}=\frac{i(-1)^{(\nu+1)/2}}{(\nu-2)\xi_0^{\nu-2}}-\mu_0 J_{0,\nu} \qquad (7.2.17b)$$

$$pJ_{\mu,\nu}=-2\mu_0 J_{\mu-1,\nu}+\xi_0^2 J_{\mu-2,\nu}+J_{\mu-2,\nu-2}, \quad \mu\geq 2. \qquad (7.2.17c)$$

Direct evaluation for $\mu=0$ and $\nu=1$ of (7.2.15) with (7.2.10a) and (7.1.8a,b) yields (since the quantities ξ_0, η_0, and κ_0 are real, and since, for the case of a bound state, the existence of a classically allowed region for the comparison equation on the positive φ_0-axis requires that $\kappa_0>\xi_0$)

$$J_{0,1}=\begin{cases} i[\pi/2+\arctan(\eta_0/\xi_0)], & \text{scattering state,} \\ i\ln\left(\frac{\kappa_0+\xi_0}{\kappa_0-\xi_0}\right)^{\frac{1}{2}}, & \text{bound state.} \end{cases} \qquad (7.2.18)$$

From (7.2.17a) one can successively calculate $J_{0,\nu}$ for ν equal to 3, 5, and 7; these expressions do not contain $J_{0,1}$. Using (7.2.17b), one can then obtain expressions for $J_{1,\nu}$ when ν is equal to 1, 3, 5, and 7; of these expressions $J_{1,1}$ contains a term proportional to $J_{0,1}$. With the aid of (7.2.17c) one can finally obtain expressions for $J_{2,\nu}$ when ν is equal to 3, 5, and 7; one can obtain expressions for $J_{3,\nu}$ when ν is equal to 3 and 7. In the expression for $J_{2,3}$, as well as in the expression for $J_{3,3}$, there appears a term proportional to $J_{0,1}$. Thus, one can obtain expressions for the quantities $J_{\mu,\nu}$ that appear in (7.2.14a,b). When these expressions are inserted into (7.2.14a) and (7.2.14b), and use is made of the fact that $p^2 = 1$; one obtains

$$-\frac{1}{2i\xi_0}\int_{\Gamma(+i0)}\left(\frac{1}{2}h - \frac{\xi_2}{\xi_0}\right)P(\varphi_0)d\varphi_0$$

$$= \frac{1}{24\xi_0^2} + \frac{p}{24(\mu_0^2 + p\xi_0^2)} + \frac{\xi_2}{\xi_0} + \frac{\xi_0\mu_2 - \xi_2\mu_0}{\xi_0^2}iJ_{0,1} \qquad (7.2.19a)$$

and

$$-\frac{1}{2i\xi_0}\int_{\Gamma(+i0)}\left(-\frac{1}{8}h^2 + \frac{S_4/T}{2P^2} - \frac{\xi_4}{\xi_0}\right)P(\varphi_0)d\varphi_0 = -\frac{1}{2880\xi_0^4} + \frac{\mu_2^2}{2(\mu_0^2 + p\xi_0^2)}$$

$$-\frac{7 + 240p\mu_0\mu_2}{2880(\mu_0^2 + p\xi_0^2)^2} + \frac{7\mu_0^2}{720(\mu_0^2 + p\xi_0^2)^3}$$

$$+\frac{\xi_2}{\xi_0}\left(-\frac{1}{24\xi_0^2} - \frac{p + 24\mu_0\mu_2}{24(\mu_0^2 + p\xi_0^2)} + \frac{p\mu_0^2}{12(\mu_0^2 + p\xi_0^2)^2}\right)$$

$$-\frac{\xi_4}{\xi_0}\frac{\mu_0^2}{\mu_0^2 + p\xi_0^2} + \frac{\xi_0\mu_4 - \xi_4\mu_0}{\xi_0^2}iJ_{0,1}. \qquad (7.2.19b)$$

Recalling (7.1.8a,b), (7.1.9a,b), and (7.2.18), we obtain from (7.2.19a) and (7.2.19b)

$$-\frac{1}{2i\xi_0}\int_{\Gamma(+i0)}\left(\frac{1}{2}h - \frac{\xi_2}{\xi_0}\right)P(\varphi_0)d\varphi_0$$

$$= \frac{1}{24\xi_0^2} + \frac{1}{24(\eta_0^2 + \xi_0^2)} + \frac{\xi_2}{\xi_0} + \frac{\xi_0\eta_2 - \xi_2\eta_0}{\xi_0^2}\left(\frac{\pi}{2} + \arctan\frac{\eta_0}{\xi_0}\right), \text{ scattering state,}$$

$$(7.2.20a)$$

$$-\frac{1}{2i\xi_0}\int_{\Gamma(+i0)}\left(\frac{1}{2}h - \frac{\xi_2}{\xi_0}\right)P(\varphi_0)d\varphi_0$$

$$= \frac{1}{24\xi_0^2} - \frac{1}{24(\kappa_0^2 - \xi_0^2)} + \frac{\xi_2}{\xi_0} - \frac{\xi_0\kappa_2 - \xi_2\kappa_0}{\xi_0^2}\ln\left(\frac{\kappa_0 + \xi_0}{\kappa_0 - \xi_0}\right)^{\frac{1}{2}}, \text{ bound state,}$$

$$(7.2.20b)$$

and

$$-\frac{1}{2i\xi_0}\int_{\Gamma(+i0)}\left(-\frac{1}{8}h^2+\frac{S_4/T}{2P^2}-\frac{\xi_4}{\xi_0}\right)P(\varphi_0)d\varphi_0$$

$$=-\frac{1}{2880\xi_0^4}+\frac{\eta_2^2}{2(\eta_0^2+\xi_0^2)}-\frac{7+240\eta_0\eta_2}{2880(\eta_0^2+\xi_0^2)^2}+\frac{7\eta_0^2}{720(\eta_0^2+\xi_0^2)^3}$$

$$+\frac{\xi_2}{\xi_0}\left(-\frac{1}{24\xi_0^2}-\frac{1+24\eta_0\eta_2}{24(\eta_0^2+\xi_0^2)}+\frac{\eta_0^2}{12(\eta_0^2+\xi_0^2)^2}\right)$$

$$-\frac{\xi_4}{\xi_0}\frac{\eta_0^2}{\eta_0^2+\xi_0^2}+\frac{\xi_0\eta_4-\xi_4\eta_0}{\xi_0^2}\left(\frac{\pi}{2}+\arctan\frac{\eta_0}{\xi_0}\right),\quad\text{scattering state,}\quad(7.2.21a)$$

$$-\frac{1}{2i\xi_0}\int_{\Gamma(+i0)}\left(-\frac{1}{8}h^2+\frac{S_4/T}{2P^2}-\frac{\xi_4}{\xi_0}\right)P(\varphi_0)d\varphi_0$$

$$=-\frac{1}{2880\xi_0^4}+\frac{\kappa_2^2}{2(\kappa_0^2-\xi_0^2)}+\frac{240\kappa_0\kappa_2-7}{2880(\kappa_0^2-\xi_0^2)^2}+\frac{7\kappa_0^2}{720(\kappa_0^2-\xi_0^2)^3}$$

$$+\frac{\xi_2}{\xi_0}\left(-\frac{1}{24\xi_0^2}+\frac{1-24\kappa_0\kappa_2}{24(\kappa_0^2-\xi_0^2)}-\frac{\kappa_0^2}{12(\kappa_0^2-\xi_0^2)^2}\right)$$

$$-\frac{\xi_4}{\xi_0}\frac{\kappa_0^2}{\kappa_0^2-\xi_0^2}-\frac{\xi_0\kappa_4-\xi_4\kappa_0}{\xi_0^2}\ln\left(\frac{\kappa_0+\xi_0}{\kappa_0-\xi_0}\right)^{\frac{1}{2}},\quad\text{bound state.}\quad(7.2.21b)$$

In order to check (7.2.21a,b) we have also derived the integrals by direct evaluation of the corresponding indefinite integrals. These calculations are not described in the present chapter.

In the phase-integral approximation up to the fifth order the behavior close to the origin of the solution of the auxiliary differential equation (6.2.4) is obtained from (7.1.13a,b) with the use of (7.1.14), (7.1.15), (7.2.5a,b), (7.2.8a,b), (7.2.20a,b), (7.2.21a,b), and (7.1.10a,b), and by taking into account the definition of L according to Eq. (6.3.4). Particularizing this result by letting $\lambda = 1$, which yields $L = \ell$ according to Eq. (6.3.4), one obtains an expression for the behavior of the solution ψ of the original radial Schrödinger equation (7.1.1) in the neighborhood of the origin. For a scattering state of the comparison equation, this wave function is represented by (7.1.16) well to the right of the turning point t_1 in Figure 6.1. For a bound state of the comparison equation, it is normalized according to (7.1.20) when A is given by (7.1.21) with (7.1.22a,b,c) or (7.1.23). In the important situation when we specifically choose $\xi_0 = \ell + \frac{1}{2}\ (> 0)$, we have $\xi_2 = \xi_4 = 0$ according to (7.2.1a,b), and the formulas obtained from (7.1.13a,b) as described above simplify considerably.

We shall now recall the original radial Schrödinger equation (7.1.1), which means that we let $\lambda = 1$; we shall consider a solution corresponding to a scattering state (for the comparison equation) in the first-order phase-integral approximation, when $\xi_0 = \ell + \frac{1}{2}$ is positive and much larger than unity, and η_0 is real and to its absolute value at most of the order of magnitude unity. For this purpose we first note that, under these assumptions, it follows from Eq. (4) on p. 32 in [3], with $z = \ell + \frac{1}{2}$ and $a = \frac{1}{2} + i\eta_0$, that

$$|\Gamma(\ell+1+i\eta_0)| = (2\pi)^{\frac{1}{2}} \left(\ell + \frac{1}{2}\right)^{\ell+\frac{1}{2}} \exp\left[-\left(\ell+\frac{1}{2}\right)\right] [1+o(1)], \quad (7.2.22a)$$

$$\arg \Gamma(\ell+1+i\eta_0) = \eta_0 \ln\left(\ell + \frac{1}{2}\right) + o(1). \quad (7.2.22b)$$

From Eq. (4) on p. 32 in [3], with $z = 2\ell + 1$ and $a = 1$, it follows that

$$\Gamma(2\ell+2) = (2\pi)^{\frac{1}{2}} (2\ell+1)^{2\ell+3/2} \exp[-(2\ell+1)] [1+o(1)]. \quad (7.2.22c)$$

Here the symbols $o(1)$ denote quantities which are at most of the order of magnitude $1/(\ell + \frac{1}{2})$. By means of (7.2.22a) and (7.2.22c) we obtain

$$\frac{2^{\ell}|\Gamma(\ell+1+i\eta_0)|}{\Gamma(2\ell+2)} = \frac{1}{2(\ell+\frac{1}{2})^{\frac{1}{2}}} \left(\frac{e}{2\ell+1}\right)^{\ell+\frac{1}{2}} [1+o(1)]. \quad (7.2.22d)$$

Next, we note that, under the same assumptions as above, it follows from (7.2.5a) that

$$\left(\lim_{z\to 0} \frac{\varphi_0}{z}\right)^{\ell+\frac{1}{2}} = \left(\frac{2\ell+1}{e}\right)^{\ell+\frac{1}{2}}$$

$$\times \exp\left[\lim_{z\to+i0}\left(-i\int_{t_1}^{z} Q(z)dz - \left(\ell+\frac{1}{2}\right)\ln z + \eta_0 \pi/2\right)\right] [1+o(1)].$$

$$(7.2.23)$$

Considering the first-order approximation, we obtain from (7.1.13a), (7.1.14) with $\lambda = 1$, (7.2.22d), and (7.2.23) the approximate formula

$$\lim_{z\to 0} \frac{\psi}{z^{\ell+1}} = \frac{1}{2} A \left(\ell+\frac{1}{2}\right)^{-\frac{1}{2}} \exp\left[\lim_{z\to+i0}\left(-i\int_{t_1}^{z} Q(z)dz - \left(\ell+\frac{1}{2}\right)\ln z\right)\right],$$

$$(7.2.24)$$

where the relative error is at most of the order of magnitude $1/(\ell + \frac{1}{2})$. For $\lambda = 1$ and $\xi_0 = \ell + \frac{1}{2}$ one obtains from Eqs. (6.3.4), (6.3.13), (6.3.26), and (6.3.35a) in the first-order approximation, when use is made of (7.2.22b), $\Delta = \pi/4 + o(1)$. With this expression for Δ we can obtain from (7.1.16) the approximate formula

$$\psi = Aq^{-\frac{1}{2}}(z) \cos\left(\frac{1}{2}\int_{\Gamma(z)} q(z)\,dz - \frac{\pi}{4}\right) \quad (7.2.25)$$

for the wave function in the classically allowed region to the right of the turning point t_1 in Figure 6.1. The formulas (7.2.24) and (7.2.25) are in agreement with what one obtains with the aid of one of the usual connection formulas.

7.3 $\xi_0 = 0$, $\mu_0 \neq 0$

In this case it follows from (7.1.3) and (7.1.7) that $Q^2(z)$ has a first-order pole at $z = 0$ and that $P^2(\varphi_0)$ has a corresponding first-order pole at $\varphi_0 = 0$. For $Q^2(z)$ we introduce the expansion

$$Q^2(z) = \alpha_{-1} z^{-1} + \alpha_0 + \alpha_1 z + \dots, \quad \alpha_{-1} \neq 0, \tag{7.3.1}$$

and for $P^2(\varphi_0)$ we obtain from (7.1.7) the expression

$$P^2(\varphi_0) = \frac{2\mu_0}{\varphi_0} + p, \quad \mu_0 \neq 0. \tag{7.3.2}$$

According to Eqs. (2.2.72) and (2.2.73), we then have

$$\lim_{z \to 0} \frac{\varphi_0}{z} = \frac{\alpha_{-1}}{2\mu_0} \tag{7.3.3}$$

and

$$\lim_{\varphi_0 \to 0} \frac{\varphi_{2N}}{\varphi_0} = 2 \lim_{\varphi_0 \to 0} I_{2N}, \quad N > 0. \tag{7.3.4}$$

The quantity I_2 is given by Eq. (2.2.33); that is,

$$I_2 = \frac{1}{2}(\varepsilon_0 - h), \tag{7.3.5}$$

from which $\lim I_2$ in (7.3.4) can be obtained. The quantity $\lim I_4$ in (7.3.4) can then be obtained from Eq. (2.2.80), with $t'_j = 0$ and $\tau'_j = \varphi_0(t'_j) = 0$; that is,

$$\lim_{\varphi_0 \to 0} I_4 = -\frac{\text{Res}(S_4/T)}{2\,\text{Res}\,P^2} - \left[\lim\left(h - \frac{\text{Res}\,h}{\varphi_0}\right)\right]$$
$$+\frac{\text{Res}\,h}{3} \lim\left(\frac{P^2}{\text{Res}\,P^2} - \frac{1}{\varphi_0}\right)\right] \lim I_2 - \left[\frac{1}{8\,\text{Res}\,P^2} + \frac{2}{3}\text{Res}\,h\right] \lim \frac{dI_2}{d\varphi_0}, \tag{7.3.6}$$

where Res denotes the residue of the function in question at $\varphi_0 = 0$, and lim denotes the limit as $\varphi_0 \to 0$.

To be able to obtain an expression for I_2 as a function of φ_0 for small values of $|\varphi_0|$ we shall first expand z in powers of φ_0. To achieve this, we shall make use of Eq. (2.2.24), with the contour Λ enclosing the origin and the turning point t_1 in the z-plane (τ_1 in the φ_0-plane), in addition to (7.1.6), obtaining

$$\int_0^{\varphi_0} P(\varphi_0) d\varphi_0 = \int_0^z Q(z)\,dz. \tag{7.3.7}$$

From (7.3.7) with (7.3.1) and (7.3.2) we obtain, making use of the fact that $p^2 = 1$,

$$z = \frac{2\mu_0}{\alpha_{-1}} \varphi_0 + \left(\frac{p}{3\alpha_{-1}} - \frac{4\mu_0^2 \alpha_0}{3\alpha_{-1}^3}\right) \varphi_0^2$$

$$+ \left(\frac{88\mu_0^3\alpha_0^2}{45\alpha_{-1}^5} - \frac{8\mu_0^3\alpha_1}{5\alpha_{-1}^4} - \frac{4p\mu_0\alpha_0}{9\alpha_{-1}^3} - \frac{1}{90\mu_0\alpha_{-1}} \right) \varphi_0^3 + \ldots . \tag{7.3.8}$$

At $z = 0$ the function $R(z)$ in the radial Schrödinger equation (7.1.1) is assumed to be regular or to have a first- or second-order pole, whereas $Q^2(z)$, as already mentioned, is assumed to have a pole of the first order there. Recalling (7.1.2), we can therefore write

$$R(z) - Q^2(z) = -\ell(\ell+1)z^{-2} + \gamma_{-1}z^{-1} + \gamma_0 + \ldots , \tag{7.3.9}$$

and, according to (7.1.11) with $\xi_0 = 0$ and $\beta = 1$, we have

$$\frac{S_2}{T} = \frac{2\mu_2}{\varphi_0} - \frac{\ell(\ell+1)}{\varphi_0^2}. \tag{7.3.10}$$

Recalling that $p^2 = 1$, we obtain from (7.1.5), (7.3.1), (7.3.9), and (7.3.8)

$$\varepsilon_0 = -\frac{3 + 16\ell(\ell+1)}{32\mu_0\varphi_0} + \ell(\ell+1) \left(\frac{2\alpha_0}{3\alpha_{-1}^2} + \frac{p}{12\mu_0^2} \right) + \frac{p}{64\mu_0^2} + \frac{\gamma_{-1}}{\alpha_{-1}}$$

$$+ \left[\ell(\ell+1) \left(\frac{8\mu_0\alpha_1}{5\alpha_{-1}^3} - \frac{26\mu_0\alpha_0^2}{15\alpha_{-1}^4} - \frac{1}{60\mu_0^3} \right) + \frac{4\mu_0\alpha_0^2}{5\alpha_{-1}^4} - \frac{6\mu_0\alpha_1}{5\alpha_{-1}^3} \right.$$

$$\left. - \frac{2\mu_0\alpha_0\gamma_{-1}}{\alpha_{-1}^3} + \frac{2\mu_0\gamma_0}{\alpha_{-1}^2} - \frac{1}{320\mu_0^3} \right] \varphi_0 + \ldots \tag{7.3.11}$$

and from (7.1.12), (7.3.2), and (7.3.10)

$$h = -\frac{3 + 16\ell(\ell+1)}{32\mu_0\varphi_0} + \left(p\frac{1 + 16\ell(\ell+1)}{64\mu_0^2} + \frac{\mu_2}{\mu_0} \right)$$

$$+ \left(\frac{3 - 8\ell(\ell+1)}{64\mu_0^3} - \frac{p\mu_2}{2\mu_0^2} \right) \varphi_0 + \ldots . \tag{7.3.12}$$

Inserting (7.3.11) and (7.3.12) into (7.3.5), we obtain

$$I_2 = \ell(\ell+1) \left(\frac{\alpha_0}{3\alpha_{-1}^2} - \frac{p}{12\mu_0^2} \right) + \frac{\gamma_{-1}}{2\alpha_{-1}} - \frac{\mu_2}{2\mu_0}$$

$$+ \left[\ell(\ell+1) \left(\frac{4\mu_0\alpha_1}{5\alpha_{-1}^3} - \frac{13\mu_0\alpha_0^2}{15\alpha_{-1}^4} + \frac{13}{240\mu_0^3} \right) \right.$$

$$\left. + \frac{2\mu_0\alpha_0^2}{5\alpha_{-1}^4} - \frac{\mu_0(5\alpha_0\gamma_{-1} + 3\alpha_1)}{5\alpha_{-1}^3} + \frac{\mu_0\gamma_0}{\alpha_{-1}^2} - \frac{1}{40\mu_0^3} + \frac{p\mu_2}{4\mu_0^2} \right] \varphi_0 + \ldots . \tag{7.3.13}$$

Hence,

$$\lim_{\varphi_0 \to 0} I_2 = \ell(\ell+1) \left(\frac{\alpha_0}{3\alpha_{-1}^2} - \frac{p}{12\mu_0^2} \right) + \frac{\gamma_{-1}}{2\alpha_{-1}} - \frac{\mu_2}{2\mu_0} \tag{7.3.14a}$$

and

$$\lim_{\varphi_0 \to 0} dI_2/d\varphi_0 = \ell(\ell+1)\left(\frac{4\mu_0\alpha_1}{5\alpha_{-1}^3} - \frac{13\mu_0\alpha_0^2}{15\alpha_{-1}^4} + \frac{13}{240\mu_0^3}\right)$$

$$+ \frac{2\mu_0\alpha_0^2}{5\alpha_{-1}^4} - \frac{\mu_0(5\alpha_0\gamma_{-1} + 3\alpha_1)}{5\alpha_{-1}^3} + \frac{\mu_0\gamma_0}{\alpha_{-1}^2} - \frac{1}{40\mu_0^3} + \frac{p\mu_2}{4\mu_0^2}. \tag{7.3.14b}$$

To be able to calculate the limit of I_4 as $\varphi_0 \to 0$, we need (7.3.14a,b) as well as expressions for the other limiting values that appear in (7.3.6), which are obtained as follows. The use of (7.1.11) with $\beta = 2$ and (7.3.2) yields

$$\frac{\text{Res}(S_4/T)}{2\,\text{Res}\,P^2} - \frac{\mu_4}{2\mu_0}. \tag{7.3.15}$$

The use of (7.3.2) and (7.3.12) yields

$$\lim\left(h - \frac{\text{Res}\,h}{\varphi_0}\right) + \frac{\text{Res}\,h}{3}\lim\left(\frac{P^2}{\text{Res}\,P^2} - \frac{1}{\varphi_0}\right) = \frac{p\ell(\ell+1)}{6\mu_0^2} + \frac{\mu_2}{\mu_0} \tag{7.3.16}$$

and

$$\frac{1}{8\,\text{Res}\,P^2} + \frac{2}{3}\,\text{Res}\,h = -\frac{\ell(\ell+1)}{3\mu_0}. \tag{7.3.17}$$

Inserting (7.3.14a,b)–(7.3.17) into (7.3.6), and noting that $p^2 = 1$ according to (7.1.8a,b); we obtain

$$\lim_{\varphi_0 \to 0} I_4 = [\ell(\ell+1)]^2\left(\frac{4\alpha_1}{15\alpha_{-1}^3} - \frac{13\alpha_0^2}{45\alpha_{-1}^4} - \frac{p\alpha_0}{18\mu_0^2\alpha_{-1}^2} + \frac{23}{720\mu_0^4}\right)$$

$$+ \ell(\ell+1)\left[\frac{2\alpha_0^2}{15\alpha_{-1}^4} - \frac{5\alpha_0\gamma_{-1} + 3\alpha_1}{15\alpha_{-1}^3} + \frac{\gamma_0}{3\alpha_{-1}^2} - \frac{p\gamma_{-1}}{12\mu_0^2\alpha_{-1}} - \frac{1}{120\mu_0^4}\right.$$

$$\left.+ \mu_2\left(\frac{p}{4\mu_0^3} - \frac{\alpha_0}{3\mu_0\alpha_{-1}^2}\right)\right] - \frac{\gamma_{-1}\mu_2}{2\mu_0\alpha_{-1}} + \frac{\mu_2^2}{2\mu_0^2} - \frac{\mu_4}{2\mu_0}. \tag{7.3.18}$$

Recalling (7.1.8a,b) and (7.1.9a,b), we obtain from (7.3.3)

$$\lim_{z \to 0}\frac{\varphi_0}{z} = -\frac{\alpha_{-1}}{2\eta_0}, \quad \text{scattering state}, \tag{7.3.19a}$$

$$\lim_{z \to 0}\frac{\varphi_0}{z} = \frac{\alpha_{-1}}{2\kappa_0}, \quad \text{bound state}, \tag{7.3.19b}$$

and from (7.3.4), (7.3.14a), (7.3.18), (7.1.8a,b), and (7.1.9a,b)

$$\lim_{\varphi_0 \to 0}\frac{\varphi_2}{\varphi_0} = \ell(\ell+1)\left(\frac{2\alpha_0}{3\alpha_{-1}^2} - \frac{1}{6\eta_0^2}\right) + \frac{\gamma_{-1}}{\alpha_{-1}} - \frac{\eta_2}{\eta_0}, \quad \text{scattering state}, \tag{7.3.20a}$$

$$\lim_{\varphi_0 \to 0}\frac{\varphi_2}{\varphi_0} = \ell(\ell+1)\left(\frac{2\alpha_0}{3\alpha_{-1}^2} + \frac{1}{6\kappa_0^2}\right) + \frac{\gamma_{-1}}{\alpha_{-1}} - \frac{\kappa_2}{\kappa_0}, \quad \text{bound state}. \tag{7.3.20b}$$

and

$$
\lim_{\varphi_0 \to 0} \frac{\varphi_4}{\varphi_0} = [\ell(\ell+1)]^2 \left(\frac{8\alpha_1}{15\alpha_{-1}^3} - \frac{26\alpha_0^2}{45\alpha_{-1}^4} - \frac{\alpha_0}{9\eta_0^2\alpha_{-1}^2} + \frac{23}{360\eta_0^4} \right)
$$

$$
+ \ell(\ell+1) \left[\frac{4\alpha_0^2}{15\alpha_{-1}^4} - \frac{10\alpha_0\gamma_{-1} + 6\alpha_1}{15\alpha_{-1}^3} + \frac{2\gamma_0}{3\alpha_{-1}^2} - \frac{\gamma_{-1}}{6\eta_0^2\alpha_{-1}} - \frac{1}{60\eta_0^4} \right.
$$

$$
\left. + \eta_2 \left(\frac{1}{2\eta_0^3} - \frac{2\alpha_0}{3\eta_0\alpha_{-1}^2} \right) \right] - \frac{\gamma_{-1}\eta_2}{\eta_0\alpha_{-1}} + \frac{\eta_2^2}{\eta_0^2} - \frac{\eta_4}{\eta_0}, \quad \text{scattering state,} \quad (7.3.21a)
$$

$$
\lim_{\varphi_0 \to 0} \frac{\varphi_4}{\varphi_0} = [\ell(\ell+1)]^2 \left(\frac{8\alpha_1}{15\alpha_{-1}^3} - \frac{26\alpha_0^2}{45\alpha_{-1}^4} + \frac{\alpha_0}{9\kappa_0^2\alpha_{-1}^2} + \frac{23}{360\kappa_0^4} \right)
$$

$$
+ \ell(\ell+1) \left[\frac{4\alpha_0^2}{15\alpha_{-1}^4} - \frac{10\alpha_0\gamma_{-1} + 6\alpha_1}{15\alpha_{-1}^3} + \frac{2\gamma_0}{3\alpha_{-1}^2} + \frac{\gamma_{-1}}{6\kappa_0^2\alpha_{-1}} - \frac{1}{60\kappa_0^4} \right.
$$

$$
\left. - \kappa_2 \left(\frac{1}{2\kappa_0^3} + \frac{2\alpha_0}{3\kappa_0\alpha_{-1}^2} \right) \right] - \frac{\gamma_{-1}\kappa_2}{\kappa_0\alpha_{-1}} + \frac{\kappa_2^2}{\kappa_0^2} - \frac{\kappa_4}{\kappa_0}, \quad \text{bound state.} \quad (7.3.21b)
$$

7.4   Summary of the Results Obtained in the Present Chapter and Discussion of Results Obtained by Previous Authors

Assuming that the function $R(z)$ in the radial Schrödinger equation (7.1.1) fulfills the condition (7.1.2), we have expressed the radial wave function ψ by means of the phase-integral approximation of arbitrary order generated from the unspecified base function $Q(z)$, which is assumed to fulfill the condition (7.1.3).

For a *scattering state* of the comparison equation, the regular wave function close to the origin is given by (7.1.13a), (7.1.14), (7.1.15), and (7.1.10a) with $\lambda = 1$ and $L = \ell$. When $\xi_0 > 0$ the limit of φ_0/z as $z \to 0$ is given by (7.2.5a); whereas the limits of φ_2/φ_0 and φ_4/φ_0 as $z \to 0$ are given by (7.2.8a) and (7.2.8b), respectively, together with (7.2.1a,b), (7.2.20a) and (7.2.21a) with (7.1.10a). When $\xi_0 = 0$ the functions $Q^2(z)$ and $R(z) - Q^2(z)$ are assumed to be expanded according to (7.3.1) and (7.3.9); the limit of φ_0/z as $z \to 0$ is then given by (7.3.19a) with (7.1.10a), whereas the limits of φ_2/φ_0 and φ_4/φ_0 as $z \to 0$ are given by (7.3.20a) and (7.3.21a) with (7.1.10a). In the classically allowed region, to the right of either the turning point t_1 on the positive part of the real axis (Figure 6.1 for $\xi_0 > 0$ and Figure 6.2 for $\xi_0 = 0$) or the origin (Figure 6.3 for $\xi_0 = 0$), the wave function is given by (7.1.16), (7.1.17), and (7.1.18) with $\lambda = 1$ and $L = \ell$.

For a *bound state* of the comparison equation, the wave function characterized by the quantum number k in the quantization condition (7.1.19), and normalized according to (7.1.20), is, close to the origin, given by

(7.1.13b) and (7.1.14) with $\lambda = 1$ and $L = \ell$ together with (7.1.21), and (7.1.22a,b,c) or (7.1.23). When $\xi_0 > 0$ the limit of φ_0/z as $z \to 0$ is given by (7.2.5b) with (7.1.10b), whereas the limits of φ_2/φ_0 and φ_4/φ_0 as $z \to 0$ are given by (7.2.8a) and (7.2.8b), respectively, together with (7.2.1a,b), (7.2.20b) and (7.2.21b) with (7.1.10b). When $\xi_0 = 0$ the functions $Q^2(z)$ and $R(z) - Q^2(z)$ are assumed to be expanded according to (7.3.1) and (7.3.9); the limit of φ_0/z as $z \to 0$ is then given by (7.3.19b) with (7.1.10b); whereas the limits of φ_2/φ_0 and φ_4/φ_0 as $z \to 0$ are given by (7.3.20b) and (7.3.21b) with (7.1.10b).

Formulas for the normalized wave function of the radial Schrödinger equation at the origin have previously been obtained both without the use of comparison equation technique [4–9] and with the use of a first-order comparison equation method [10, 11] (see also further references to earlier work given in these papers). With regard to [10] the results in Tables I and II in that work show that, for $\ell = 0$ the authors have achieved a substantial improvement in comparison with ordinary WKB, while for $\ell \neq 0$ the improvement is moderate. However, for $\ell = 0$ simple, more accurate, formulas can be obtained without comparison equation technique with the use of the arbitrary-order phase-integral approximation generated from an unspecified base function [5], as one can see by comparing the results for $\ell = 0$ in Table I in [10] with the results in Table II in [5]. Thus, one finds that the error in the first-order approximation of the phase-integral results in Table II of [5] is about the same as the error in Table I of [10]; whereas the error in the third- and fifth-order approximations of the phase-integral results is considerably smaller. Fröman and Fröman [7] have also derived formulas for the normalized wave function of the radial Schrödinger equation close to the origin that are very accurate for small values of the angular momentum quantum number ℓ. Durand and Durand [11] found that these formulas are, in certain contexts, not useful for large values of ℓ and have derived alternative formulas by means of a first-order comparison equation method.

Acknowledgement. We would like to thank Dr. Finn Karlsson for valuable cooperation at the initial stage of the present work.

References

[1] Fröman, N. and Fröman, P.O., Technique of the comparison equation adapted to the phase-integral method. This is Chapter 2 in the present monograph.

[2] Fröman, N., Fröman, P.O., and Linnaeus, S., Phase-integral formulas for the regular wave function when there are turning points close to a pole of the potential. This is Chapter 6 in the present monograph.

[3] Luke, Y.L., *The Special Functions and Their Approximations*, Vol. I, Academic Press, New York and London, 1969.

[4] Fröman, N. and Fröman, P.O., *Phys. Rev.* **A6** (1972), 2064–2067.

[5] Fröman, N., *J. Math. Phys.* **19** (1978), 1141–1146.

[6] Thidé, B., *J. Math. Phys.* **21** (1980), 1408–1415.

[7] Fröman, N. and Fröman, P.O., *J. Physique* (Paris) **42** (1981), 1491–1504.

[8] Linnaeus, S. and Düring, M., *Ann. Phys.* (N.Y.) **164** (1985), 506–515.

[9] Yngve, S., *Phys. Rev.* **A33** (1986), 96–104.

[10] Durand, B. and Durand, L., *Phys. Rev.* **A33** (1986), 2887–2898.

[11] Durand, B. and Durand, L., *Phys. Rev.* **A33** (1986), 2899–2906.

8
Phase-Amplitude Method Combined with Comparison Equation Technique Applied to an Important Special Problem

Per Olof Fröman, Anders Hökback, and Nanny Fröman

Abstract. We show how comparison equation technique can be used to overcome a difficulty that arises in the neighborhood of the origin in the numerical integration of a Schrödinger-like differential equation by means of the phase-amplitude method, when the effective potential behaves as $1/(4z^2)$ close to the origin. These results are applied to the calculation of the energy eigenvalues of a two-dimensional anharmonic oscillator.

8.1   Introduction

For a one-dimensional Schrödinger-like differential equation, pertaining to an effective potential that is real on the real axis, there is, on the real axis, a simple connection between phase and amplitude for every complex solution that has a nonconstant phase on this axis. Therefore, on the real axis, every real solution of the differential equation in question can be written (in a nonunique way) as the product of a z-dependent local amplitude and the sine of a z-dependent local phase such that the above connection between phase and amplitude prevails. This fact has been used by several authors for numerical solution of the Schrödinger equation as far back as in the thirties [1–5]. Young [3, 4] pointed out its fundamental importance in quantum mechanics, and Wheeler [5], adding to Milne's [1] work, showed the great practical advantage of what he called "the amplitude-phase method" for efficient numerical solution of the radial Schrödinger equation. The reason for the success of this method is that, by starting the numerical integration judiciously, one can achieve the result that the local amplitude and the local phase are nonoscillating functions.

This chapter deals with a special case, in which a serious difficulty in the integration procedure appears close to the origin; *viz.* the case when the effective potential behaves as $1/(4z^2)$ close to the origin, where z is the independent variable. For the radial Schrödinger equation this situation corresponds to the nonphysical value $\ell = -1/2$ of the angular momentum quantum number. On separation of the Schrödinger equation for a two-dimensional oscillator (in polar coordinates), for a hydrogen atom in a homogeneous electric field (in parabolic coordinates), and for the two-center Coulomb problem (in elliptic coordinates), the case in question occurs when the magnetic quantum number is equal to zero. These are but a few concrete examples of a complicated situation that one may encounter by the solution of physical problems described by differential equations of the Schrödinger type. The purpose of this chapter is to show how this difficulty can be overcome with the use of the phase-amplitude method in combination with comparison equation technique.

We consider the quantization condition for a bound-state problem within the framework of the phase-amplitude method. This condition is derived in §8.2. In §8.3 comparison equation solutions, expressed in terms of Coulomb wave functions, are used to obtain an approximate formula useful in the neighborhood of the origin in order to overcome the difficulty mentioned above and to make numerical integration of high accuracy possible and efficient. An application to a two-dimensional anharmonic oscillator is described in §8.4, where numerical results for a number of energy eigenvalues are given with 12 digits.

8.2   Quantization Condition

Consider the differential equation

$$\frac{d^2\psi}{dz^2} + R(z)\psi = 0, \tag{8.2.1}$$

where it is assumed that $R(z)$ is real on the real z-axis and that

$$\lim_{z \to 0} z^2 R(z) = -\ell(\ell+1); \tag{8.2.2}$$

see Eqs. (6.2.1) and (6.2.3) in [6]. Except for an arbitrary constant factor, the solution of (8.2.1) that is equal to zero for $z = 0$, can be written as

$$\psi = q^{-1/2}(z) \sin\left(\int_0^z q(z)\,dz\right), \tag{8.2.3}$$

where $q(z)$ is a convenient solution of the q-equation (1.3.3) in [7], that is,

$$q^{-3/2}\frac{d^2}{dz^2}q^{-1/2} + \frac{R(z)}{q^2} - 1 = 0. \tag{8.2.4}$$

To obtain a nonoscillatory solution $q(z)$ we start the numerical integration of (8.2.4) in the middle of a region where $R(z)$ is positive by using the phase-integral formula (1.3.19) for $q(z)$, that is,

$$q(z) = Q(z) \sum_{n=0}^{N} Y_{2n}, \qquad (8.2.5)$$

with convenient choices of $Q(z)$ and N. When $R(z)$ corresponds to what in quantal language is a single well potential, then one can obtain a function $q(z)$ that is positive and nonoscillatory in the whole interval $0 < z < \infty$. This is the function $q(z)$ which is to be used in (8.2.3). Requiring that $\psi(z)$ approaches zero as $z \rightarrow +\infty$, one obtains from (8.2.3) the quantization condition

$$\int_{0}^{\infty} q(z)\,dz = s\pi, \qquad (8.2.6)$$

where s is a positive integer. The calculation of the integral in the left-hand member of (8.2.6) is rather easily performed except when $\ell = -1/2$, in which case $q(z)$ approaches zero so slowly as $z \rightarrow 0$ that the integral cannot be calculated numerically with sufficient accuracy by conventional methods. To master this difficulty we shall in the following section use comparison equation technique to obtain for the integral of $q(z)$ from 0 to z an approximate analytic formula that is very accurate when z is sufficiently small.

8.3   Solution of the Difficulty at the Origin by Means of Comparison Equation Solutions Expressed in Terms of Coulomb Wave Functions

According to Eq. (6.3.5), and the corresponding formula with F_L replaced by G_L, we have the two linearly independent comparison equation solutions

$$\psi_1 = \left(\frac{d\rho}{dz}\right)^{-\frac{1}{2}} F_L(\eta, \rho), \qquad (8.3.1a)$$

$$\psi_2 = \left(\frac{d\rho}{dz}\right)^{-\frac{1}{2}} G_L(\eta, \rho), \qquad (8.3.1b)$$

where ρ, η, and L are given by Eqs. (6.3.2), (6.3.3), and (6.3.4). The Wronskian of the solutions (8.3.1a,b) is

$$W = \psi_1 d\psi_2/dz - \psi_2 d\psi_1/dz$$

$$= F_L(\eta, \rho) dG_L(\eta, \rho)/d\rho - G_L(\eta, \rho) dF_L(\eta, \rho)/d\rho = -1. \qquad (8.3.2)$$

Inserting (8.3.1a,b) and (8.3.2) into Eq. (1.5.14), we find that the general solution of the q-equation is given by the formula

$$\frac{d\rho/dz}{q(z)} = \alpha F_L^2(\eta, \rho) + \beta G_L^2(\eta, \rho) \pm 2(\alpha\beta - 1)^{1/2} F_L(\eta, \rho) G_L(\eta, \rho), \quad (8.3.3)$$

where α and β are arbitrary constants. Using the asymptotic formulas for the Coulomb wave functions, one finds that sufficiently far away from the origin, in the classically allowed region, (8.3.3) yields

$$\frac{d\rho/dz}{q(z)} \sim \frac{1}{2}(\alpha + \beta) - \frac{1}{2}(\alpha - \beta)\cos(2\theta) \pm (\alpha\beta - 1)^{1/2}\sin(2\theta), \quad (8.3.4)$$

where

$$\theta = \rho - \eta \ln(2\rho) - L\pi/2 + \arg(L + 1 + i\eta). \quad (8.3.5)$$

The function on the right-hand side of (8.3.4) is rapidly oscillating unless the coefficients of $\cos(2\theta)$ and $\sin(2\theta)$ are equal to zero; that is, unless

$$\alpha = \beta = \pm 1. \quad (8.3.6)$$

With these values of α and β we obtain from (8.3.3)

$$q(z) = \pm\frac{d\rho/dz}{F_L^2(\eta, \rho) + G_L^2(\eta, \rho)}, \quad (8.3.7)$$

that is, with the aid of (8.3.2),

$$q(z) = \pm\frac{d}{dz}\arctan\frac{F_L(\eta, \rho)}{G_L(\eta, \rho)}. \quad (8.3.7')$$

From (8.3.7') we obtain

$$\int_0^z q(z)\,dz = \pm\arctan\frac{F_L(\eta, \rho)}{G_L(\eta, \rho)}. \quad (8.3.8)$$

When $\ell = -1/2$ and $\lambda = 1$ we find from Eq. (6.3.4) that $L = -1/2$. For this particular value of L we have (when ρ is small) the formulas

$$F_{-1/2}(\eta, \rho) = \left(\frac{\pi}{1 + \exp(2\pi\eta)}\right)^{1/2} \rho^{1/2}[1 + O(\rho)] \quad (8.3.9)$$

and

$$\frac{G_{-1/2}(\eta, \rho)}{F_{-1/2}(\eta, \rho)} = -\frac{1 + \exp(2\pi\eta)}{\pi}\left[\ln(2\rho) + 2\gamma\right.$$
$$\left. + \frac{1}{2}\left(\frac{\Gamma'(1/2 + i\eta)}{\Gamma(1/2 + i\eta)} + \frac{\Gamma'(1/2 - i\eta)}{\Gamma(1/2 - i\eta)}\right) + O(\rho)\right], \quad (8.3.10)$$

where $\gamma = -\Gamma'(1)/\Gamma(1) = 0.5772\ldots$, which is Euler's constant. For sufficiently small values of z the function $\rho(z)$ is approximately proportional to z, and with the aid of (8.3.9) and (8.3.10) we obtain from (8.3.7), (8.3.8), and (8.3.7'), where we choose the upper signs, the approximate formula

$$q(z) = \frac{B}{z[1 + B^2(C - \ln z)^2]} \tag{8.3.11}$$

and

$$\int_0^z q(z)\, dz = \arctan\frac{1}{B(C - \ln z)}, \tag{8.3.12}$$

where

$$B = \frac{1}{\pi}[1 + \exp(2\pi\eta)] \tag{8.3.13}$$

and

$$C = -\left[\ln\left(\lim_{z\to 0}\frac{\rho}{z}\right) + \ln 2 + 2\gamma + \frac{1}{2}\left(\frac{\Gamma'(1/2 + i\eta)}{\Gamma(1/2 + i\eta)} + \frac{\Gamma'(1/2 - i\eta)}{\Gamma(1/2 - i\eta)}\right)\right]. \tag{8.3.14}$$

In particular we have

$$B = \frac{2}{\pi} \quad \text{when} \quad \eta = 0 \tag{8.3.13'}$$

and, since $\Gamma'(1/2)/\Gamma(1/2) = -(\gamma + 2\ln 2)$,

$$C = -\left[\ln\left(\lim_{z\to 0}\frac{\rho}{z}\right) + \gamma - \ln 2\right] \quad \text{when} \quad \eta = 0. \tag{8.3.14'}$$

The parameter η and the limit of ρ/z as $z \to 0$ can be determined as described in Chapter 6. If the resulting expressions are inserted into (8.3.13) and (8.3.14), one can obtain comparison equation expressions for B and C. We shall, however, proceed in a simpler way that is more convenient for our purposes here. From (8.3.11) it follows that

$$B(C - \ln z) = -z\frac{d}{dz}\left(\frac{1}{2zq(z)}\right). \tag{8.3.15}$$

With the aid of this formula we can write (8.3.12) as

$$\int_0^z q(z)\, dz = \arctan\frac{2zq^2(z)}{\frac{d}{dz}[zq(z)]}, \tag{8.3.16}$$

that is

$$\int_0^z q(z)\, dz = \arctan\frac{2zq(z)}{1 + \frac{z}{q(z)}\frac{dq(z)}{dz}}. \tag{8.3.16'}$$

These are approximate formulas that are very accurate when z is sufficiently small. One gains some accuracy by rewriting the q-equation as a differential equation for the function $zq(z)$ and using the formula (8.3.16), instead of using the differential equation for the function $q(z)$ and using the formula (8.16′).

With the aid of (8.3.16) or (8.3.16′) we can rewrite the quantization condition (8.2.6) as

$$\arctan\frac{2zq^2(z)}{\frac{d}{dz}[zq(z)]} + \int_z^\infty q(z)\,dz = s\pi, \quad s = 1, 2, 3, \ldots, \qquad (8.3.17)$$

or

$$\arctan\frac{2zq(z)}{1 + \frac{z}{q(z)}\frac{dq(z)}{dz}} + \int_z^\infty q(z)\,dz = s\pi, \quad s = 1, 2, 3, \ldots, \qquad (8.3.17′)$$

where we choose z to be small enough so that the expression in the left-hand member of (8.3.17) or (8.3.17′) remains constant (within the prescribed degree of numerical accuracy) when z decreases further.

8.4   Application to a Two-Dimensional Anharmonic Oscillator

Using obvious notation, we may write the Schrödinger equation for the motion of a quantal particle in a plane as

$$\left[-\frac{\hbar^2}{2\mu}\Delta + V(r, \varphi)\right]\Psi = E\Psi, \qquad (8.4.1)$$

where Δ is the two-dimensional Laplacian, and r and φ are polar coordinates in the plane. The Schrödinger equation (8.4.1) can be written as

$$\frac{\partial^2(r^{1/2}\Psi)}{\partial r^2} + \frac{r^{1/2}\Psi}{4r^2} + \frac{1}{r^2}\frac{\partial^2(r^{1/2}\Psi)}{\partial \varphi^2} + \frac{2\mu}{\hbar^2}[E - V(r,\varphi)]r^{1/2}\Psi = 0. \quad (8.4.2)$$

Assuming the potential to be independent of φ and inserting

$$\Psi = r^{-1/2}\psi(r)e^{im\varphi}, \quad m = 0, \pm 1, \pm 2, \ldots, \qquad (8.4.3)$$

into (8.4.2), we obtain

$$\frac{d^2\psi}{dr^2} + \left\{\frac{2\mu}{\hbar^2}[E - V(r)] - \frac{m^2 - 1/4}{r^2}\right\}\psi = 0. \qquad (8.4.4)$$

Since the energy levels corresponding to the same value of $|m| + 2(s-1)$ but different values of $|m|$ coincide for the harmonic oscillator and lie rather close together for an anharmonic oscillator; it is convenient to introduce the quantum number

$$v = |m| + 2(s - 1), \quad s = 1, 2, 3, \ldots, \qquad (8.4.5)$$

instead of the quantum number s in the quantization condition. The quantum number v can assume the values $|m|, |m| + 2, |m| + 4, \ldots$.

If the oscillator is purely quartic, and if we introduce a conveniently chosen dimensionless radial coordinate z instead of r, the Schrödinger equation (8.4.4) may be written in the form (8.2.1) with

$$R(z) = A - z^4 - \frac{(|m| - 1/2)(|m| + 1/2)}{z^2}, \qquad (8.4.6)$$

where A is a dimensionless constant. Thus, we may choose the parameter ℓ in (8.2.2) to be

$$\ell = |m| - 1/2. \qquad (8.4.7)$$

In order to use the phase-amplitude method for numerical determination of the eigenvalues of A, we begin to solve the q-equation (8.2.4) at a point z_0 in the middle of the potential well, where for $q(z)$ we use the phase-integral formula (8.2.5) with $Q^2(z) = R(z)$ or $Q^2(z) = R(z) - 1/(4z^2)$. It turns out that it is sufficient to use the first-order approximation; i.e., to set $N = 0$ in (8.2.5). First, one solves numerically the q-equation as z increases from z_0 towards $+\infty$; at the same time, one calculates the integral of $q(z)$ from z_0 to $+\infty$. Then one solves the q-equation as z decreases from z_0 towards zero, at the same time calculating the integral of $q(z)$ from 0 to z_0. In practice one performs these calculations of integrals by solving the differential equation for $\int_{z_0}^{z} q(z)dz$. For all integer values of m, except for $m = 0$, the phase-amplitude method, with the quantization condition (8.2.6), gives excellent results and yields the eigenvalues of A with at least 12 significant digits. The case $m = 0$ must, however, be given special treatment, because the integral of $q(z)$ from 0 to z approaches zero very slowly as $z \to 0$. This difficulty is not removed even if one introduces $\ln z$ as a new independent variable in the q-equation. For $m = 0$ one first calculates the integral of $q(z)$ from z_0 to $+\infty$ (as described above); then one calculates the integral of $q(z)$ from a small value z (> 0), so far unspecified, to z_0. By adding these integrals, one obtains the integral of $q(z)$ from the small value z (> 0) to $+\infty$. This integral is then inserted into the quantization condition (8.3.17) or (8.3.17′), where z is chosen to be so small that the expression in the left-hand member of the quantization condition remains constant (within the prescribed numerical accuracy) when z decreases further. The eigenvalues of A are then obtained from (8.3.17) or (8.3.17′) by iteration.

The eigenvalues of the two-dimensional quartic oscillator, calculated with 12 significant digits by the method described above, are shown in Table 8.1 for even values of $|m|$ and in Table 8.2 for odd values of $|m|$. The same eigenvalues have previously been calculated by Bell, Davidson, and Warsop [8] (see their Tables 2 and 3, where the quantum number ℓ is the same as our quantum number $|m|$). Comparison with our results shows that the values given in Tables 2 and 3 of [8] are inaccurate by at the most a few units in the penultimate digit.

TABLE 8.1. Eigenvalues of the Two-Dimensional Quartic Oscillator for Even Values of $|m|$

$|m|$ v	0	2	4	6	8
0	1.47714975358				
2	6.00338608331	5.62433934939			
4	11.8024335951	11.5347494634	10.7582651654		
6	18.4588187041	18.2454190471	17.6161535360	16.5993938148	
8	25.7917923785	25.6114884883	25.0761859502	24.2012853355	23.0085828705
10	33.6942798766	33.5366321542	33.0669774618	32.2946108625	31.2333713758
12	42.0938077108	41.9528155033	41.5319464543	40.8372588340	39.8780530766
14	50.9374043246	50.8092644590	50.4262843401	49.7926458577	48.9148971678
16	60.1843312661	60.0664640544	59.7138936277	59.1296371320	58.3184829832
18	69.8020966749	69.6926625364	69.3651267735	68.8217436896	68.0661253945
20	79.7640658245	79.6617014855	79.3551956010	78.8462808852	78.1377532669
22	90.0479326379	89.9515964682	89.6630486547	89.1836524902	88.5156195083
24	100.634691871	100.543567970	100.270565135	99.8167772264	99.1839860015
26	111.507919998	111.421355958	111.161964279	110.730637383	110.128833622
28	122.653255555	122.570720546	122.323363847	121.911924043	121.337611242
30	134.058013393	133.979069038	133.742443849	133.348756758	132.799023985
32	145.710891761	145.635171345	145.408186051	145.030460244	144.502856379
34	157.601745507	157.528938792	157.310669055	156.947384852	156.439825034
36	169.721407483	169.651249404	169.440904860	169.090760875	168.601455731
38	182.061545832	181.993807915	181.790706873	181.452579222	180.979980528
40	194.614548434	194.549032194	194.352582111	194.025492825	193.568251203
42	207.373428272	207.309960051	207.119642263	206.802734509	206.359666091
44	220.331745104	220.270171994	220.085529623	219.778048031	219.348107896
46	233.483540040	233.423726620	233.244354888	232.945629764	232.527890550
48	246.823280419	246.765106112	246.590644520	246.300079053	245.893713570

TABLE 8.2. Eigenvalues of the Two-Dimensional Quartic Oscillator for Odd Values of $|m|$

$|m|$	1	3	5	7	9
v					
1	3.39815017612				
3	8.70045381407	8.09066777036			
5	14.9778083724	14.5086752577	13.6008780592		
7	21.9996010346	21.6113406074	20.8495182161	19.7391730543	
9	29.6348795564	29.2999027995	28.6381966888	27.6646126668	26.3981184227
11	37.7983477541	37.5015281726	36.9130077468	36.0422964436	34.9020995078
13	46.4293001915	46.1613712780	45.6289255711	44.8384252120	43.7987560635
15	55.4819391193	55.2367731055	54.7488394606	54.0227380241	53.0649178994
17	64.9202624005	64.6935745854	64.2419551090	63.5688015252	62.6789427480
19	74.7150890157	74.5037533478	74.0824102998	73.4536457785	72.6211717923
21	84.8422004757	84.6438586523	84.2482064951	83.6572614570	82.8739411694
23	95.2811144947	95.0939386790	94.7204057303	94.1621225628	93.4214264008
25	106.014240271	105.836781852	105.482526840	104.952777861	104.249438063
27	117.026275995	116.857365248	116.520087226	116.015513959	115.345217293
29	128.303766543	128.142443797	127.820250664	127.338081033	126.697249357
31	139.834770810	139.680237864	139.371552961	138.909470050	138.295100441
33	151.608606253	151.460191249	151.163685461	150.719731277	150.129277439
35	163.615649199	163.472780665	163.187322058	162.759825655	162.191108474
37	175.847176286	175.709363614	175.433979399	175.021502285	174.472641358
39	188.295236822	188.162055329	187.895902687	187.497196739	186.966557240
41	200.952548783	200.823628115	200.565971493	200.179946077	199.666097012
43	213.812413129	213.687428145	213.437621364	213.063317336	212.564998374
45	226.868642540	226.747306054	226.141346962	226.141346962	225.657441796
47	240.115501599	239.997558545	239.761801862	239.408489184	238.938003952
49	253.547656189	253.432878879	253.203440349	252.859571766	252.401617391

References

[1] Milne, W.E., *Phys. Rev.* **35** (1930), 863–867.

[2] Wilson, H.A., *Phys. Rev.* **35** (1930), 948–956.

[3] Young, L.A., *Phys. Rev.* **38** (1931), 1612–1614.

[4] Young, L.A., *Phys. Rev.* **39** (1932), 455–457.

[5] Wheeler, J.A., *Phys. Rev.* **52** (1937), 1123–1127.

[6] Fröman, N., Fröman, P.O., and Linnaeus, S., Phase-integral formulas for the regular wave function when there are turning points close to a pole of the potential. This is Chapter 6 in the present monograph.

[7] Fröman, N. and Fröman, P.O., Phase-integral approximation of arbitrary order generated from an unspecified base function. This is Chapter 1 in the present monograph.

[8] Bell, S., Davidson, R., and Warsop, P.A., *J. Phys.* **B3** (1970), 113–122.

9
Improved Phase-Integral Treatment of the Combined Linear and Coulomb Potential

Staffan Linnaeus

Abstract. The combined linear and Coulomb potential $V(r) = ar - b/r$ is considered. Previously given phase-integral formulas for energy levels and expectation values of integer powers of r (from which the probability density at the origin can also be calculated) are improved by means of results from comparison equation theory. All formulas are expressed analytically in terms of complete elliptic integrals.

9.1   Introduction

In previous work [1, 2], analytical phase-integral formulas for energy levels and expectation values associated with the radial Schrödinger equation containing a linear potential plus a Coulomb potential

$$V(r) = ar - b/r \qquad (9.1.1)$$

were derived. These formulas were found to give very accurate results for sufficiently large values of the angular momentum quantum number ℓ, whereas less satisfactory results were obtained for small values of ℓ. One can understand this as being due to a transition zero on the negative part of the real axis, which draws closer to the origin as the angular momentum diminishes. The purpose of this chapter is to include into the phase-integral formulas the effect of this transition zero as well as of a transition zero on the positive part of the real axis which, like the transition zero on the negative part of the real axis, approaches the origin when the energy increases. We achieve this result by means of a new formula for the wave function derived with comparison equation technique [3]. In this way, the accuracy of the phase-integral formulas can be increased dramatically.

The quantization condition is treated in §9.2. In §9.3 we deal with the expectation values of integer powers of r, which also give the probability density at the origin. All results are expressed analytically in terms of complete elliptic integrals in the Appendix. As this chapter is a continuation of [1] and [2], we use definitions and results from these papers.

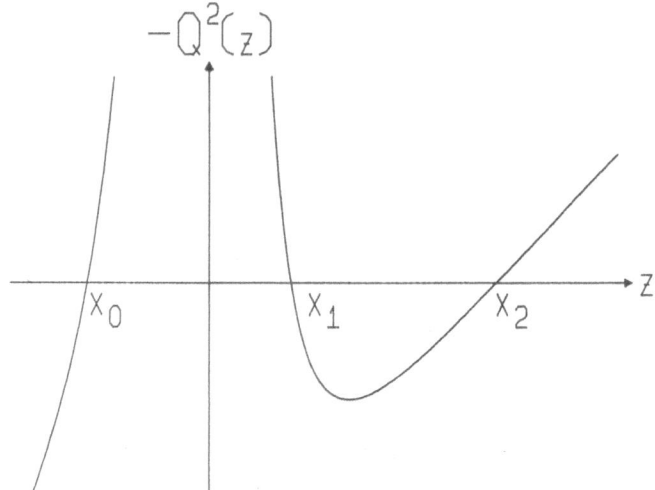

FIGURE 9.1. Qualitative behavior of the function $-Q^2(z)$. The three zeros x_0, x_1 and x_2 of $Q^2(z)$ are indicated.

9.2   Energy Levels

Introducing a dimensionless variable z and dimensionless parameters A and B as in [1], we may write the radial Schrödinger equation with the potential (9.1.1) as

$$\frac{d^2\psi}{dz^2} + R(z)\psi = 0, \qquad (9.2.1)$$

where

$$R(z) = A - z + \frac{B}{z} - \frac{\ell(\ell+1)}{z^2}. \qquad (9.2.2)$$

The phase-integral *base function* $Q(z)$ is chosen as it is in [1]; that is,

$$Q^2(z) = R(z) - \frac{1}{4z^2} = A - z + \frac{B}{z} - \frac{(\ell+\frac{1}{2})^2}{z^2}. \qquad (9.2.3)$$

As illustrated in Figure 9.1, the squared base function has one negative zero, x_0; and two positive zeros, x_1 and x_2. As the energy increases, x_0 and x_1 approach the origin, while x_2 recedes from the origin. We shall allow x_0 and x_1 to be arbitrarily close to the origin, whereas x_2 should be regarded as well isolated from x_1. Thus, we are primarily concerned with excited states, although our treatment turns out to work fairly well even for the ground state.

Let ψ_1 and ψ_2 be two solutions of the Schrödinger equation (9.2.1) for the combined linear and Coulomb potential such that

$$\psi_1(0) = 0 \qquad (9.2.4a)$$

$$\psi_2(z) \to 0, \quad z \to +\infty. \tag{9.2.4b}$$

These solutions are uniquely determined, apart from a constant factor for each one of them. They are linearly independent, unless the energy assumes an eigenvalue.

According to Eq. (6.3.50) we have

$$\psi_1(z) = C_1 q^{-\frac{1}{2}}(z) \sin\left[\frac{1}{2}\int_{\Gamma_1(z)} q(z)\, dz + \Delta\right], \quad x_1 < z < x_2, \tag{9.2.5}$$

where C_1 is an arbitrary constant, $q(z)$ is given by Eq. (2.5) in [1], and $\Gamma_1(z)$ is the contour shown in Figure 9.2(a). The quantity Δ accounts for the effect of the transition zeros x_0 and x_1, and is in the $(2N+1)$th-order approximation given by Eqs. (6.3.36) and (6.3.37a–c) that, with our choice of the base function, can be written as

$$\Delta = \pi/4 + \arg\Gamma\left(\frac{1}{2} + \xi + i\eta\right) - \xi\arctan(\eta_0/\xi)$$

$$- \frac{1}{2}\eta\ln(\xi^2 + \eta_0^2) + \sum_{n=0}^{N}\Delta^{(2n+1)}, \tag{9.2.6}$$

where

$$\Delta^{(1)} = \eta_0 \tag{9.2.7a}$$

$$\Delta^{(3)} = -\frac{\eta_0}{24(\xi^2 + \eta_0^2)} \tag{9.2.7b}$$

$$\Delta^{(5)} = -\frac{\eta_2 + 12\eta_0\eta_2^2}{24(\xi^2 + \eta_0^2)} + \frac{7\eta_0 + 80\eta_0^2\eta_2}{960(\xi^2 + \eta_0^2)^2} - \frac{7\eta_0^3}{720(\xi^2 + \eta_0^2)^3} \tag{9.2.7c}$$

with

$$\xi = \xi_0 = -i\operatorname{Res}_{z=0}Q(z) = \ell + \frac{1}{2} \tag{9.2.8}$$

and, according to Eqs. (6.3.39) and (6.3.9),

$$\eta = \sum_{n=0}^{N}\eta_{2n} = \sum_{n=0}^{N}\frac{i}{2\pi}\int_{\Gamma_\eta} Y_{2n}Q(z)\, dz. \tag{9.2.9}$$

Γ_η is the contour of integration shown in Figure 9.2(b).

In order to obtain an approximation of $\psi_2(z)$ in the classically allowed region we use the phase-integral connection formula (21) in [4], which yields

$$\psi_2(z) = C_2 q^{-\frac{1}{2}}(z)\cos\left(-\frac{1}{2}\int_{\Gamma_2(z)} q(z)\, dz - \frac{\pi}{4}\right), \quad x_1 < z < x_2, \tag{9.2.10}$$

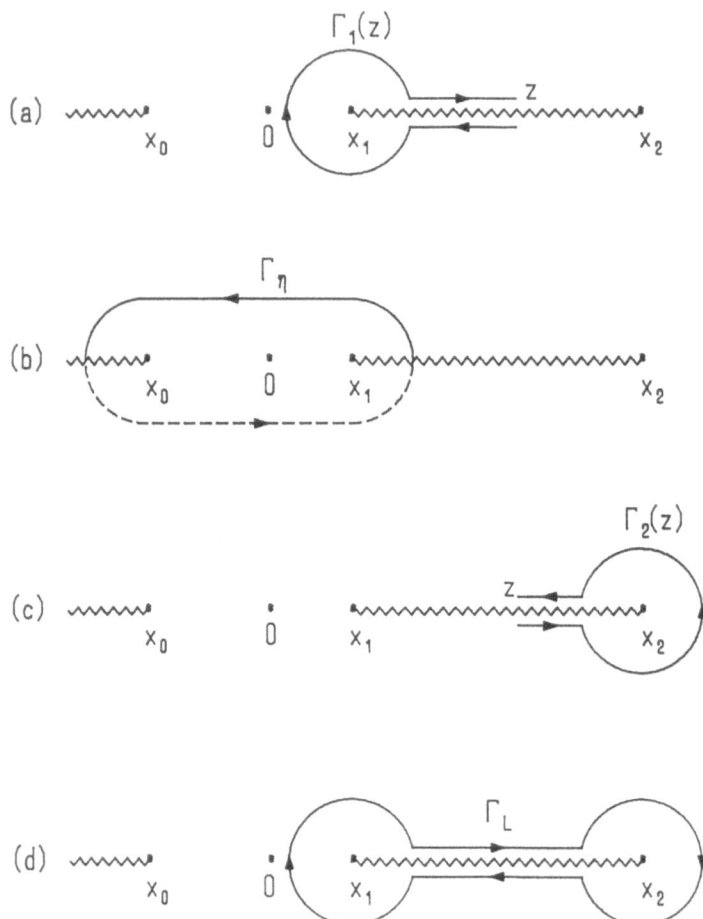

FIGURE 9.2. Contours of integration in the complex z-plane. Wavy lines indicate branch cuts. The dashed part of the contour Γ_η lies on a Riemann sheet adjacent to the complex z-plane under consideration. The base function is chosen positive on the upper lip of the branch cut between x_1 and x_2.

where C_2 is an arbitrary constant and $\Gamma_2(z)$ is the contour shown in Figure 9.1(c). Equivalently, we can write

$$\psi_2(z) = C_2 q^{-\frac{1}{2}}(z) \sin\left(\frac{1}{2}\int_{\Gamma_1(z)} q(z)\,dz + \frac{3\pi}{4} - L\right), \quad x_1 < z < x_2,$$
(9.2.11)

where (cf. Eqs. [2.5] and [2.10] in [1])

$$L = \sum_{n=0}^{N} L^{(2n+1)} = \sum_{n=0}^{N} \frac{1}{2}\int_{\Gamma_L} Y_{2n} Q(z)\,dz.$$
(9.2.12)

Γ_L is the contour shown in Figure 9.2(d).

Requiring that ψ_1 and ψ_2 be the same function at the eigenenergies, we obtain from (9.2.5) and (9.2.11) the quantization condition

$$L + \Delta = \left(s + \frac{3}{4}\right)\pi, \quad s = 0, 1, 2, \ldots.$$
(9.2.13)

Note that Eq. (2.9) in [1] is recovered from this formula by the approximation $\Delta \approx \pi/4$.

The integrals $L^{(2n+1)}$ were evaluated in terms of complete elliptic integrals in [1] for $n = 0$, 1 and 2 (see Eqs. [3.6] and [3.6a–c] in that paper). The formula for $L^{(2n+1)}$, and the corresponding formula for η_{2n}, are given in the Appendix of this chapter.

In Figure 9.3 we illustrate the error of the approximate quantization condition (9.2.13). By comparing these diagrams with Figures 4a–d in [1] we shall see that in most cases a considerable improvement of the accuracy has been gained by the treatment set forth here.

9.3   Expectation Values

The quantization condition (9.2.13) can also be used to calculate expectation values of multiplicative operators by the method devised by Fröman and Fröman [5]. Formula (6) in [5] gives (cf. Eqs. [9.2.1], [9.2.2], and [9.2.8])

$$\langle z^{-1} \rangle = \frac{\partial(L + \Delta)}{\partial B} \Big/ \frac{\partial(L + \Delta)}{\partial A}$$
(9.3.1a)

$$\langle z^{-2} \rangle = -\frac{1}{2\xi}\frac{\partial(L + \Delta)}{\partial \xi} \Big/ \frac{\partial(L + \Delta)}{\partial A}.$$
(9.3.1b)

From (9.2.6) with (9.2.7a–c) we obtain

$$\frac{\partial \Delta}{\partial \kappa} = \left[\operatorname{Re}\psi\left(\frac{1}{2} + \xi + i\eta\right) - \frac{1}{2}\ln(\xi^2 + \eta_0^2)\right]\frac{\partial \eta}{\partial \kappa} - \frac{\xi^2 + \eta\eta_0}{\xi^2 + \eta_0^2}\frac{\partial \eta_0}{\partial \kappa}$$

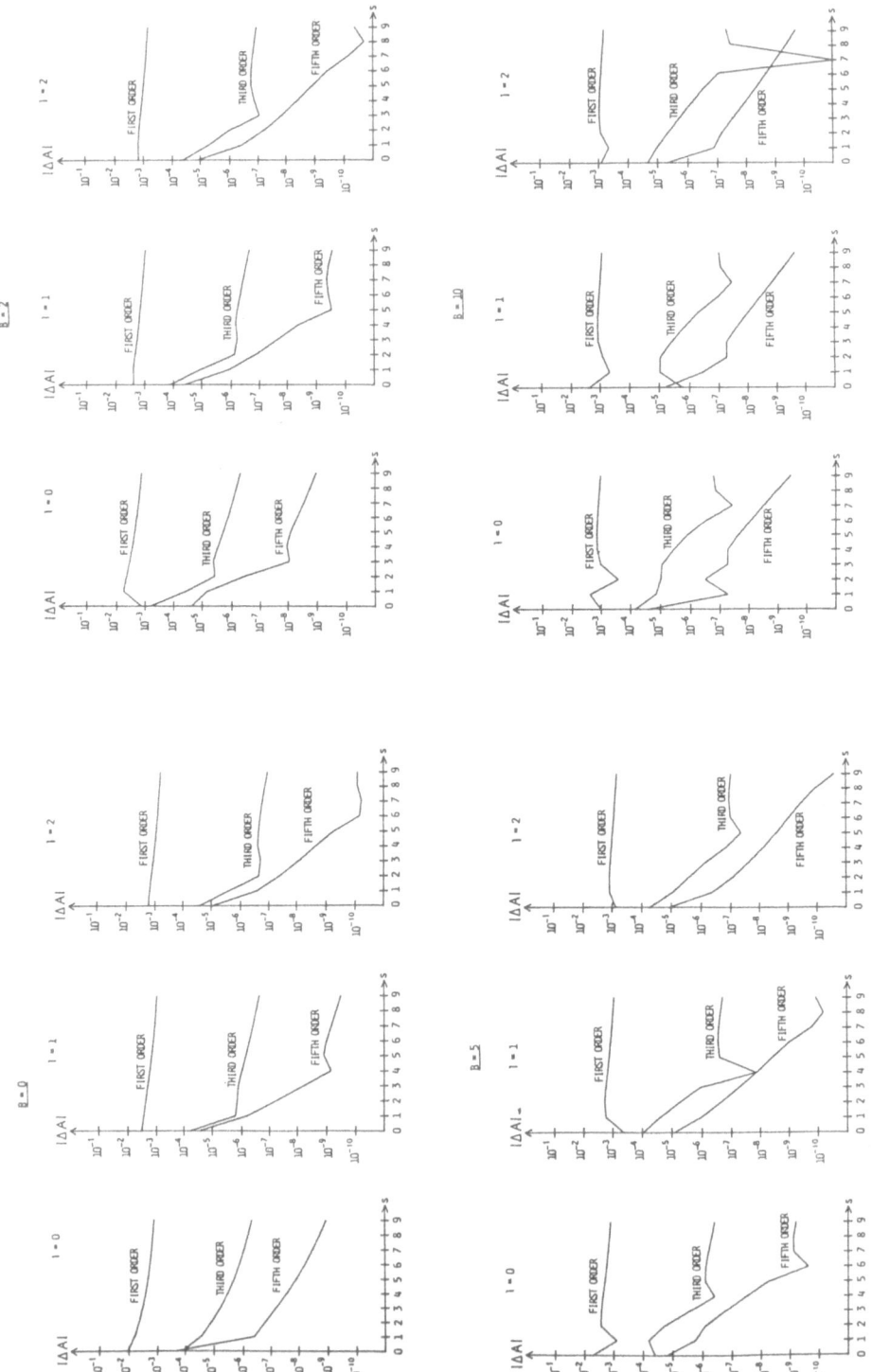

FIGURE 9.3. Error of the approximate energy levels obtained from (9.2.13). The quantity ΔA is defined as $A_{\text{phase-integral}} - A_{\text{exact}}$.

$$+ \left[\operatorname{Im} \psi \left(\frac{1}{2} + \xi + i\eta \right) - \arctan \frac{\eta_0}{\xi} + \frac{\xi(\eta_0 - \eta)}{\xi^2 + \eta_0^2} \right] \frac{\partial \xi}{\partial \kappa} + \sum_{n=0}^{N} \frac{\partial \Delta^{(2n+1)}}{\partial \kappa}$$

$$(9.3.2)$$

with

$$\frac{\partial \Delta^{(1)}}{\partial \kappa} = \frac{\partial \eta_0}{\partial \kappa} \tag{9.3.3a}$$

$$\frac{\partial \Delta^{(3)}}{\partial \kappa} = \frac{\eta_0^2 - \xi^2}{24(\xi^2 + \eta_0^2)^2} \frac{\partial \eta_0}{\partial \kappa} + \frac{\xi\eta_0}{12(\xi^2 + \eta_0^2)^2} \frac{\partial \xi}{\partial \kappa} \tag{9.3.3b}$$

$$\frac{\partial \Delta^{(5)}}{\partial \kappa} = \left(-\frac{\eta_2^2}{2(\xi^2 + \eta_0^2)} + \frac{7 + 240\eta_0\eta_2 + 960\eta_0^2\eta_2^2}{960(\xi^2 + \eta_0^2)} - \frac{7\eta_0^2 + 40\eta_0^3\eta_2}{120(\xi^2 + \eta_0^2)^3} \right.$$

$$\left. + \frac{7\eta_0^4}{120(\xi^2 + \eta_0^2)^4} \right) \frac{\partial \eta_0}{\partial \kappa} + \left(-\frac{1 + 24\eta_0\eta_2}{24(\xi^2 + \eta_0^2)} + \frac{\eta_0^2}{12(\xi^2 + \eta_0^2)^2} \right) \frac{\partial \eta_2}{\partial \kappa}$$

$$+ \left(\frac{\eta_2 + 12\eta_0\eta_2^2}{12(\xi^2 + \eta_0^2)^2} - \frac{7\eta_0 + 80\eta_0^2\eta_2}{240(\xi^2 + \eta_0^2)^3} + \frac{7\eta_0^3}{120(\xi^2 + \eta_0^2)^4} \right) \xi \frac{\partial \xi}{\partial \kappa}, \tag{9.3.3c}$$

where κ is an arbitrary parameter and $\psi(\xi + \frac{1}{2} + i\eta)$ (not to be confused with the wave function) is the logarithmic derivative of the gamma function. Explicit expressions for the derivatives of $L^{(2n+1)}$ and η_{2n} in (9.3.1a,b) are given in the Appendix. From the expectation values $\langle z^{-1} \rangle$ and $\langle z^{-2} \rangle$ one can calculate the expectation value of any nonnegative power of z by means of the exact recurrence relation (3.1) in [2]; similarly one can calculate all existing expectation values of negative powers of z by means of the exact recurrence relation (3.2) in [2]. Furthermore, one may obtain the probability density close to the origin through the exact formula (4.1) in [2]. (There is, however, a misprint in that formula. To obtain the correct formula, $\langle z^{-2\ell-1} \rangle$ in the last term should be replaced by $\langle z^{-2\ell-2} \rangle$.

In Tables 9.1 and 9.2a,b we compare the values of $\langle z^{-1} \rangle$ and $\langle z^{-2} \rangle$ obtained from our phase-integral formulas with results from exact numerical calculations (second order difference approximation with Richardson extrapolation). It may be seen that the fifth-order phase-integral values are extremely accurate especially for excited states, even when $\ell = 0$.

Acknowledgment. I wish to thank Professor N. Fröman, Professor P.O. Fröman, and Dr. S. Yngve for much helpful advice during the course of this work.

Table 9.1. Expectation values of z^{-1} for $B = 0$

$\ell = 0$

s	0	1	2	3
First order	0.834 7	0.581 7	0.472 1	0.407 0
Third order	0.834 4	0.582 1	0.472 310 6	0.407 144 95
Fifth order	0.835 1	0.582 171	0.472 312 02	0.407 145 110
Numerical	0.834 87	0.582 169	0.472 311 94	0.407 145 115

$\ell = 1$

s	0	1	2	3
First order	0.512 4	0.411 00	0.352 84	0.313 88
Third order	0.512 19	0.411 013	0.352 875 9	0.313 912 0
Fifth order	0.512 24	0.411 017 2	0.352 876 72	0.313 912 242
Numerical	0.512 23	0.411 016 9	0.352 876 69	0.313 912 235

$\ell = 2$

s	0	1	2	3
First order	0.387 70	0.330 664	0.293 131	0.266 014
Third order	.0.387 621	0.330 654 7	0.293 136 5	0.266 022 71
Fifth order	0.387 630 3	0.330 655 88	0.293 136 890	0.266 022 840
Numerical	0.387 628 5	0.330 655 83	0.293 136 881	0.266 022 838

Table 9.2a. Expectation values of z^{-2} for $B = 0$

$\ell = 0$				
s	0	1	2	3
First order	1.107	0.817	0.694	0.622
Third order	1.124 9	0.823 77	0.698 35	0.624 493
Fifth order	1.125 2	0.823 671	0.698 317	0.624 481 7
Numerical	1.124 8	0.823 680	0.698 319	0.624 482 1

$\ell = 1$				
s	0	1	2	3
First order	0.313 3	0.256 0	0.224 5	0.203 8
Third order	0.314 12	0.256 540	0.224 833 5	0.204 038 3
Fifth order	0.314 150	0.256 538 33	0.224 832 375	0.204 037 726
Numerical	0.314 136	0.256 538 17	0.224 832 376	0.204 037 730

$\ell = 2$				
s	0	1	2	3
First order	0.167 7	0.144 3	0.129 53	0.119 07
Third order	0.167 859	0.144 482 75	0.129 636 02	0.119 150 23
Fifth order	0.167 864 3	0.144 482 87	0.129 635 916	0.119 150 132
Numerical	0.167 862 6	0.144 482 83	0.129 635 913	0.119 150 132

Table 9.2b. Expectation values of z^{-2} for $B = 1.4$

$\ell = 0$

s	0	1	2	3
First order	2.458	1.286	0.989	0.843
Third order	2.457	1.292 701	0.993 36	0.845 432
Fifth order	2.460 03	1.292 707	0.993 341 7	0.845 422 0
Numerical	2.459 90	1.292 698	0.993 341 8	0.845 422 1

$\ell = 1$

s	0	1	2	3
First order	0.413 0	0.311 9	0.263 6	0.234 2
Third order	0.413 60	0.312 476 8	0.264 013 7	0.234 477 0
Fifth order	0.413 68	0.312 478 5	0.264 013 07	0.234 476 481
Numerical	0.413 66	0.312 478 1	0.264 013 050	0.234 476 481

$\ell = 2$

s	0	1	2	3
First order	0.195 20	0.162 67	0.143 29	0.130 13
Third order	0.195 342	0.162 807 1	0.143 399 38	0.130 216 22
Fifth order	0.195 351	0.162 807 6	0.143 399 356	0.130 216 151
Numerical	0.195 348	0.162 807 53	0.143 399 349	0.130 216 150

Appendix: Expressions for Phase-Integral Quantities in Terms of Complete Elliptic Integrals

Recalling that the square of the base function $Q(z)$ has three real zeros x_0, x_1, and x_2 (see Figure 9.1), we define the three independent quantities (cf. Eqs. (3.5a–c) in [1]):

$$d = (x_2 - x_0)^{\frac{1}{2}} \tag{9.A.1a}$$

$$k = \left(\frac{x_2 - x_1}{x_2 - x_0}\right)^{\frac{1}{2}} \tag{9.A.1b}$$

$$\alpha^2 = \frac{x_2 - x_1}{x_2} \tag{9.A.1c}$$

and also introduce the notations (cf. Eqs. [3.5d–e] in [1])

$$k' = (1 - k^2)^{\frac{1}{2}} \tag{9.A.1d}$$

$$\alpha'^2 = \frac{\alpha^2(1-k^2)}{\alpha^2 - k^2}. \tag{9.A.1e}$$

The integrals $L^{(2n+1)}$ and η_{2n} appearing in the quantization condition given in §9.2 can be evaluated using the standard technique [1], which yields (cf. Eq. (3.6) in [1])

$$L^{(2n+1)} = d^{3-6n}\left[f_{2n+1}(k^2,\alpha^2)\frac{K(k)-E(k)}{k^2} + (-1)^n f_{2n+1}(k'^2,\alpha'^2)\frac{E(k)}{k'^2} \right.$$
$$\left. + \frac{2k^2 k'^2(\alpha^2-1)}{\alpha^2 \alpha'^2}\delta_{n,0}\Pi(\alpha^2,k) \right] \tag{9.A.2a}$$

and

$$\eta_{2n} = -\frac{1}{\pi}d^{3-6n}\left(f_{2n+1}(k^2,\alpha^2)\frac{E(k')}{k^2} + (-1)^n f_{2n+1}(k'^2,\alpha'^2)\frac{K(k')-E(k')}{k'^2} \right.$$
$$\left. + \frac{2k^2 k'^2(\alpha'^2-1)}{\alpha^2\alpha'^2}\delta_{n,0}\Pi(\alpha'^2,k') \right), \tag{9.A.2b}$$

where $K(k)$, $E(k)$, and $\Pi(\alpha^2,k)$ are the complete elliptic integrals of the first, second, and third kind, and the functions f_{2n+1} for $n = 0$, 1, and 2 are given by Eqs. (3.6a–c) in [1].

We shall also give the parameter derivatives of $L^{(2n+1)}$ and η_{2n}, which are necessary in the calculation of the expectation values of z^{-1} and z^{-2} according to the phase-integral formulas (9.3.1a,b). The results can be concisely notated as with the convention

$$\frac{\partial}{\partial \kappa_\nu} = \begin{cases} \frac{\partial}{\partial A} & \nu = 0 \\ \frac{\partial}{\partial B} & \nu = -1 \\ \frac{\partial}{\partial(-\xi^2)} = -\frac{1}{2\xi}\frac{\partial}{\partial\xi} & \nu = -2. \end{cases} \tag{9.A.3}$$

Then we have

$$\frac{\partial L^{(2n+1)}}{\partial \kappa_\nu} = \frac{1}{2}d^{2\nu-6n+1}\left(g_{2n+1}^\nu(k^2,\alpha^2)\frac{K(k)-E(k)}{k^2} \right.$$
$$\left. - (-1)^{\nu+n}g_{2n+1}^\nu(k'^2,\alpha'^2)\frac{E(k)}{k'^2} + \frac{2\alpha^2}{k^2}\delta_{\nu,-2}\delta_{n,0}\Pi(\alpha^2,k) \right), \tag{9.A.4a}$$

$$\frac{\partial \eta_{2n}}{\partial \kappa_\nu} = -\frac{1}{2\pi}d^{2\nu-6n+1}\left(g_{2n+1}^\nu(k^2,\alpha^2)\frac{E(k')}{k^2} \right.$$
$$\left. - (-1)^{\nu+n}g_{2n+1}^\nu(k'^2,\alpha'^2)\frac{K(k')-E(k')}{k'^2} \right.$$
$$\left. - \frac{2\alpha'^2}{k'^2}\delta_{\nu,-2}\delta_{n,0}\Pi(\alpha'^2,k') \right), \quad \nu = 0,-1,-2, \tag{9.A.4b}$$

where the functions we need, g_{2n+1}^ν, are given by Eqs. (2.2a–i) in [2].

References

[1] Thidé, B. and Linnaeus, S., *Ann. Phys.* (N.Y.) **164** (1985), 495–505.

[2] Linnaeus, S. and Düring, M., *Ann. Phys.* (N.Y.) **164** (1985), 506–515.

[3] Fröman, N., Fröman, P.O., and Linnaeus, S., Phase-integral formulas for the regular wave function when there are turning points close to a pole of the potential. This is Chapter 6 in the present monograph.

[4] Fröman, N., *Ann. Phys.* (N.Y.) **61** (1970), 451–464.

[5] Fröman, N. and Fröman, P.O., *Ann. Phys.* (N.Y.) **163** (1985), 215–226.

10
High-Energy Scattering from a Yukawa Potential

Staffan Linnaeus

Abstract. Phase shifts and probability densities at the origin for a nonrelativistic particle in a Yukawa potential are calculated by means of arbitrary-order phase-integral formulas, obtained from a comparison equation treatment. Numerical calculations show that the formulas are very accurate even for the lowest partial waves, provided that the energy is sufficiently high.

10.1   Introduction

The arbitrary-order phase-integral method [1–5] is, in general, very convenient for calculating phase shifts and other quantities, but for low values of the angular momentum, it is sometimes less satisfactory. A possibility for overcoming the difficulties associated with the lowest partial waves was devised by N. Fröman, P.O. Fröman, and S. Linnaeus [6], who derived a new connection formula that takes account of the turning points on each side of the origin by means of comparison equation technique. The supplementary quantities obtained in this way make it possible to handle all partial waves accurately if the energy is sufficiently high.

In this chapter we shall apply the results from the treatment in Chapter 6 to the nonrelativistic scattering problem for the Yukawa potential

$$V(r) = -Be^{-\lambda r}/r, \qquad (10.1.1)$$

where B and λ are positive constants. The phase shifts produced by this potential are considered in §10.2, where the new phase-integral formula is tested against exact calculations and found to be very accurate for reasonably high energies, as expected. In §10.3 we perform the corresponding investigation for the probability density at the origin. The numerical method used for calculating the exact reference values is described in the Appendix.

10.2   Phase Shifts

The radial Schrödinger equation can be written

$$d^2 u_\ell(r)/dr^2 + R(r)u_\ell(r) = 0 \qquad (10.2.1)$$

where

$$R(r) = k^2 - 2mV(r)/\hbar^2 - \ell(\ell+1)/r^2. \qquad (10.2.2)$$

According to the new connection formula, given in §6.3.2, the wave function that is regular at the origin is (when r lies well to the right of the positive turning point t_+ in Figure 10.1) given by the approximate formula

$$u_\ell(r) = \text{const } q^{-\frac{1}{2}}(r) \sin\left(\frac{1}{2}\int_{\Gamma(r)} q(r)dr + \Delta\right), \qquad (10.2.3)$$

where $\Gamma(r)$ is the contour shown in Figure 10.1(a) and, in the $(2N+1)$th order of approximation,

$$q(r) = Q(r) \sum_{n=0}^{N} Y_{2n}. \qquad (10.2.4)$$

$Q(r)$ is a function we have not yet specified, called the *base function*. The first few functions Y_{2n} are

$$Y_0 = 1 \qquad (10.2.5a)$$

$$Y_2 = \frac{1}{2}\varepsilon_0 \qquad (10.2.5b)$$

$$Y_4 = -\frac{1}{8}\varepsilon_0^2 - \frac{1}{8}\varepsilon_2 \qquad (10.2.5c)$$

with

$$\varepsilon_0 = \frac{R(r) - Q^2(r)}{Q^2(r)} + Q^{-\frac{3}{2}}(r)\frac{d^2}{dr^2}Q^{-\frac{1}{2}}(r) \qquad (10.2.6)$$

and

$$\varepsilon_n = \left(\frac{1}{Q(r)}\frac{d}{dr}\right)^n \varepsilon_0. \qquad (10.2.7)$$

Choosing

$$Q^2(r) = R(r) - 1/(4r^2) = k^2 - 2mV(r)/\hbar^2 - \left(\ell + \frac{1}{2}\right)^2/r^2, \qquad (10.2.8)$$

we obtain from §6.3.2

$$\Delta = \frac{\pi}{4} + \arg\Gamma\left(\frac{1}{2} + \xi + i\eta\right) - \frac{1}{2}\eta\ln(\xi^2 + \eta_0^2)$$

$$- \xi \arctan(\eta_0/\xi) + \sum_{n=0}^{N} \Delta^{(2n+1)}, \tag{10.2.9}$$

where

$$\Delta^{(1)} = \eta_0 \tag{10.2.10a}$$

$$\Delta^{(3)} = -\frac{\eta_0}{24(\xi^2 + \eta_0^2)} \tag{10.2.10b}$$

$$\Delta^{(5)} = -\frac{\eta_2 + 12\eta_0\eta_2^2}{24(\xi^2 + \eta_0^2)} + \frac{7\eta_0 + 80\eta_0\eta_2^2}{960(\xi^2 + \eta_0^2)^2} - \frac{7\eta_0^3}{720(\xi^2 + \eta_0^2)} \tag{10.2.10c}$$

with

$$\xi = \xi_0 = \ell + \frac{1}{2} \tag{10.2.11}$$

$$\eta = \sum_{n=0}^{N} \eta_{2n} = \sum_{n=0}^{N} \frac{i}{2\pi} \int_{\Gamma} Q(z) Y_{2n} dz; \tag{10.2.12}$$

Γ is the closed contour shown in Figure 10.1(c).

The phase shift δ_ℓ is defined from the asymptotic formula

$$u_\ell(r) \sim \text{const } \sin(kr - \ell\pi/2 + \delta_\ell), \quad r \to +\infty. \tag{10.2.13}$$

By comparing (10.2.13) with (10.2.3) we obtain the following approximate formula for the phase shift:

$$\delta_\ell = \lim_{r \to +\infty} \left(\frac{1}{2} \int_{\Gamma(r)} q(r) dr - kr \right) + \frac{\ell\pi}{2} + \Delta. \tag{10.2.14}$$

The accuracy of (10.2.14) in the first, third, and fifth order of phase-integral approximation is illustrated in Figure 10.2. It may be seen that the third- and fifth-order values become extremely accurate for large k (i.e., high energy) and small λ (i.e., weak screening).

10.3   Probability Density at the Origin

The radial probability density is given by

$$\rho(r) = |u_\ell(r)|^2/r^2, \tag{10.3.1}$$

where $u_\ell(r)$ is a suitably normalized wave function; in the present section we require that $u_\ell(r)$ has unit amplitude at infinity. Close to the origin, we have

$$\rho(r) \sim \rho_0 r^{2\ell}, \tag{10.3.2}$$

where ρ_0 is a constant that must be calculated.

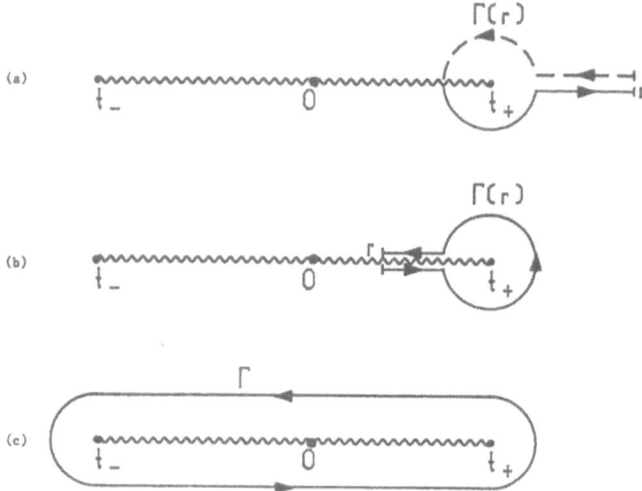

FIGURE 10.1. Contours of integration in the complex r-plane. The positive and the negative turning points are indicated as t_+ and t_-, respectively. A branch cut, marked as a wavy line, is introduced so as to make $Q(r)$ single-valued. The contour $\Gamma(r)$ is shown both for $r < t_+$ and for $r > t_+$. The dashed part of $\Gamma(r)$ for $r > t_+$ lies on a Riemann sheet adjacent to the complex r-plane under consideration.

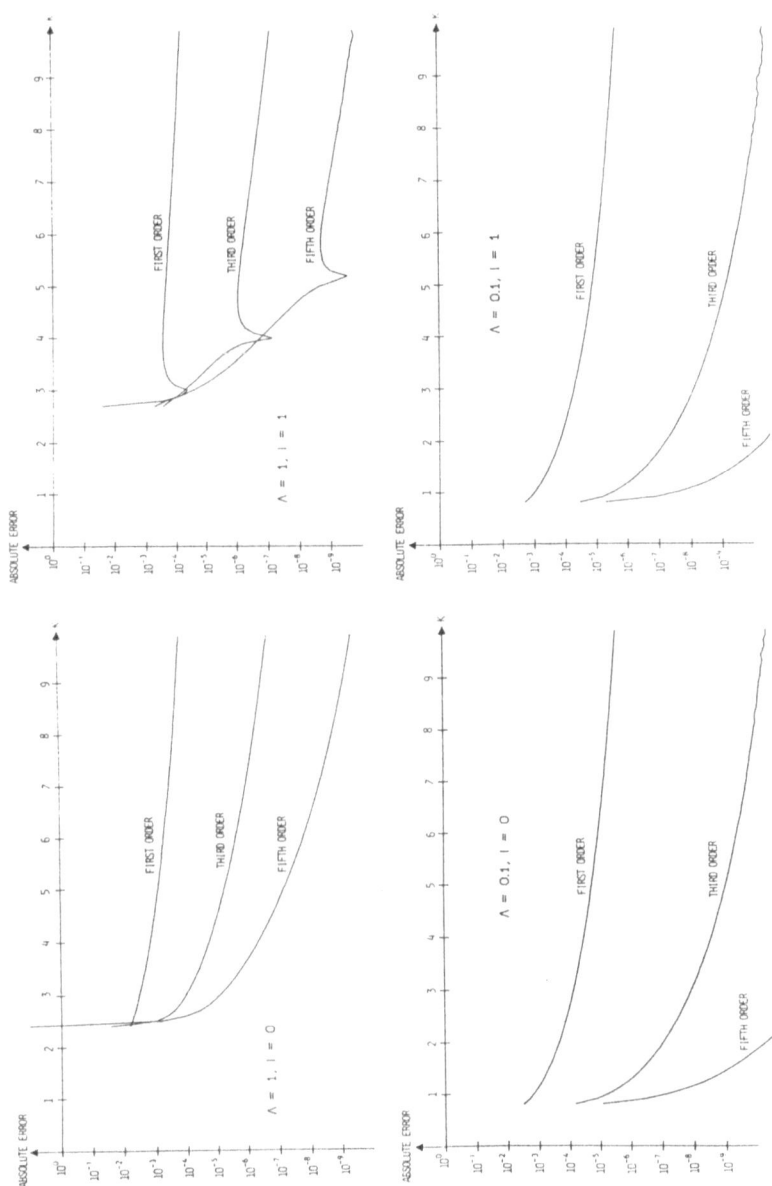

FIGURE 10.2. Absolute value of the error of the phase shift formula (10.2.14) when applied to the Yukawa potential (10.1.1). The dimensionless quantities K and Λ are defined as $K = k\hbar^2/(mB)$ and $\Lambda = \lambda\hbar^2/(mB)$.

Recalling the treatment in [7], we obtain from Eqs. (7.1.13a), (7.1.16), (7.2.5a), (7.2.8a,b), (7.2.20a), and (7.2.21a) the approximate formula

$$\rho_0 = 2^{2\ell} e^{-\pi\eta} \left(\frac{|\Gamma(\ell+1+i\eta)|}{(2\ell+1)!} \right)^2 k\,\alpha^{2\ell+1}, \qquad (10.3.3)$$

where, in the $(2N+1)$th-order approximation,

$$\alpha = \sum_{n=0}^{N} \alpha_{2n}, \qquad (10.3.4)$$

with

$$\alpha_0 = \frac{2\xi^2}{(\xi^2+\eta_0^2)^{\frac{1}{2}}} \exp\left[-\frac{\gamma_0}{\xi} + \frac{\eta_0}{\xi}\left(\arctan\frac{\eta_0}{\xi} + \frac{\pi}{2} \right) - 1 \right], \qquad (10.3.4a)$$

$$\alpha_2 = \left[-\frac{\gamma_2}{\xi} + \frac{1}{24\xi^2} + \frac{1}{24(\xi^2+\eta_0^2)} + \frac{\eta_2}{\xi}\left(\arctan\frac{\eta_0}{\xi} + \frac{\pi}{2} \right) \right]\alpha_0, \qquad (10.3.4b)$$

$$\alpha_4 = \left[-\frac{\gamma_4}{\xi} - \frac{1}{2880\xi^4} + \frac{\eta_2^2}{2(\xi^2+\eta_0^2)} - \frac{7+240\eta_0\eta_2}{2880(\xi^2+\eta_0^2)^2} + \frac{7\eta_0^2}{720(\xi^2+\eta_0^2)^3} \right.$$
$$\left. + \frac{\eta_4}{\xi}\left(\arctan\frac{\eta_0}{\xi} + \frac{\pi}{2} \right) \right]\alpha_0 + \frac{\alpha_2^2}{2\alpha_0}. \qquad (10.3.4c)$$

The quantities γ_{2n} are defined by

$$\gamma_0 = \lim_{r\to 0} \left(\frac{i}{2} \int_{\Gamma(r)} Q(r)\,dr + \xi\ln r \right), \qquad (10.3.5a)$$

$$\gamma_{2n} = \frac{i}{2} \int_{\Gamma(0)} Q(r)Y_{2n}dr, \quad n > 0; \qquad (10.3.5b)$$

$\Gamma(r)$ is the contour shown in Figure 10.1(b). For the practical calculation of γ_0 we note that

$$\gamma_0 = \int_0^{t_+} \left[|Q(r)| - \xi\frac{(1-r/t_+)^{\frac{1}{2}}}{r} \right] dr + \xi[\ln(4t_+) - 2], \qquad (10.3.6)$$

where t_+ is the positive turning point (see Figure 10.1). The integral in the right-hand member is conveniently calculated by means of quadrature according to 25.4.34 in [8].

The accuracy of formula (10.3.3) is illustrated in Figure 10.3. As the phase shift formula (10.2.14), formula (10.3.3) provides very accurate values in the third and the fifth order of approximation for sufficiently high energy, especially for weak screening.

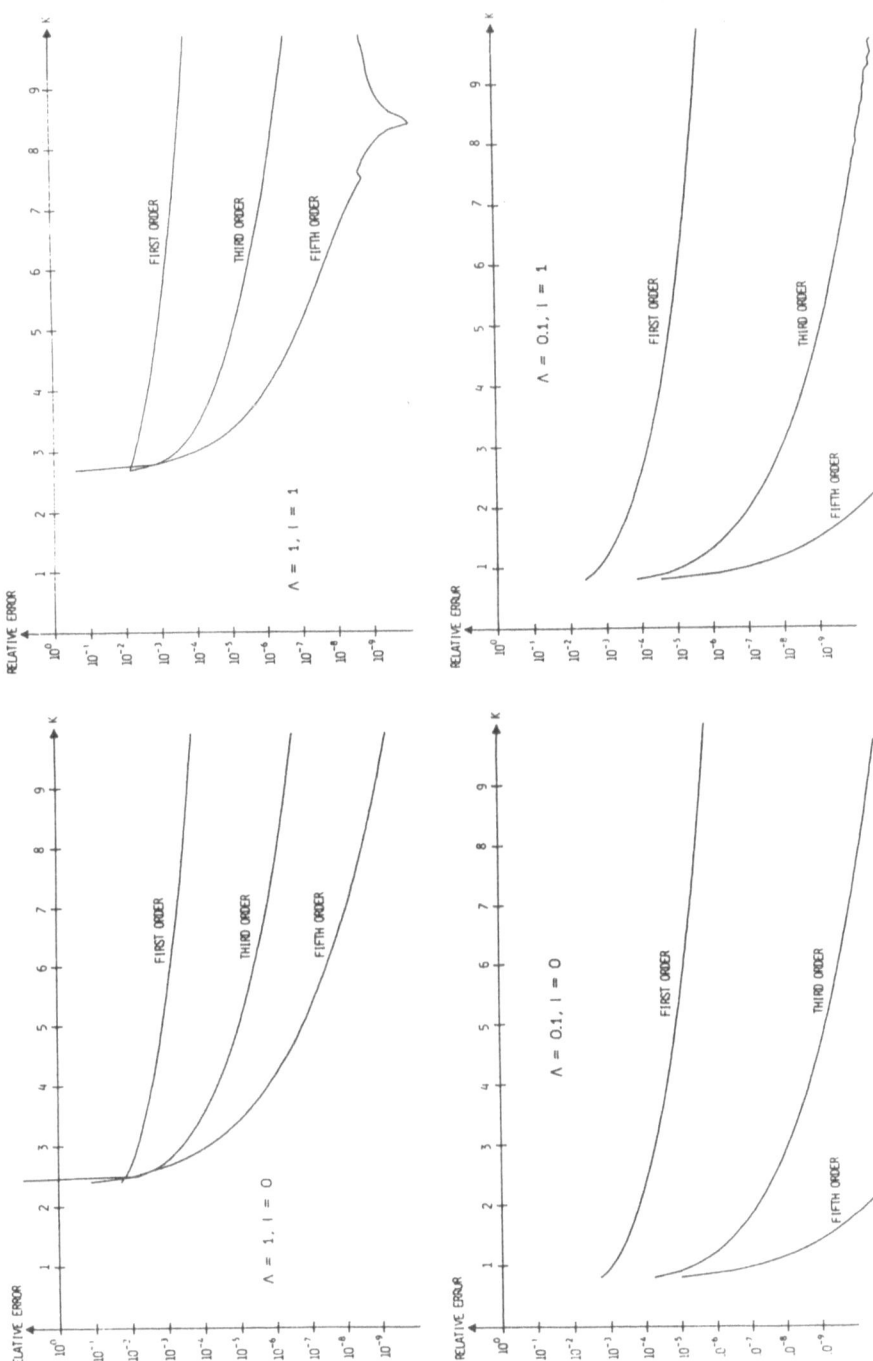

FIGURE 10.3. Absolute value of the relative error of ρ_0 according to (10.3.3) when applied to the Yukawa potential (10.1.1). The dimensionless quantities K and Λ are defined as $K = k\hbar^2/(mB)$ and $\Lambda = \lambda\hbar^2/(mB)$.

Appendix: Numerical Solution of the Schrödinger Equation

We define two functions $a_\ell(r)$ and $b_\ell(r)$ by requiring that

$$u_\ell(r) = a_\ell(r)f_\ell(r) + b_\ell(r)g_\ell(r) \tag{10.A.1a}$$

$$u_\ell'(r) = a_\ell(r)f_\ell'(r) + b_\ell(r)g_\ell'(r), \tag{10.A.1b}$$

where $f_\ell(r)$ and $g_\ell(r)$ are the regular and irregular free particle wave functions:

$$f_\ell(r) = kr\, j_\ell(kr) \tag{10.A.2a}$$

$$g_\ell(r) = -kr\, y_\ell(kr). \tag{10.A.2b}$$

j_ℓ and y_ℓ are spherical Bessel functions (see 10.1 in [8]). From the Schrödinger equation we get two coupled, first-order differential equations for $a_\ell(r)$ and $b_\ell(r)$:

$$a_\ell'(r) = 2mV(r)[a_\ell(r)f_\ell(r) + b_\ell(r)g_\ell(r)]g_\ell(r)/(\hbar^2 k), \tag{10.A.3a}$$

$$b_\ell'(r) = -2mV(r)[a_\ell(r)f_\ell(r) + b_\ell(r)g_\ell(r)]f_\ell(r)/(\hbar^2 k). \tag{10.A.3b}$$

At the origin we have

$$f_\ell(r) \sim (kr)^{\ell+1}/(2\ell+1)!! \tag{10.A.4a}$$

$$g_\ell(r) \sim \begin{cases} 1 & \text{when } \ell = 0 \\ (2\ell-1)!!(kr)^{-\ell} & \text{when } \ell > 0. \end{cases} \tag{10.A.4b}$$

Here we will use another boundary condition for $u_\ell(r)$, when r approaches zero, than we did in §10.3. Thus, specifying

$$\lim_{r\to 0}[u_\ell(r)/r^{\ell+1}] = 1, \tag{10.A.5}$$

we obtain from (10.A.1a,b) and (10.A.4a,b) the following boundary conditions for $a_\ell(r)$ and $b_\ell(r)$:

$$a_\ell(0) = (2\ell+1)!!/k^{\ell+1} \tag{10.A.6a}$$

$$\lim_{r\to 0}\frac{b_\ell(r)}{r^{2\ell+1}} = 0. \tag{10.A.6b}$$

The values of $a_\ell'(r)$ and $b_\ell'(r)$ at the origin are obtained from (10.A.3a,b), (10.A.4a,b), and (10.A.6a,b):

$$a_\ell'(0) = \frac{(2\ell-1)!!}{k^{\ell+1}}\frac{2m}{\hbar^2}\lim_{r\to 0} r\, V(r) \tag{10.A.7a}$$

$$b_\ell'(0) = 0. \tag{10.A.7b}$$

The equations (10.A.3a,b) can be solved by standard methods; we used the routine DO2CAF from the NAG library. Recalling (10.2.13), and using in (10.A.1a) the asymptotic relations

$$f_\ell(kr) \sim \sin(kr - \ell\pi/2), \quad r \to +\infty, \tag{10.A.8a}$$

$$g_\ell(kr) \sim \cos(kr - \ell\pi/2), \quad r \to +\infty, \tag{10.A.8b}$$

we find that the phase shift is obtained from

$$\tan\delta_\ell = b_\ell(+\infty)/a_\ell(+\infty). \tag{10.A.9}$$

Multiplying the function $u_\ell(r)$ in this appendix by $[a_\ell^2(\infty) + b_\ell^2(\infty)]^{-\frac{1}{2}}$, we obtain the function $u_\ell(r)$ in §10.3. In this fashion we find that the quantity ρ_0, defined in (10.3.2), is given by

$$\rho_0 = 1/[a_\ell^2(+\infty) + b_\ell^2(+\infty)]. \tag{10.A.10}$$

References

[1] Fröman, N. and Fröman, P.O., *JWKB Approximation, Contributions to the Theory*, North-Holland, Amsterdam, 1965. Russian translation: *MIR*, Moscow, 1967.

[2] Fröman, N., *Ark. Fys.* **32** (1966), 541–548.

[3] Fröman, N., *Ann. Phys.* (N.Y.) **61** (1970), 451–464.

[4] Fröman, N. and Fröman, P.O., *Ann. Phys.* (N.Y.) **83** (1974), 103–107.

[5] Fröman, N. and Fröman, P.O., Phase-integral approximation of arbitrary order generated from an unspecified base function. Review article in: *Forty More Years of Ramifications: Spectral Asymptotics and Its Applications*, edited by S.A. Fulling and F.J. Narcowich, Discourses in Mathematics and Its Applications, No. 1, Department of Mathematics, Texas A & M University, College Station, Texas, 1991, pp. 121–159. After minor changes reprinted as Chapter 1 in the present monograph.

[6] Fröman, N., Fröman, P.O., and Linnnaeus, S., Phase-integral formula for the regular wave function when there are turning points close to a pole of the potential. This is Chapter 6 in the present monograph.

[7] Fröman, N., Fröman, P.O., Walles, E., and Linnaeus, S., Phase-integral formulas for the normalized wave function of the radial Schrödinger equation close to the origin. This is Chapter 7 in the present monograph.

[8] Abramowitz, M. and Stegun, I.A., Editors, Handbook of Mathematical Functions, National Bureau of Standards, Applied Mathematics Series, 55. Seventh printing, May 1968, with corrections.

11
Probabilities for Transitions Between Bound States in a Yukawa Potential, Calculated with Comparison Equation Technique

Staffan Linnaeus

Abstract. Transition probabilities between the six lowest states in a Yukawa potential are calculated in the dipole approximation. The required matrix elements of r are evaluated from a phase-integral formula not involving wave functions, as well as from the definition of a matrix element with the wave functions obtained by means of comparison equation technique. Comparison with results from numerical calculations show that the comparison equation treatment in particular is very accurate.

11.1 Introduction

The calculation of transition probabilities, which may be reduced to the calculation of certain nondiagonal matrix elements, is an important problem in atomic physics. Since there has been a great interest in the study of highly excited states in recent years, it seems natural to apply asymptotic methods. Fröman and Fröman [1] derived an elegant, arbitrary-order phase-integral formula in which the wave functions do not appear. This formula is, however, less accurate for small angular momenta, in which case the centrifugal barrier is very thin. In order to deal with such cases one can resort to comparison equation technique and thus get approximate wave functions valid through the turning points. This approach was used by Fröman, Fröman, and Linnaeus in order to derive new connection formulas for the phase-integral approximation [2]. The purpose of this chapter is to demonstrate that it is also possible to use the wave functions obtained from the comparison equation treatment to calculate transition probabilities directly from the definition of a matrix element.

We shall consider the Yukawa potential

$$V(r) = -\frac{2\hbar^2}{ma_0} \frac{e^{-\lambda r}}{r}, \tag{11.1.1}$$

where a_0 is the Bohr radius. In the dipole approximation, the probability per unit of time for a transition from the state (n, ℓ) to the state (n', ℓ'), where $E_{n'\ell'} < E_{n\ell}$ and $\ell' = \ell \pm 1$, is given by (see §59 and §60 in [3])

$$A_{n\ell}^{n'\ell'} = \frac{4}{3m\hbar^2 a_0 c^3}(E_{n\ell} - E_{n'\ell'})^3 \frac{\max(\ell, \ell')}{2\ell + 1}\langle n\ell|r|n'\ell'\rangle^2; \qquad (11.1.2)$$

$E_{n\ell}$ is the energy of the state (n, ℓ). Numerical calculations of the transition probabilities in the dipole approximation between the six lowest states for various values of λ have been performed by Roussel and O'Connell [4]. For a selection of λ-values we have also calculated the matrix elements of r numerically with higher accuracy. We then calculated the same quantities by means of the phase-integral formula in [1] up to the fifth-order approximation; we also calculated these quantities up to the third-order approximation by using comparison equation wave functions in the integral defining the matrix element. Our approximate results are in good agreement with the numerical ones. The comparison equation treatment is usually superior to the phase-integral formula, at least if values pertaining to the same order of approximation are compared.

11.2   Phase-Integral Formulas

The radial Schrödinger equation reads

$$\frac{d^2}{dr^2}u_{n\ell}(r) + R(r)u_{n\ell}(r) = 0, \qquad (11.2.1)$$

where

$$R(r) = 2m[E_{n\ell} - V(r)]/\hbar^2 - \ell(\ell + 1)/r^2.$$

In the $(2N+1)$th-order phase-integral approximation there appears a function $q(r)$ given in [5] as

$$q(r) = Q(r)\sum_{i=0}^{N} Y_{2i}, \qquad (11.2.2)$$

where $Q(r)$ is an as yet unspecified function, referred to as the *base function*; the first few functions Y_{2i} are

$$Y_0 = 1 \qquad (11.2.3a)$$

$$Y_2 = \frac{1}{2}\varepsilon_0 \qquad (11.2.3b)$$

$$Y_4 = -\frac{1}{8}\varepsilon_0^2 - \frac{1}{8}\left(\frac{1}{Q(r)}\frac{d}{dr}\right)^2\varepsilon_0 \qquad (11.2.3c)$$

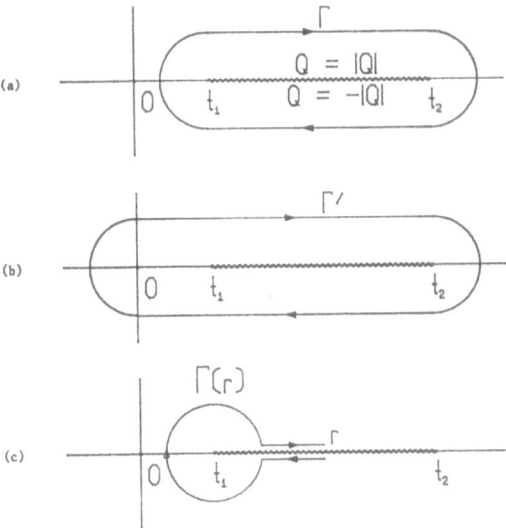

FIGURE 11.1. Contours of integration in the complex r-plane. The wavy line indicates a branch cut.

with
$$\varepsilon_0 = \frac{R(r) - Q^2(r)}{Q^2(r)} + Q^{-\frac{3}{2}}(r)\frac{d^2}{dr^2}Q^{-\frac{1}{2}}(r). \tag{11.2.4}$$

We choose the base function as follows:
$$Q^2(r) = R(r) - \frac{1}{4r^2}. \tag{11.2.5}$$

For bound states, $Q^2(r)$ has a pole of the second order at the origin and two positive zeros, t_1 and t_2 (the generalized classical turning points).

The energy levels are obtained from the quantization condition [6]
$$\frac{1}{2}\int_\Gamma q(r)\,dr = \left(k + \frac{1}{2}\right)\pi, \quad k = 0, 1, 2, \ldots, \tag{11.2.6}$$

where Γ is a contour encircling t_1 and t_2 as shown in Figure 11.1(a), and k is the number of nodes of the radial wave function. Since the residue of $q(r)$ at the origin is $(\ell + \frac{1}{2})i$, this phase-integral quantization condition can be rewritten as
$$\frac{1}{2}\int_{\Gamma'} q(r)\,dr = n\pi, \tag{11.2.7}$$

where n is the principal quantum number, related to k by
$$n = k + \ell + 1, \tag{11.2.8}$$

and the contour Γ' encircles the origin in addition to t_1 and t_2 [see Figure 11.1(b)].

The phase-integral calculation of nondiagonal matrix elements was considered by Fröman and Fröman [1], who gave the following formula:

$$\langle n\ell|f(r)|n'\ell'\rangle = \frac{\int_\Gamma \frac{f(r)\exp\{i[w(r)-w'(r)]\}dr}{[q(r)q'(r)]^{1/2}}}{\left[\int_\Gamma \frac{dr}{q(r)} \int_\Gamma \frac{dr}{q'(r)}\right]^{1/2}}, \qquad (11.2.9)$$

where the unprimed and primed quantities pertain to the initial and final state, respectively; and where

$$w(r) = \frac{1}{2}\int_{\Gamma(r)} q(z)\,dz; \qquad (11.2.10)$$

$\Gamma(r)$ is the contour shown in Figure 10.1(c), and Γ is the contour shown in Figure 10.1(a).

We note that (11.2.9) cannot be expected to give accurate results if the innermost turning point of either of the states lies too close to the origin; this occurs if n is much larger than ℓ. No such restriction exists for the validity of the quantization condition (11.2.7).

11.3   Comparison Equation Formulas

In §6.4, approximate wave functions, valid through the origin and both turning points, were derived for the bound state case. The result (with the expansion parameter equal to unity) can be written

$$u_{n\ell}(r) = N_{n\ell}\left(\frac{d\varphi}{dz}\right)^{-\frac{1}{2}} W_{\kappa,\ell+\frac{1}{2}}(2\varphi), \qquad (11.3.1)$$

where $N_{n\ell}$ is a normalization factor, $W_{\kappa,\ell+\frac{1}{2}}(2\varphi)$ is the Whittaker function [7], and, in the $(2N+1)$th-order approximation,

$$\kappa = \sum_{i=0}^{N} \kappa_{2i} = \sum_{i=0}^{N} \frac{1}{2\pi}\int_{\Gamma'} Q(r)Y_{2i}(r)\,dr, \qquad (11.3.2)$$

$$\varphi = \sum_{i=0}^{N} \varphi_{2i}. \qquad (11.3.3)$$

The first two functions φ_{2i} are obtained from the formulas

$$\int_{\kappa_0-[\kappa_0^2-(\ell+\frac{1}{2})^2]^{\frac{1}{2}}}^{\varphi_0} P(\varphi_0)d\varphi_0 = \int_{t_1}^{r} Q(r)\,dr, \qquad (11.3.4a)$$

$$\varphi_2 = \frac{1}{P(\varphi_0)} \int_{\kappa_0 - [\kappa_0^2 - (\ell + \frac{1}{2})^2]^{\frac{1}{2}}}^{\varphi_0} P(\varphi_0) \left(\frac{1}{2}\varepsilon_0 - \frac{1}{2}h \right) d\varphi_0, \qquad (11.3.4b)$$

where

$$P(\varphi_0) = \left(-1 + \frac{2\kappa_0}{\varphi_0} - \frac{(\ell + \frac{1}{2})^2}{\varphi_0^2} \right)^{\frac{1}{2}}, \qquad (11.3.5)$$

$$h(\varphi_0) = \frac{2\kappa_2}{\varphi_0 P^2(\varphi_0)} + \frac{1}{4\varphi_0^2 P^2(\varphi_0)} + P^{-\frac{3}{2}}(\varphi_0) \frac{d^2}{d\varphi_0^2} P^{-\frac{1}{2}}(\varphi_0). \qquad (11.3.6)$$

As shown in §6.4, the comparison equation treatment leads to the quantization condition

$$\kappa = n, \qquad (11.3.7)$$

which is equivalent to the phase-integral quantization condition (11.2.7). Hence, the comparison equation treatment yields no correction to the phase-integral energy levels.

A closed phase-integral expression for the normalization factor $N_{n\ell}$ in (11.3.1) can be obtained from Eq. (6.4.25). With the usual convention

$$\int_0^\infty u_{n\ell}^2(r)\, dr = 1 \qquad (11.3.8)$$

we obtain $N_{n\ell}$ from the normalization factor A in Chapter 6 (with the expansion parameter equal to unity) by setting $N_{n\ell} = A\Gamma(2\ell + 2)/\Gamma(\kappa + \ell + 1)$ and using the quantization condition (11.3.7). The result is

$$N_{n\ell} = \left[(n + \ell)!(n - \ell - 1)! \frac{1}{2\pi} \int_\Gamma \sum_{i=0}^N C_{2i} \frac{dr}{Q(r)} \right]^{-\frac{1}{2}}, \qquad (11.3.9)$$

where

$$C_0 = 1 \qquad (11.3.10a)$$

$$C_2 = -\frac{1}{2}\varepsilon_0. \qquad (11.3.10b)$$

Using the normalized comparison equation wave functions, we can evaluate the matrix elements of r needed for the calculation of the transition probabilities according to (11.1.2). The results are shown in Tables 11.1–11.4 along with the corresponding results obtained from the phase-integral formula (11.2.9) and the numerical values given in [4]. In Tables 11.5–11.6 we compare the matrix elements of r/a_0, as obtained from the two approximation schemes, with our own numerically exact results. Tables 11.5–11.6, where the numerically exact values are reliable, show that the comparison equation method gives more accurate values than the phase-integral formula in the same order of approximation; not even the fifth-order phase-integral approximation is as accurate as the result obtained from the third-order comparison equation treatment. In Tables 11.1–11.4 the numerical values from [4] are not accurate enough to allow us to draw any safe conclusions; however, it seems that the results in these tables support the conclusions obtained from Tables 11.5–11.6.

TABLE 11.1. Transition probabilities for the Yukawa potential with $\lambda = 0.01/a_0$.

Transition	Phase-Integral			Comparison Eq.		Ref [4]	Unit
	first order	third order	fifth order	first order	third order		
$2p \to 1s$	6.512	6.302	6.256	6.253	6.253	6.26	$10^8 s^{-1}$
$2p \to 2s$	2.051	2.124	2.164	2.146	2.145	2.14	$10^{-2} s^{-1}$
$3s \to 2p$	5.658	6.292	6.175	6.226	6.228	6.23	$10^6 s^{-1}$
$3p \to 1s$	1.568	1.680	1.642	1.656	1.655	1.66	$10^8 s^{-1}$
$3p \to 2s$	2.212	2.211	2.209	2.209	2.209	2.21	$10^7 s^{-1}$
$3p \to 3s$	1.096	1.118	1.119	1.120	1.119	1.12	$10^{-1} s^{-1}$
$3d \to 2p$	6.524	6.394	6.387	6.387	6.386	6.39	$10^7 s^{-1}$
$3d \to 3p$	6.672	6.695	6.822	6.767	6.764	6.77	$10^{-1} s^{-1}$

TABLE 11.2. Transition probabilities for the Yukawa potential with $\lambda = 0.025/a_0$.

Transition	Phase-Integral			Comparison Eq.		Ref [4]	Unit
	first order	third order	fifth order	first order	third order		
$2p \to 1s$	6.435	6.225	6.180	6.177	6.176	6.18	$10^8 s^{-1}$
$2p \to 2s$	4.254	4.410	4.493	4.461	4.455	4.47	$10^0 s^{-1}$
$3s \to 2p$	5.291	5.887	5.777	5.817	5.826	5.83	$10^6 s^{-1}$
$3p \to 1s$	1.490	1.596	1.560	1.573	1.572	1.57	$10^8 s^{-1}$
$3p \to 2s$	2.045	2.044	2.042	2.041	2.042	2.04	$10^7 s^{-1}$
$3p \to 3s$	1.887	1.928	1.929	1.935	1.929	1.94	$10^1 s^{-1}$
$3d \to 2p$	6.139	6.006	5.998	6.001	5.998	6.01	$10^7 s^{-1}$
$3d \to 3p$	1.188	1.196	1.219	1.212	1.209	1.21	$10^2 s^{-1}$

TABLE 11.3. Transition probabilities for the Yukawa potential with $\lambda = 0.05/a_0$.

Transition	Phase-Integral			Comparison Eq.		Ref [4]	Unit
	first order	third order	fifth order	first order	third order		
$2p \rightarrow 1s$	6.179	5.973	5.930	5.928	5.926	5.93	$10^8 s^{-1}$
$2p \rightarrow 2s$	2.123	2.209	2.251	2.244	2.232	2.24	$10^2 s^{-1}$
$3s \rightarrow 2p$	4.261	4.751	4.659	4.670	4.699	4.71	$10^6 s^{-1}$
$3p \rightarrow 1s$	1.256	1.343	1.313	1.325	1.323	1.32	$10^8 s^{-1}$
$3p \rightarrow 2s$	1.570	1.570	1.568	1.565	1.568	1.57	$10^7 s^{-1}$
$3p \rightarrow 3s$	7.017	7.203	7.207	7.306	7.206	7.21	$10^2 s^{-1}$
$3d \rightarrow 2p$	4.951	4.817	4.810	4.816	4.808	4.82	$10^7 s^{-1}$
$3d \rightarrow 3p$	4.944	5.049	5.147	5.184	5.102	5.11	$10^3 s^{-1}$

TABLE 11.4. Transition probabilities for the Yukawa potential with $\lambda = 0.1/a_0$.

Transition	Phase-Integral			Comparison Eq.		Ref [4]	Unit
	first order	third order	fifth order	first order	third order		
$2p \rightarrow 2s$	5.283	5.097	5.059	5.057	5.053	5.06	$10^8 s^{-1}$
$2p \rightarrow 2s$	8.718	9.183	9.362	9.513	9.280	9.30	$10^3 s^{-1}$
$3s \rightarrow 2p$	1.655	1.869	1.824	1.788	1.840	1.84	$10^6 s^{-1}$
$3p \rightarrow 1s$	4.982	5.334	5.216	5.252	5.239	5.26	$10^7 s^{-1}$
$3p \rightarrow 2s$	4.224	4.298	4.283	4.219	4.271	4.29	$10^6 s^{-1}$
$3p \rightarrow 3s$	1.231	1.315	1.320	1.546	1.309	1.32	$10^4 s^{-1}$

TABLE 11.5. Matrix elements of r/a_0 for the Yukawa potential with $\lambda = 0.01/a_0$.

Transition	Phase-Integral			Comparison Eq.		Exact
	first order	third order	fifth order	first order	third order	
$2p \rightarrow 1s$	1.32	1.295	1.2899	1.28958	1.289557	1.289557
$2p \rightarrow 2s$	5.09	5.716	5.225	5.2030	5.202634	5.202634
$3s \rightarrow 2p$	0.90	0.945	0.936	0.93973	0.939822	0.939822
$3p \rightarrow 1s$	0.501	0.519	0.513	0.51484	0.514810	0.514810
$3p \rightarrow 2s$	3.062	3.061	3.05981	3.05981	3.059768	3.059768
$3p \rightarrow 3s$	12.66	12.790	12.7934	12.795	12.792917	12.792917
$3d \rightarrow 2p$	4.80	4.747	4.7445	4.7447	4.744374	4.744374
$3d \rightarrow 3p$	10.06	10.07	10.17	10.128	10.126810	10.126809

TABLE 11.6. Matrix elements of r/a_0 for the Yukawa potential with $\lambda = 0.05/a_0$.

Transition	Phase-Integral			Comparison Eq.		Exact
	first order	third order	fifth order	first order	third order	
$2p \rightarrow 1s$	1.30	1.280	1.2749	1.2749	1.274470	1.274470
$2p \rightarrow 2s$	5.20	5.31	5.36	5.350	5.342254	5.342253
$3s \rightarrow 2p$	0.93	0.979	0.970	0.972	0.973773	0.973767
$3p \rightarrow 1s$	0.465	0.481	0.475	0.4778	0.477344	0.477345
$3p \rightarrow 2s$	2.953	2.950	2.94867	2.94848	2.948613	2.948607
$3p \rightarrow 3s$	13.94	14.175	14.1789	14.23	14.17834	14.178415
$3d \rightarrow 2p$	4.72	4.650	4.6467	4.653	4.645996	4.646005
$3d \rightarrow 3p$	11.37	11.53	11.65	11.64	11.59636	11.596455

References

[1] Fröman, N. and Fröman, P.O., *J. Math. Phys.* **18** (1977), 903–906.

[2] Fröman, N., Fröman, P.O., and Linnaeus, S., Phase-integral formulas for the regular wave function when there are turning points close to a pole of the potential. This is Chapter 6 in the present monograph.

[3] Bethe, H.A. and Salpeter, E.E., *Quantum Mechanics of One- and*

Two-Electron Systems, Handbuch der Physik, Vol. XXXV, Springer-Verlag, Berlin, Göttingen, Heidelberg, 1957.

[4] Roussel, K.M. and O'Connell, R.F., *Phys. Rev.* **A9** (1974), 52–56.

[5] Fröman, N. and Fröman, P.O., Phase-integral approximation of arbitrary order generated from an unspecified base function. Review article in: *Forty More Years of Ramifications: Spectral Asymptotics and Its Applications*, edited by S.A. Fulling and F.J. Narcowich, Discourses in Mathematics and Its Applications, No. 1, Department of Mathematics, Texas A & M University, College Station, Texas, 1991, pp. 121–159. After minor changes reprinted as Chapter 1 in the present monograph.

[6] Fröman, N. and Fröman, P.O., *J. Math. Phys.* **19** (1978), 1830–1837.

[7] Buchholz, H., *The Confluent Hypergeometric Function*, Springer-Verlag, Berlin, Heidelberg, New York, 1969.

Author Index

Subject Index

Springer Tracts in Natural Philosophy

(continued after index)

Springer Tracts in Natural Philosophy

(Continued from page ii)